Making a Semiconductor Superpower

This book provides real stories about the South Korean semiconductor community. It explores the lives and careers of six influential semiconductor engineers who all studied at Korea Advanced Institute of Science and Technology (KAIST) under the mentorship of Dr. Kim Choong-Ki, the most influential semiconductor professor in South Korea during the last quarter of the twentieth century. Kim's students became known as "Kim's Mafia" because of the important positions they went on to hold in industry, government, and academia. This book will be of interest to semiconductor engineers and electronics engineers, historians of science and technology, and scholars and students of East Asian studies.

"They were called 'Kim's Mafia.' Kim Choong-Ki himself wouldn't have put it that way. But it was true what semiconductor engineers in South Korea whispered about his former students: They were everywhere. … Kim was the first professor in South Korea to systematically teach semiconductor engineering. From 1975, when the nation had barely begun producing its first transistors, to 2008, when he retired from teaching, Kim trained more than 100 students, effectively creating the first two generations of South Korean semiconductor experts." (Source: *IEEE Spectrum*, October, 2022.)

Making a Semiconductor Superpower

The Seven Engineers from KAIST Who Shaped the Chip Industry

Dong-Won Kim

CRC Press
Taylor & Francis Group
Boca Raton London New York

CRC Press is an imprint of the
Taylor & Francis Group, an **informa** business

Designed cover image: www.shutterstock.com

First edition published 2024
by CRC Press
6000 Broken Sound Parkway NW, Suite 300, Boca Raton, FL 33487-2742

and by CRC Press
4 Park Square, Milton Park, Abingdon, Oxon, OX14 4RN

CRC Press is an imprint of Taylor & Francis Group, LLC

ISBN: 978-1-032-40604-6 (hbk)
ISBN: 978-1-032-40292-5 (pbk)
ISBN: 978-1-003-35391-1 (ebk)

DOI: 10.1201/9781003353911

Typeset in Times
by Newgen Publishing UK

Dedication

Dedicated to South Korean Semiconductor Engineers
Who Made a Miracle

Contents

PART I   *Mentor*

PART II   *Mentees*

Abbreviations

μm	micrometer
ADD	Agency for Defense Development
AMOLED	active matrix organic light-emitting diode
CAD	computer-aided design
CCD	charge-coupled device
CHiPS	Center for High-Performance Integrated Systems
CMOS	complementary metal-oxide semiconductor
CPU	central processing unit
DDR	double-data-rate
DRAM	dynamic random-access memory
EEPROM	electrically erasable programmable random-only memory
ETRI	Electronics and Telecommunications Research Institute
Gb	gigabit
GB	gigabyte
HD	high definition
IC	integrated circuit
ICT	information and communications technology
IDEC	Integrated Circuit Design Education Center
IEDM	International Electron Device Meeting
IEEE	Institute of Electrical and Electronics Engineers
IPS	in-plane switching
ISPP	incremental step programming pulse
KAIS	Korea Advanced Institute of Science (1971–1980)
KAIST	Korea Advanced Institute of Science and Technology (1981–)
Kb	kilobit
KIET	Korea Institute of Electronics Technology (later ETRI)
KIPRIS	Korea Intellectual Property Rights Information Service
KIST	Korea Institute of Science and Technology
KIT	Korea Institute of Technology
LAN	local area network
LCD	liquid crystal display
LDI	LCD drive (or driver) IC
LG	"Life's Good" from 1995 on (formerly Lucky-GoldStar)
LPCVD	low-pressure chemical vapor deposition
LSI	large-scale integration
LTPO	low temperature polycrystalline oxide
LTPS	low temperature polycrystalline silicon
LWIR	long-wave infrared
M3D	monolithic 3D integration
Mb	megabit
MHz	megahertz
MOS	metal-oxide-semiconductor

MOSFET	metal-oxide-semiconductor field-effect transistor
MROM	mask (or masked) random-only memory
NAND	"not and" (a Boolean operator and logic gate)
nm	nanometer
NNFC	National NanoFab Center
NOR	"not or" (a Boolean operator and logic gate)
NUS	National University of Singapore
NVM	nonvolatile memory
OLED	organic light-emitting diode
PCIe	peripheral component interconnect express
PMOS	p-type metal-oxide-semiconductor
pOLED	plastic OLED, or plastic organic light-emitting diode
poly-Si	polycrystalline silicon
POSTECH	Pohang University of Science and Technology
QUXGA	quad ultra-extended graphic array
QVGA	quarter video graphics array
RAM	random access memory
RF	radio frequency
ROM	random-only memory
SDRAM	synchronous dynamic random-access memory
SK	Sunkyong conglomerate
SLC	single-level cell
SNDL	Silicon Nano Device Laboratory
SNU	Seoul National University
SoC (or SOC)	system-on-a-chip
SOI	silicon-on-insulator
SRAM	static random-access memory
SSD	solid-state device
SWIR	short-wave infrared
TFT	thin-film transistor
TFT-LCD	thin-film-transistor liquid-crystal display
TLC	triple-level-cell
TSMC	Taiwan Semiconductor Manufacturing Company
UMC	United Microelectronics Corporation (in Taiwan)
USB	universal serial bus
VA	vertical alignment
VGA	video graphics array
VLSI	very large-scale integration
VR	virtual reality

Note on Romanization of Korean

In general, all Korean names and words are spelled according to the Revised Romanization of Korean rule rather than the McCune-Reischauer rule. The Revised Romanization rule was issued by the South Korean Ministry of Culture in 2000 and does not use any special characters as the McCune-Reischauer system does.

Korean names follow the East Asian word order of surname first, followed by given name. Most Western media now follow this rule (for example, Xi Jinping or Lee Jay-Yong). For works published in English, authors' names follow the English word order of given name first, followed by surname. For works published in Korean, the surname appears first, followed by the given name.

Many Koreans and Korean companies do not strictly follow the Revised Romanization rule but adopt slightly different spellings: for example, Kyung Chong-Min rather than Gyeong Jong-Min, Samsung Electronics rather than Samseong Electronics, and Hankuk Kyungje rather than Hankuk Gyeongje. I honor their choices herein.

Titles of books or articles in Korean are romanized as they are pronounced. For selected important books and articles, I add a translation after the romanized Korean title: for example, *Uri Kim Choong-Ki Seonsaengnim* (*Our Teacher, Kim Choong-Ki*), *Samsungjeonja 30nyeonsa* (*Thirty-Year History of Samsung Electronics*). If the books have their own English titles, these appear after their romanized titles. Most Korean professional journals, especially engineering journals, require that authors supply English titles and summaries of their articles. In those cases, I use their English titles.

For technical terms in the titles of Korean books, papers, and newspaper articles, I use the original English words: for example, "flash memory" instead of "plaesi memori."

Preface

I had yet to even imagine this book project when my article "Transfer of 'Engineer's Mind': Kim Choong-Ki and the Semiconductor Industry in South Korea" was published in *Engineering Studies* in the summer of 2019. Kim Choong-Ki is one of my first cousins, and my mother had told me a lot about her talented nephew when I was young. I had, however, paid little attention to his career as a semiconductor engineer but had instead concentrated on my own choice—great physicists in history. My first two books were therefore a history of the Cavendish Laboratory under J. J. Thomson and a biography of Yoshio Nishina (who is often called the "Father of Modern Physics in Japan"). In both works, I particularly emphasized J. J. Thomson's and Nishina's roles as successful mentors. The principal and common question I posed was, "How could a teacher train so many talented students so successfully that they eventually changed their science community?" Kim Choong-Ki became my third subject within this line of research.

My interest in Kim Choong-Ki began to increase once I became an assistant professor at KAIST in the spring of 1994. Kim had received the coveted Ho-Am Prize for Science and Engineering the year before, and his name began to appear in the newspapers. While I was writing the history of KAIST in 1994–1996, I also learned more about my renowned cousin, who had earned the deep respect of KAIST faculty members in both his own department and other departments as well. It was, however, *Uri Kim Choong-Ki Seonsaengnim* (*Our Teacher, Kim Choong-Ki*) that first prompted me to consider writing an article about him. This small book was privately printed to celebrate Kim's 60th birthday in 2002 and is filled with 67 short essays by his former students. Most of them are recollections of their student days at KAIST in which—like the memoirs of J. J. Thomson's and Nishina's former students—they happily remember how their mentor influenced their future lives and work and often recount interesting anecdotes about or favorite dicta of their mentor.

There were, however, three obstacles to my writing about Kim Choong-Ki. The first was a possible conflict of interest, since both he and I were professors at KAIST. The second was that I had never worked on the history of technology, and my knowledge of semiconductors was very limited. The third potential obstacle was that I was not sure I could gather sufficient data to write an article because South Korean scientists/engineers are not good at keeping records, and companies are ultrasensitive in guarding their information. So I gave up the idea.

In 2005, I resigned my professorship at KAIST and moved permanently to the United States with my two daughters. I briefly returned to KAIST between 2009 and 2012 to serve as dean in the humanities and social science division. Between 2012 and 2017, I taught the history of science at Harvard University, Johns Hopkins University, the National University of Singapore, and the University of Pennsylvania, as either a visiting professor or a lecturer. These diverse experiences in different countries and institutions during the 2010s led me to pay more attention to South Korean science and technology. The first result was an article on science fiction in South and North Korea, which was well received around the world except in South Korea. After publishing that article, I became notorious among some South Korean scholars and Koreanists, who considered me eccentric. After I failed to find a suitable publisher for a paper on the state of the science museum in South Korea that included strong criticism of these museums' culture, I realized that, given my unique views, there was little future chance of publishing any works on South Korean popular images of science and technology.

Thus, in 2017, I reconsidered Kim Choong-Ki's career as a subject of research and found that it suited my interests perfectly. First, although South Korea had been a primary actor in the world semiconductor market since the 1990s, few academic writings had been published on this phenomenon and almost nothing about the engineers who had actually shaped the chip industry; second, Kim's case study could attract the interest of engineers as well as historians of science and technology; and, third, I was now free to write about Kim because I had not been on the faculty at KAIST since 2012. Having collected data and information about the South Korean semiconductor industry since the early 2000s, I was able to complete the first draft of an article rather quickly. Suzanne Moon kindly advised me to contact *Engineering Studies* for its publication. After some revisions in response to feedback from the journal's editor Cyrus Mody and three anonymous reviewers, "Transfer of 'Engineer's Mind'" was finally uploaded to the internet on July 29, 2019. I sent a copy of the article and my apology to Kim, whom I had not informed about my work until then. He generously forgave me.

The response to the 2019 article surprised me. I received many encouraging emails and comments from engineers not only in South Korea but also in other countries. Two of them especially surprised and pleased me. The first was an inquiry from the *IEEE Spectrum*. Just a few days after the article on Kim Choong-Ki came out, Jean Kumagai, senior editor of *IEEE Spectrum* at that time, contacted me to ask whether I would write a short biographical essay on Kim for the journal. After some thought, I accepted the challenge, though the project was delayed until the spring of 2022, partly due to the COVID-19 pandemic. The other piece of mail that surprised me greatly was from Dr. Edward S. Yang, Kim's thesis advisor at Columbia University. He recounted interesting anecdotes and provided valuable information about his former student, some of which I have quoted in the *IEEE Spectrum* article as well as in this book. All this unexpected encouragement after the publication of the *Engineering Studies* article encouraged me to take on another, much bolder challenge—writing a book on Kim and his former students.

I have been extremely lucky to receive a great deal of support from so many people during the project. First of all, I would like to express my deepest gratitude to the seven protagonists of the book—Kim Choong-Ki, Kyung Chong-Min, Lim Hyung-Kyu, Cho Byung-Jin, Ha Yong-Min, Park Sung-Kye, and Chung Han. They each allowed me to conduct two or more extensive interviews with them, during the COVID-19 pandemic period. They also provided valuable photos and other necessary data. I was privileged to meet and interview several other semiconductor engineers as well, many of whom had studied under Kim. Among them, I am especially grateful to Kim Oh-Hyun, Jung Hee-Buhm, and Yoon Nan-Young, all of whom could well have been profiled in a chapter of their own, had time and space permitted.

Instead, the information they so generously provided is interwoven throughout the book. Several casual conversations with Kook Yoon-Jae about Kim Choong-Ki and his former students also allowed me to add some interesting and important stories to the book. Several other people whom I contacted during the process preferred to remain anonymous, and I have faithfully kept my word on this. I thank them equally.

The editors and editorial staff of both *Engineering Studies* and *IEEE Spectrum* deserve special thanks as well. Cyrus Mody, the editor of *Engineering Studies*, carefully read my draft article on Kim and provided important feedback for revising it further. Without his assistance, the article might have been less interesting and far less popular. Likewise, the staff of *IEEE Spectrum* helped me greatly to write and revise the draft of "Godfather of South Korea's Chip Industry: How Kim Choong-Ki Helped the Nation Become a Semiconductor Superpower." Since I had no prior experience writing a journalistic article in such a short space and time, their ongoing assistance and feedback between April and August of 2022 were most helpful. It was a new experience for me, and I really learned a lot during the process. I especially thank Ariel Bleicher, who was my primary contact and editor of the article, as well as Samuel K. Moore, senior editor.

CRC Press kindly accepted my book proposal and decided to publish *Making a Semiconductor Superpower*. I am deeply grateful to Marc Gutierrez, who was my primary contact during the process. He was very patient with my continuous inquiries and provided much valuable advice, including on the title of the book as well as the design of its cover. I also thank the designer of the book cover who realized Marc's and my ideas wonderfully. Victoria R. M. Scott, an independent editor, kindly took up the task of editing the final draft, and I thank her for her unflagging efforts.

Some parts of Chapter 1 and Chapter 2 are based on my early article with Stuart W. Leslie, "Winning Markets or Winning Nobel Prizes: KAIST and the Challenges of Late Industrialization" (*Osiris* 1998), and my recent article "Transfer of 'Engineer's Mind': Kim Choong-Ki and the Semiconductor Industry in South Korea" (*Engineering Studies* 2019). I thank the University of Chicago Press and Taylor & Francis, respectively, for permission to use the articles. Chang Hea-Ja (Mrs. Kim Choong-Ki) provided many valuable photos for the book.

Last but never least, I would like to thank my two daughters, Da-Ye and Jean-Sol, who have been the source of my strength from the beginning to the end of this book project.

Dong-Won Kim
Lexington, Massachusetts
February 2023

About the Author

Dong-Won Kim is a historian of science and technology. After receiving his Ph.D. in the history of science from Harvard University in 1991, he was a professor and dean at the Korea Advanced Institute of Science and Technology (KAIST). He has also taught at Johns Hopkins, Harvard, the National University of Singapore, and the University of Pennsylvania as a visiting professor or lecturer. His major research fields are the history of physics and the history of science and technology in Korea and Japan. He has published several papers on these subjects, as well as two monographs: *Leadership and Creativity: A History of the Cavendish Laboratory, 1871–1919* (Springer, 2002) and *Yoshio Nishina: Father of Modern Physics in Japan* (Taylor & Francis, 2007). His next project is on cosmic ray research in the first half of the twentieth century.

Introduction

By the mid-2000s, South Korea had become a dominant power in the world semi-conductor market, and by the mid-2010s, its worldwide market share of memory had climbed to more than 60%.[1] In 2017, Samsung Electronics, South Korea's flagship electronics company, became the number one semiconductor manufacturer in the world in terms of revenue, and another South Korean semiconductor company, SK Hynix, was number three.[2] Since the early 2000s, two South Korean electronics companies, Samsung Display and LG Display, had become the major players in the thin-film-transistor liquid-crystal display (TFT-LCD) market and then in the organic light-emitting diode (OLED) market as well. In 2018, semiconductors made up 21% of South Korea's entire exports and 6.7% of its GDP, and its share increased in the early 2020s.[3] This is a truly amazing achievement for a nation that started assembling its first radio sets only in 1959 and that began to manufacture outdated memory chips only in the mid-1980s.

Numerous books and articles have been published to explain how South Korea's semiconductor industry accomplished such brilliant success within such a short period. These works usually emphasize the roles and contributions of the South Korean government and individual companies, but they almost completely neglect those of the South Korean academy. Instead, they focus on how the South Korean government made effective plans for the growth of the industry, how the heads of South Korean conglomerates made correct decisions to enter or support the semiconductor sector, or how well South Korea's semiconductor companies were organized and operated for successful results.[4] In addition, whether written by economic analysts or professors of business management, these works consider US-trained South Korean engineers in the 1980s and 1990s to have been convenient and important conduits through which state-of-the-art knowledge and technology flowed smoothly from the United States to South Korea.[5] Almost no studies, however, clearly explain *what* was actually brought from the United States to South Korea or *how* that transferred technology and knowledge were successfully disseminated to and implemented in the South Korean semiconductor industry. The research and development and manufacturing of semiconductor devices required more than a handful of highly educated and experienced US-trained star engineers: they required dozens or even hundreds of rank-and-file engineers as well.

DOI: 10.1201/9781003353911-1

A more serious problem is previous studies' neglect of the contributions of individual semiconductor engineers. The names and achievements of these engineers rarely appear in print or other media except when they are promoted to high management posts or receive coveted prizes. I often compare the situation to that of ceramic artisans in premodern Korea. These artisans left many magnificent works, some of which are displayed in prestigious museums around the world, yet we don't know who they were because there are no records of their names. Likewise, although South Korean semiconductor engineers have made great contributions to the success of the industry since the mid-1970s, few of their names are known to the general public. Even those few may be forgotten sooner or later.

This book aims to fill these two gaps in the history of semiconductors in South Korea. First, it analyzes the role of academia in the development of semiconductors in South Korea by focusing on Kim Choong-Ki, the most influential teacher and mentor during the last quarter of the twentieth century (see Chapter 2). Second, it offers six case studies of semiconductor specialists who earned their advanced degrees under Kim at the Korea Advanced Institute of Science and Technology (KAIST) and who have contributed greatly to the growth of the semiconductor community over the last four decades.

This book does not intend to cover the entire history of the South Korean semiconductor industry but instead concentrates on the lives and works of seven semiconductor engineers who have greatly contributed to shaping South Korea's semiconductor chip industry. The choice of engineers from KAIST is obvious: the training of semiconductor specialists in South Korea started there in the spring of 1975, when Kim Choong-Ki began to teach the subject in its Department of Electrical Engineering, and the semiconductor has been a core of KAIST's educational program ever since. As semiconductors became a major South Korean industry from the mid-1980s on, the importance of training semiconductor specialists rose quickly within KAIST. By the early twenty-first century, more than one-third of the faculty members of the Department of Electrical Engineering at KAIST were working on semiconductors, and many members of other departments in the College of Engineering and College of Natural Sciences were also involved in teaching and doing research on semiconductors. The South Korean semiconductor industry particularly welcomed KAIST graduates in semiconductors because they could handle both theory and practice. The success of KAIST's semiconductor program also stimulated other South Korean universities to follow suit.

Kim Choong-Ki was at the center of this development from the mid-1970s to the end of the 2000s, training the first two generations of semiconductor specialists for South Korean industry and academia. Educated at Columbia University in the late 1960s and trained at Fairchild Camera and Instrument in the early 1970s, Kim was the ideal engineer to teach both the theory and practice of semiconductors. In the 1990s and 2000s, members of the South Korean semiconductor community dubbed Kim's former students "Kim's Mafia" because they filled many of the most important positions in industry, government research institutes, and academia. The South Korean media took note and began to call Kim the "godfather" of the country's semiconductor community. Accordingly, Part I, "Mentor," consists solely of Chapter 2, "Godfather of the South Korean Semiconductor Community: Kim Choong-Ki," which systematically analyzes how Kim, a meticulous introvert, became an inspiring university professor and the most influential mentor of the South Korean semiconductor community.

Part II, "Mentees," consists of six chapters that profile a half dozen of Kim's former students at KAIST. From about a hundred of Kim's former students, I carefully selected two from academia and four from industry. Some of these members of "Kim's Mafia" went on to surpass their teacher in terms of achievements as well as fame. Although Samsung Electronics' share in the development of South Korea's semiconductor industry has been dominant, and many of Kim's talented former students have worked there, I deliberately selected only one from Samsung. I instead profile two engineers each from LG Display and SK Hynix, and another who started his own business after graduation. These six engineers are profiled in roughly chronological order in order to show the changing environment in Kim's laboratory from the mid-1970s to the mid-1990s, as well as corresponding changes in the South Korean semiconductor community from the early 1980s on.

In South Korea, the fame of engineers has often been determined by how frequently they appear in the mass media, how high they climb in the company's or university's hierarchy, or even how much they earn. Some of these six engineers are therefore less known to the public than others and are recognized only by their peers in the semiconductor community. Since my goal was neither to select the most important semiconductor engineers in South Korea nor to judge which of Kim's students have contributed the most to the development of semiconductors, I freely but carefully chose six with different backgrounds, career paths, and contributions to South Korea's semiconductor community. I regret not being able to include Jung Hee-Bum, who has worked at the Electronics and Telcommunications Research Institute (ETRI) for many decades, and Yoon Nan-Young, the only female student under Kim's supervision during his entire career.[6]

The diversity of these men's characters, lives, and works are reflected in the diverse sources for each chapter. I conducted at least two interviews with each engineer profiled in the book and interviewed other semiconductor engineers as well. Since none kept diaries or similar records, the descriptions of their private lives are heavily dependent on their own memories. *Uri Kim Choong-Ki Seonsaengnim* (*Our Teacher, Kim Choong-Ki*), a collective memoir of Kim's former students, published in 2002 to celebrate Kim's 60th birthday, was particularly helpful in providing vivid descriptions of their student years at KAIST.[7] Research papers, newspaper articles, government documents, and companies' official documents were also thoroughly studied and used. Some sensitive contents from the point of view of specific companies are often mentioned or described by referring to newspaper articles instead. I did my best to use correct technological terms and explanations, but some mistakes may be inevitable. Also, I am the only person responsible for any mistakes and wrong interpretations of the main engineers profiled in the book.

Let's start our journey to the South Korean semiconductor community.

NOTES

1 Ministry of Trade, Industry and Energy, "Bandoche, Display Saneop Donghyang," *e-Narajipyo* (September 14, 2022), www.index.go.kr/unity/potal/main/EachDtlPageDetail.do?idx_cd=1155 (searched on October 6, 2022).

2 Frost and Sullivan, "Samsung Surpasses Intel in 2017-Q2 Revenue, Grabs the Pole Position! But Will the Lead Continue?" www.frost.com/frost-perspectives/samsung-surpasses-intel-in-2017-q2-revenue-grabs-the-pole-position-but-will-the-lead-continue/; and Business Wire, "With Its Highest Growth Rate in 14 Years, the Global Semiconductor Industry Topped $429 Billion in 2017, HIS Markit Says," *Business Wire* (March 28, 2018), www.businesswire.com/news/home/20180328006092/en/With-its-Highest-Growth-Rate-in-14-Years-the-Global-Semiconductor-Industry-Topped-429-Billion-in-2017-IHS-Markit-Says (searched March 2, 2022).

3 Tongsang, "Bandochesaneop: dashi oneun Chohohwangki, gyeonjohan Heureumse Jeonmang," *Tongsang, 104* (January 2021), https://tongsangnews.kr/webzine/2101/sub2_2.html (searched on September 6, 2022). *Tongsang* is a monthly web-magazine issued by the Ministry of Trade, Industry and Energy.

4 A good example is Kim Su-Yeon, Baik You-Jin, and Park Young-Ryeol, "Hankuk Bandochesaneopui Seongjangsa: Memory Bandochereul jungsimeuro [The Historical Review of the Semiconductor Industry]," *Gyeongyeongsahak, 30:3* (2015), 145–166. All three authors belong to the School of Business in Yonsei University. Other examples are Linsu Kim, *Imitation to Innovation: The Dynamics of Korea's Technological Learning* (Boston: Harvard Business School Press, 1997); Youngil Lim, *Technology and Productivity: The Korean Way of Learning and Catching Up* (Cambridge, Mass.: The MIT Press, 1999); and Myung Oh and James F. Larson, *Digital Development in Korea* (New York: Routledge, 2011).

5 Examples include Linsu Kim, *Imitation to Innovation*, 153–167, and Youngil Lim, *Technology and Productivity*, 83–84, 110–113.

6 I conducted long interviews with both Jung Hee-Bum and Yoon Nan-Young in May 2022 and finished the first drafts of profiles of them but finally decided not to include them. Some of the contents of their interviews, however, appear in the book.

7 Park Sang-In et al., *Uri Kim Choong-Ki Seonsaengnim* [*Our Teacher, Kim Choong-Ki*] (Daejeon: Privately printed, 2002).

1 Historical Background

Science and Technology in South Korea, KAIST, and the Semiconductor Industry

This chapter provides the historical and cultural background of South Korean science and technology in the twentieth century. Since all seven semiconductor engineers in the book worked or studied at Korea Advanced Institute of Science and Technology (KAIST), a short history of KAIST is also provided. This chapter also explains the unique triangular relationship among the South Korean government, industry, and academia, which became the primary cause of the country's successful development of the semiconductor industry over the past 50 years.

SCIENCE AND TECHNOLOGY IN SOUTH KOREA TO THE END OF THE TWENTIETH CENTURY

Premodern Korea had not provided science and technology with a favorable environment. Until the very end of the nineteenth century, few Koreans paid serious attention to science and technology. Neo-Confucianism, the official ideology during the Chosun dynasty (1392–1910), emphasized the importance of the humanities, and Chosun elites "undervalued work, especially manual labor."[1] Only a few Korean intellectuals studied astronomy, for calendar-making or for its philosophical connection with the Confucian worldview, and some wrote books on natural history, but their activities were usually considered marginal.[2] A recent Korean history textbook nicely summarizes this point: "Education trained the cultivated generalists. There was disdain for the specialist and for technical training that prevailed into recent times. ... Education was basically of a nonspecialized, literary nature, which has remained the preference of most Koreans."[3] This prejudice against manual labor and against subjects not included in the humanities created a negative image of science and technology during the Chosun dynasty that continued even into the late twentieth century.

The best example of this general trend during the premodern period is the fate of Korean ceramic artisans, who produced very fine ceramics during both the Koryo (918–1392) and Chosun dynasties.[4] Korean ceramic objects, which "are perhaps only the best-known example of Korean creativity within the Sinitic cultural world [in the premodern period]," have long represented Korean culture at the British Museum, the Louvre, the New York Metropolitan Museum, and other bastions of high art.[5] However, the artisans who made these magnificent artifacts left no names in Korean

DOI: 10.1201/9781003353911-2

history because they were among the lowest social strata in Korean society: "They had only heavy burdens to carry out as ceramic artisans. According to the *Gyeongkuk Daejeon* [*Great Code of Administration*, Chosun's code of laws, acts, customs, and ordinances], ceramic artisans had to participate in manufacturing ceramics for their whole lives, and their task was inherited by their offspring. All ceramic artisans were registered to the government, and they could not be freed from their bondage. ... If there were any food crises, it was ceramic artisans who first died of hunger."[6]

There are some exceptions to this history of anonymous Korean ceramic artisans. During the Japanese invasions of Korea (1592–1598, known as the Wanli Korean campaign, or *Bunroku no eki*), many Korean ceramic artisans were forcibly or voluntarily moved to Japan. They were treated well there and contributed greatly to the development of Japanese ceramics. The two most famous cases are Shim Soo-Kwan (Chin Jukan) and Yi Sam-Pyeong (Ri San-Pei): Shim settled in Kagoshima on Kyushu Island, where he and fifteen generations of his descendants produced fine porcelains, while Yi became the father of Imari (Arita) porcelain and is honored in the Sueyama (Tozan) Shrine in Arita on Kyushu.[7] It is therefore no wonder that when the Chosun government demanded that Korean ceramic artisans return after the end of the war, most of them refused to do so because of their miserable living conditions at home.[8]

Although some Confucian values—such as the emphasis on education, the meritocratic tradition, and the importance of the group over the individual—contributed greatly to the rapid economic development of late twentieth-century South Korea, this deeply rooted prejudice against science and technology nonetheless continued.[9] For example, until the late 1990s, engineering students were often ridiculed as *gongdori*, a disrespectful epithet for factory workers.[10] In 1971, the founders of Korea Advanced Institute of Science (KAIS) deliberately omitted the word "technology" from its title because they feared that if it were included, most Koreans would mistake this graduate-only institute for a vocational school.[11] The prejudice against science and technology seems to have weakened by the end of the 1980s as heavy industry became the mainstream economy in South Korea, only to revive in the early 1990s with the new civilian governments and the rise of stronger nationalism. When *Joong-Ang Ilbo* and *Daedeok Net* conducted a survey in the summer of 2012, a majority of the South Korean scientists and engineers who responded pointed to deep-rooted social and cultural discrimination against science and technology as the major obstacle blocking the further development of these fields in South Korea.[12] In November 2016, a senior sociologist, Song Bok, concluded his article on how to reform the conservatism in South Korea as follows: "We must discard the [old] idea that only *munkwa* [文科, humanities generalists] can rule the country. In other words, the idea that only those who major in humanities, social sciences, or law can rule the country must be abandoned now."[13] The English saying that "Old habits die hard" applies perfectly to Koreans' attitude toward science and technology.

It was not just a negative image of science and technology in premodern Korea but also negative policies that obstructed the development of these fields. For example, King Jeongjo (reigned 1776–1800), who is considered one of the most enlightened kings during the Chosun dynasty, officially prohibited the import of foreign books from China and labeled ideas or philosophies outside the official Neo-Confucianism as "heretical."[14] As a result, nineteenth-century Korean governments' efforts to bring

Western science and technology into the country were far less enthusiastic and systematic than those of the Japanese and even the Chinese governments. Until the end of the nineteenth century, the concepts of Western science and technology remained foreign to most Koreans. When the Korean people at last developed a fledgling interest in Western science and technology in the early years of the twentieth century, it was, tragically, too late to catch up overnight.

In 1910, Japan, which had already transformed itself into East Asia's first modern power, annexed Korea and began a 35-year occupation of that country that did not end until Japan surrendered to the Allied Forces on August 15, 1945. Japan characterized its iron-fisted rule of Korea as the desperately needed modernization of a backward nation. To further its colonial rule, the Japanese colonial government systematically introduced certain concepts of Western science and technology by establishing the Central Testing Laboratory in Seoul (1912) and opening a few schools, including Keijo (Kyungseong, presently Seoul) Technical High School (1916). In 1926, the Japanese colonial government opened Korea's first and only university, Keijo Imperial University, in Seoul.[15] Keijo Imperial University had two faculties—law and medicine—and during its first decade, neither science nor engineering instruction was systematically offered. In 1938, a combined science and engineering faculty was added to increase the supply of engineers for Japan's war against China.[16] This new faculty included a small physics department, a small chemistry department, and relatively large departments of civil, mechanical, electrical, chemical, and mining and metallurgical engineering. A strict quota was enforced of only twelve ethnic Koreans per year: by the time of the Japanese surrender in 1945, only 37 Koreans had received science and technology training at the University. Thus, Korea's only university before 1945 contributed very little to the future development of science and engineering in that nation.

During its occupation by Japan, Korea produced fewer than 400 native Korean scientists and engineers, almost all of whom earned their bachelor's or more advanced degrees outside the Korean peninsula.[17] About 200 earned their degrees in Japan, 100 in the United States, 30 in Manchuria at Japanese-established universities, 30 in Europe, and 20 in China or the Soviet Union. Only 11 Koreans earned doctorates in science or engineering: 5 in Japan, 5 in the United States, and 1 in Europe. Of these 11 doctorates, only one earned his degree in engineering. That engineer was Li Seung-Ki, who had developed a new synthetic fiber from polyvinyl alcohol and received his doctoral degree from Kyoto Imperial University in 1939. In short, in the first half of the twentieth century, the foundation of Korea's science and engineering was very fragile indeed.

For Korea, Japan's unconditional surrender to the Allied Forces in August 1945 meant liberation from Japanese occupation, but it did not mean independent sovereignty. The Allied Forces divided the Korean peninsula into two regions along the 38th parallel, and for the next three years, South Korea was occupied by the United States and North Korea by the Soviet Union. These two superpowers exerted strong influences on the development of science and technology in the divided Korea—influences that sprang from their opposing political ideologies, very different educational and research systems, and conflicting visions of Korea's future.

In South Korea, the five years following the Japanese surrender were a time of turmoil in which two occurrences significantly affected the development of science and technology. The first was the establishment of Seoul National University (SNU) in 1946 by the American Occupation Authority.[18] SNU was formed by merging the former Keijo Imperial University with several vocational schools scattered throughout Seoul. As the sole heir of Keijo Imperial University, SNU soon became South Korea's premier center of academic teaching. The second was the South's loss of many qualified scientists and engineers. Between 1945 and 1953, as many as 40% of South Korea's scientists and engineers moved to North Korea, many others emigrated to the United States or other countries, and some entered government service.[19]

When the Korean War ended in 1953, science and technology in South Korea therefore barely existed. Most of the country's science and engineering facilities had been destroyed during the war, and most of its scientific and engineering community had dispersed. Only one or two qualified representatives of each scientific or engineering field remained in South Korea, and even at SNU, professorships were often held by new graduates with bachelor's degrees. The Korean War, however, had left two positive results for the future development of science and technology in South Korea: first, in its competition with the North, the South Korean government finally began to pay attention to the development of science and technology; second, South Korea started wholeheartedly importing and absorbing American science and engineering. Both the introduction of an experimental nuclear reactor from the United States and the US-government-sponsored Minnesota Project symbolized the changing attitude after the end of the Korean War.[20] It was also after the end of the Korean War that many South Koreans went privately to the United States to study either science or engineering.

Meanwhile, South Korea began to rebuild its industry over the ruins. Its major industries during the 1950s were agriculture and fishery, but it also had small consumer-oriented or labor-intensive businesses such as the textile and food industries. The Samsung, LG, and SK conglomerates, whose companies would become major players in the world's electronics market half a century later, started out as modest suppliers of daily consumer products such as flour and sugar (Samsung), toothpaste (LG), and textiles (SK). Heavy industry barely existed, and almost all electronic devices were imported from abroad. The country's electricity supply was meager, and the communication lines were unstable until the end of the 1950s. The most popular electronics item during the 1950s were US-made radio sets such as the Zenith Trans-Oceanic, sold at a price equivalent to "fifty bags of rice."[21] In 1947, the number of telephones in the country was about 45,000, and in 1960, it was only about 90,000, even though South Korea's population had reached about 30 million by then.[22]

An important event for the development of the electronics industry occurred in October 1958, when Koo In-Hwoi, the founder of the future LG conglomerate, decided to launch a new company, GoldStar, to manufacture radio sets. On November 15, 1959, the company presented its "first domestic radio A-501" to the market.[23] A year after that, it succeeded in developing its first transistor radio, the T-701. Though GoldStar successfully designed its first radio sets and localized some less important parts, it remained heavily dependent on the import of core parts, such as vacuum tubes, transistors, and speakers. Moreover, in the beginning "market demand was very limited due to the fact that most people in Korea were on low incomes, and those who could afford radios preferred foreign brands. In 1961, GoldStar had to seriously consider whether to keep or abandon the company."[24] This changed dramatically after May 16, 1961, when a military junta led by Major General Park Chung-Hee mounted a coup and ushered in a twenty-year program of modernization by fiat. The new military government began to distribute radio sets to rural areas and also strongly encouraged the populace to buy goods "made in Korea." By the end of 1962, GoldStar had sold more than a million radios and had begun to export them to the world. This not only made GoldStar the front-runner in electronics in the domestic market but also stimulated others, such as Taihan Cable, to enter the electronics business.

Under Park's strong leadership and with the support of US-trained technocrats, South Korean industrialization and its economic development proceeded rapidly during the 1960s and 1970s. Within two decades, the four Five-Year Economic Development Plans (1962–1966, 1967–1971, 1972–1976, and 1977–1981) fundamentally changed South Korea from a poor agriculture-fishery country into the most enthusiastic industrial upstart in East Asia.[25] Whereas the first five-year economic plan focused on strengthening the infrastructure and encouraging light industry for export, the next three plans concentrated on developing heavy industry, including iron and steel, machinery, shipbuilding, chemicals, and electronics. Many well-known South Korean companies—for example, Hyundai Motor Company (1967), Samsung Electronics (1969), and Hyundai Heavy Industry (1972)—were established during the second and third economic development periods. The result was truly astounding. When Park's government launched its first economic development plan in 1961, South Korea's GNP per capita was less than $100, and its major export item was tungsten ore, which made up more than 60% of all its exports.[26] By the time Park was assassinated in the fall of 1979, South Korea's GNP per capita had reached $1,680, and it had become a country that exported not only textiles and shoes but ships, automobiles, chemicals, and TV sets to the whole world.

Electronics was a key target of Park's economic plans from the time of his second five-year plan (1967–1971). In September 1967, Park's government invited Kim Wan-Hee of Columbia University to visit South Korea to draw up a blueprint for the development of electronics there.[27] Kim visited GoldStar and other electronics companies and also inspected some government research institutes. On September 16, he personally presented seven suggestions to Park, who immediately ordered the preparation of a more detailed plan. Kim and two hundred specialists worked on the "Report on the Development of Electronics" for the next eight months, which became the bible for the development of electronics during the next decade.

Most important, however, was the establishment of Samsung Electronics in 1969.[28] The founder of the Samsung conglomerate, Lee Byung-Chul, had been carefully examining both the domestic and the foreign markets for years before he decided to enter the electronics industry. To appease its rivals, who opposed Samsung's entrance into the electronics business, Samsung Electronics initially promised to export all its products. But this unreasonable shackle became meaningless once it began to produce its TV sets, refrigerators, and other electronics. Samsung Electronics aimed to manufacture all necessary parts for its end products and was eager to introduce advanced technology from abroad, especially from Japan and the United States. It soon became

GoldStar's most formidable rival in the domestic market, especially for TV sets and refrigerators. Samsung Electronics was also the first South Korean electronics company to focus on exporting its products: in 1973, for instance, it exported more than 50,000 TV sets to the world, while its domestic sales remained below 30,000.[29] In short, the entry of Samsung into electronics in 1969 provided a great stimulus to the South Korean electronics industry and paved the way for the accelerated development of the field.

South Korean industry went on to make a quantum leap during the 1980s, transforming itself from a copycat into a creative innovator. Kim Linsu's 1997 *Imitation to Innovation: The Dynamics of Korea's Technological Learning* nicely summarizes and analyzes how South Korea made such a qualitative change during the 1980s and 1990s by acquiring and digesting the necessary technology.[30] Kim identifies the smooth technology transfer from the advanced countries—such as the United States and Japan—to South Korea during this period as an important key to South Korea's success: in the case of the semiconductor industry, for example, he argues that South Korean engineers who had been trained at American universities and satellite research institutes in Silicon Valley became major conduits for the necessary advanced technology during the 1980s.[31] By the mid-1990s, South Korea had become "second in the world in shipbuilding and consumer electronics, third in semiconductor memory chips, fifth in textiles, chemical fibers, petrochemicals, and electronics, and sixth in automobiles, and iron and steel."[32] The country experienced a brief setback during the economic crisis that badly hit several Asian countries in the late 1990s but survived and returned even more strongly in the early twenty-first century. By the early 2020s, South Korea had become the tenth largest economy in the world, whose electronics and automobile companies were among the top five.

KOREA ADVANCED INSTITUTE OF SCIENCE AND TECHNOLOGY (KAIST)[33]

When Park's government enthusiastically promoted its ambitious economic development plans in the beginning of the 1960s, it encountered two serious problems: South Korea possessed neither the necessary technology nor the manpower. At the outset, therefore, South Korean industry was heavily dependent on imported technology and hardware, and this dependency became worse as its exports increased: the more South Korea exported its products, the more it was forced to import necessary parts and pay license fees. The Park government desperately wanted to break this cycle, but in the 1960s there were few engineers or applied scientists in South Korea to solve this problem.

In 1966, Park established a new research institute, Korea Institute of Science and Technology (KIST), with US government support that was provided in exchange for South Korea sending troops to Vietnam.[34] The new institute was designed to become

the window through which the transfer of foreign technology to domestic industry can be made. ... It guides and counsels industries in selecting appropriate technologies for import and in modifying, improving, and adapting imported technology for application and dissemination. KIST is the bridge between domestic industry and advanced technologies of foreign countries.[35]

South Korean scientists and engineers—trained and working mostly in the United States but also in Germany, Japan, and Great Britain—were invited to join KIST with offers of higher salaries and other privileges. From its inception, KIST contributed greatly to shipbuilding, chemical engineering, the steel industry, and various other industries. Its very success, however, soon created a new problem: there were too few well trained engineers and applied scientists to work in South Korean industry or in government research institutes such as KIST. Two possible solutions were suggested: either transform existing engineering colleges so that they could produce the manpower required or establish a new institute for that specific purpose. Park chose the latter course.

In 1971, Korea Advanced Institute of Science (KAIS) was established with financial support from the United States Agency for International Development (USAID).[36] The institute became known as KAIST when "Technology" was added to its title in 1981. Frederick E. Terman, longtime provost of Stanford University and often called the "Father of Silicon Valley," was invited to draw up the blueprint of this new institute along with Thomas L. Martin, Donald L. Benedict, Franklin A. Long, and Chung Kun-Mo. In the resulting "Survey Report on the Establishment of Korea Advanced Institute of Science," Terman emphasized that this new graduate-only institute was being created to "satisfy the needs of Korean industry and Korean industrial establishments for highly trained and innovative specialists, rather than to add to the world's store of basic knowledge."[37] In line with Terman's "steeples of excellence" strategy at Stanford, KAIS would "offer instruction in only a limited number of high priority fields" with the goal of "achieving a truly outstanding program in each of these fields."[38] A few selective fields—mechanical engineering, chemical engineering and applied chemistry, electrical engineering, industrial engineering, material science, and biology—were chosen to support the government's third and fourth Five-Year Economic Development Plans (1972–1976 and 1977–1981). Close cooperation with Korean industry and government research institutes was strongly encouraged. In short, KAIS intended to win the South Korean market with its graduates rather than aiming to win Nobel Prizes.

In many respects, the establishment of KAIS in 1971 was a coup for Park Chung-Hee in the realm of higher education. It went against all the ideas of higher education in South Korea at the time. For most South Koreans, going to university meant entering an undergraduate program, whereas the new institution would have only a graduate program. Most universities had strong humanities, social sciences, law, and art departments but relatively weak science and engineering departments, whereas KAIS would focus on producing many specialists in science and engineering alone. The Ministry of Education managed all levels and all kinds of education, whereas the new institution would instead be under the jurisdiction of the Ministry of Science and Technology. At a time when only military academies required their students to live in dormitories, all KAIS students would live in dorms. At a time when fellowships were very rare but tuition and fees relatively high, all KAIS students would receive full scholarships and generous stipends, just like students at military academies. And at a time when being a professor was honorable but poorly paid, professors at the new institute would receive much higher salaries than the norm and live in free housing near the campus.

Yet even all these innovations were not necessarily enough to ensure that KAIS would survive and flourish in the very conservative South Korean society. Park therefore reluctantly exempted male KAIS students from the nation's compulsory three-year military service. This was perhaps the crucial factor that led the country's most talented science and engineering students to apply to this new graduate-only institute. During the 1970s, graduates of the science or engineering departments at Seoul National University often made up more than 70% of each year's KAIS incoming class.

Once accepted, the students were expected to apply themselves more than full-time. Kyung Chong-Min, who entered KAIS in the spring of 1975, remembered:

> After the entrance ceremony, we were "captured" by Prof. Park Song-Bai [chair of the department of electrical engineering] and led to the laboratory: we were not allowed to greet our family members or even take photos. That very night we were forced to attend a seminar until 11 p.m. where our seniors presented their cases, and Prof. Park gave us homework which was due the day after. When we were finally allowed to return to the dormitory after midnight, we wondered whether it was easier to go to the army. ... We usually went to bed around 2 or 3 a.m., and many students found themselves bleeding from the nose in the morning. After a year, we found that we had really learned a lot, and that we could do something for ourselves.[39]

It was not just the hard-working environment but the things they studied that differentiated KAIS students from the outset; at KAIS they learned not only theory but practice, to which no South Korean universities had thus far paid any serious attention. Titles of KAIS master's and doctoral theses produced during the 1970s include "Non-Holographic Spatial Frequency Filtering of the Korean Characters," "Determination of Air Pollution Level by Measuring the Reduction of Contrast due to Atmospheric Mie Scattering," "A Variable Inductance Accelerator for Electric Powered Vehicles," "A Simulation Study of the Emergency Core Cooling System for the Ko-Ri Nuclear Power Plant," "Automatic Telephone Line Testing System Using Micro-Processor," "Design and Fabrication of the Manual-Rice-Transplanter, and Its Linkage Synthesis and Analysis," "A Study on the Thermal Efficiency of the House Heating System," "Design of a Thermometer for Refrigerators," "Recovery of Uranium from Sea Water by Adsorption," "Organoleptic Description of Kimchi Quality and Its Preservation on a Compressed Form," and "Studies on the Process Optimization of Penicillin Fermentation."[40] As these titles indicate, KAIS students learned how to apply theory to practice in the real world during their student years. It is therefore no wonder that the South Korean industry welcomed graduates of KAIS from the very beginning: the head of the GoldStar (later LG Electronics) central research laboratory, for example, asked Professor Park Song-Bai of the Department of Electrical Engineering for the list of its graduates and immediately hired ten of them.[41]

The KAIS Department of Electrical Engineering had strongly supported Terman's original idea since the beginning. Park Song-Bai, the head of the department during the 1970s, was ready to resist any faculty who tried to digress from the institute's original path: he therefore engaged in some memorable scenes with Jeon Mu-Sik, the head of the Department of Chemistry, who argued that KAIS must aim to win Nobel Prizes too.[42] The Department of Electrical Engineering's most successful achievements were usually closely related either to South Korean industry or to the country's defense: between 1975 and 2000, for example, its faculty members succeeded in developing the "Adaptive Delta Modulation System for Defense Applications" (Un Chong-Kwan, 1978), "512-Bit Mask-Programmable ROM" (Kim Choong-Ki, 1981), "Turret Survo Drive System" (Youn Myung-Joong, 1984), "2-Tesla Nuclear Magnetic Resonance Imaging System" (Cho Zang-Hee, 1985), "Ultrasonic Imaging System" (Park Song-Bai, 1986), "45 Mbps Video Codec" (Kim Jae-Kyoon, 1988), "4-Legged Robot" (Bien Zeung-Nam, 1989), "[Mechanism to Detect] the Fourth Infiltration Tunnel Excavated by North Korea" (Ra Jung-Woong, 1990), "KITSAT-1 and KITSAT-2 [South Korea's first two satellites]" (Choi Soon-Dal, 1992 and 1993), "KAICUBE Hanbit-1, a 2-Gflops Parallel Computer" (Park Kyu-Ho, 1994), "Wireless IR Printer-Sharing Unit" (Shin Sang-Young, 1995), "50-MHz Pentium Chip" (Kyung Chong-Min, 1997), "Room-Temperature IR Sensor" (Hong Song-Cheol, 1998), and "Medical Diagnosis Simulator Based on 3-Dimensional Virtual Reality Image" (Ra Jong-Beom, 2000).[43]

In January 1981, the new military government forced KAIS and KIST to merge and form KAIST. This shotgun marriage between a reputable graduate-only institution of higher education and the most prestigious research institute at the time was expected to produce a synergy that would boost the development of science and technology but soon proved to be a failure. Every effort to integrate the two different cultures by means of several organizational restructurings was in vain, and from the mid-1980s on, the two institutions functioned semi-independently under the same roof. The establishment of Korea Institute of Technology (KIT) in 1985, a college intended to train undergraduates for science and technology, added to the confusion. KIT was nominally a part of KAIST but functioned independently and was located in the Daejeon area, not in Seoul. The government finally decided to solve this dilemma: KIST would become an independent institution from 1989 on, using its former title, while KAIST (formerly KAIS) would move to the new campus in Daejeon and merge with KIT.[44] Many KAIST professors objected to the transfer of their main campus from Seoul to Daejeon because they were afraid that KAIST might fail to recruit the best students if it was no longer in the capital. Nonetheless, KAIST's move to Daejeon and merger with KIT occurred as planned, and a new KAIST soon emerged as the core of the Daedok Science Town, where more than 50 research institutes gathered in the early 1990s.

Despite the organizational turmoil in the 1980s and the unwanted move from Seoul to Daejeon in the early 1990s, KAIST continued training the best applied scientists and engineers for South Korean industry as well as producing excellent research results. In the early 1980s, it adopted a special rule for its doctoral candidates: they must publish their research results in prestigious international journals before receiving their degree. Owing to this unique regulation, "students' foreign-language publications rose from 21 in 1979 to 178 in 1989 and to 407 in 1996."[45] This helped KAIST raise its standards above those of any other South Korean university and also guaranteed that its graduates were trained to be not only industry-oriented but also academically competent. It was also during this period that KAIST established its long-term cooperation with industry by offering short-term workshops, building targeted research centers or educational programs sponsored by private companies, and organizing special interuniversity educational programs for specific industries. For instance, during the 1980s and 1990s, KAIST professors regularly offered two- and three-day workshops to disseminate top-notch knowledge and technology; in the 1990s, Hyundai Motor Company, Hyundai Heavy Industries, and Samsung Electronics supported the establishment of specific research centers or educational programs for their companies; and in 1995, the Integrated-Circuit Design Education Center (IDEC) was established to provide South Korean industry with chip designers.[46]

Ironically, it was during the turbulent 1980s and 1990s that KAIST became firmly rooted in South Korean higher education and began to influence its overall approach to teaching science and engineering. During the 1970s and early 1980s, South Korean universities saw KAIST not as their equal but as an exception, mainly because KAIST has no undergraduate program; in fact, they often ridiculed KAIST as a vocational school while maintaining that they themselves were ivory towers. However, from the mid-1980s on, as KAIST's graduates began to occupy key positions in industry and research institutes, South Korean universities slowly began to emulate KAIST's educational model. It was none other than the College of Natural Sciences and the College of Engineering at SNU that were most eager to follow KAIST's example. In the late 1970s, both colleges had begun to aggressively recruit new faculty members trained in the United States and to emphasize research results both for new hires and when faculty came up for promotion. Beginning in the early 1980s, both colleges greatly strengthened their graduate programs and began to produce their own doctoral graduates. The College of Engineering even began working with specific companies to "win markets": from the beginning of the 1990s, for example, the Inter-University Semiconductor Research Center (established in 1985) began to work with Samsung Display Device, GoldStar, Hyundai Electronics, and other electronics companies; and the Institute of Advanced Machines and Design (established in 1989) organized a consortium to support about ten large companies, such as Hyundai Motor Company, Samsung Electronics, and Hyundai Heavy Industries.[47] By the early 2000s, several of these SNU departments had become true rivals of their counterparts at KAIST, in terms of both their number of publications and their cooperation with the industry. Many other universities in South Korea followed SNU's path and became eager to cooperate with industry.

The advent of the new century therefore posed a new challenge to KAIST: how could it maintain its leading role in science and engineering in South Korea, as it had done in the last three decades of the twentieth century? KAIST's solution was to hire foreigners as its presidents—an innovation as radical as Park's decision in 1971 to establish KAIS separately from the existing educational system.[48] In May 2004, KAIST's board of trustees appointed Robert B. Laughlin, a 1998 Nobel Prize Winner in physics, as its new president. Laughlin aimed to make KAIST a true leading science and engineering university "in the world," like Stanford University, where he had worked for decades. He proposed several radical changes to achieve this goal, which included enlarging its student body from 7,000 to 20,000, establishing pre-med, pre-law, and pre-MBA programs, using English as the official language on campus, employing more non-Korean faculty members, and applying stricter rules to the tenure review process.[49] Unfortunately, Laughlin soon faced strong objections from within KAIST because his proposals did not fit well with the original KAIST spirit, and some of them even went against it. His unequivocal way of expressing his opinions and his arrogant manner often made enemies rather than allies within the campus community. After struggling for two years, he stepped down in the summer of 2006.

Surprisingly, KAIST's board of trustees selected another foreigner, the Korean American Suh Nam-Pyo, as Laughlin's successor. Suh had returned to the Massachusetts Institute of Technology (MIT), his alma mater, in 1970, where he had headed the Department of Mechanical Engineering for a decade in the 1990s. He had also served as the research director responsible for engineering at the National Science Foundation under the Ronald Reagan administration during the 1980s. This highly accomplished engineering professor was chosen as a fixer to rescue KAIST from its turmoil.

Suh did not discard the radical changes that Laughlin had started, as some conservative KAIST members hoped, but instead accelerated most of them, with more resolution. He enthusiastically promoted the use of English in teaching and everyday life on campus; required that deans and department heads hire more foreigners as faculty members; radically increased the number of undergraduates and strengthened undergraduate programs; and established several new schools for bio and brain engineering, medical engineering, and green traffic engineering. Suh's most influential and controversial policy was to further strengthen the tenure review system: he demanded that those applying for tenure prove themselves to be world-class scholars in their field. If they failed, they had to leave the KAIST campus after a grace period. When the KAIST personnel committee denied tenure to some applicants in the fall of 2007, it was quite a shock to the entire university community in South Korea.[50] Opposition to Suh then began to form, both inside and outside of KAIST, and began to criticize his policies severely. Suh successfully defended himself and his reform programs in his first term, but the attack on him intensified from the beginning of his second term. He finally resigned in the spring of 2013, about a year and a half before the end of his second term.

Suh's successor was Kang Sung-Mo, another Korean American. Kang had worked as an electrical engineering professor for more than two decades at several American universities and had recently served as chancellor of the University of California, Merced. Kang focused on stabilizing KAIST while continuing the reforms that the previous two presidents had initiated. By the end of his tenure in the spring of 2017, KAIST had become more diversified in both faculty members and students, had a bigger and more vibrant undergraduate program, and was engaging in stronger, more science-based research. In short, thanks to the reforms of the three "foreign" presidents between 2004 and 2017, KAIST successfully transformed itself from a special training institute for South Korean industry into a science- and engineering-focused university like MIT or Stanford University.

TRIANGULAR RELATIONSHIP AMONG THE GOVERNMENT, INDUSTRY, AND ACADEMIA FOR THE DEVELOPMENT OF SEMICONDUCTORS[51]

To understand the rapid, almost miraculous development of the semiconductor industry in South Korea during the last half century, it is essential to analyze the country's unique framework of the collaboration among government, industry, and academia during the period. This triangular relationship was differently balanced in other countries. In the United States, government-supported R&D (particularly from the military and the space program) provided the infant semiconductor industry with important stimuli until the early 1970s, when the initiative shifted to industry, and the role of academia also increased. In Japan, industry was the major player from the beginning of the development of semiconductors, while both government and academia were largely consigned to supporting the industry.[52] The severe competition between the American and Japanese semiconductor industries during the 1980s made each government ask its industry to form a consortium for the development of very large-scale integration (VLSI) semiconductors, but this was a rare and one-time intervention.[53] In Taiwan, South Korea's rival in semiconductors, the government's initiative and dominant role quickly passed to industry, especially after the Taiwan Semiconductor Manufacturing Company (TSMC) was established in 1987, and the academy made some contributions to the process.[54]

In South Korea, the initiative and the power balance among the three actors changed dramatically between 1965 and 2000 (Figure 1.1). In the early phase between the mid-1960s and the mid-1980s, the South Korean government was the true prime mover in semiconductor development by setting specific goals, establishing related research institutes, and pushing both industry and academia, which were less enthusiastic or even passive in comparison. From 1983 on, however, it was industry that led the rapid and successful development of semiconductors, and the academy's role and contribution also steadily rose, while the government's overly dominant influence declined rapidly.

At the outset, it was the South Korean government that seriously considered semiconductors to have a promising future in the South Korean economy. As already mentioned, in 1967 the South Korean government had invited Kim Wan-Hee to draft a plan for the development of the electronics industry in the country. Following Kim's strong advice, President Park Chung-Hee had ordered preparation of the necessary laws and detailed plans, which were enacted as the Electronics Industry Promotion Act (1968) and the Eight-Year Plan for the Electronics Industry Promotion (1969). On January 12, 1973, President Park declared that his government would focus both the Third Five-Year Economic Development Plan (1972–1976) and the following Fourth Plan (1977–1981) on heavy and chemical industry, which included iron/steel, non-ferrous metals, chemicals, machinery, shipbuilding, and electronics. The oil crisis in the fall of 1973 then forced the South Korean government to pay more attention to the electronics industry, which consumed relatively less energy than other heavy industry. As a result, the final version of the Fourth Five-Year Economic Development Plan promoted electronics from a minor field under machinery in the Third Economic Plan to the second most important field under heavy industry and gave special emphasis

1967-1982

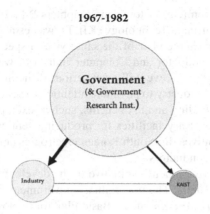

1983-2000

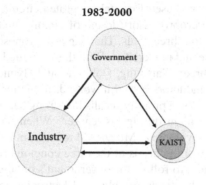

2001 –

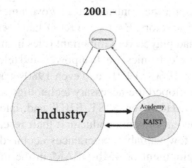

FIGURE 1.1   Change over time in the triangular relationship among government, industry, and the academy in South Korea.

Source: Dong-Won Kim, "Transfer of 'Engineer's Mind,'" 122.

to the development of semiconductors and computers.[55] In 1976, a special institute, Korea Institute of Electronics Technology (KIET), was established to accelerate the government's efforts. In September of the same year, a special report, "Electronics Promotion Plan: Semiconductor and Computer Industry," warned that "if we don't start semiconductor right now, we will meet serious difficulties soon."[56] Park's government not only forced industry to pay more attention to semiconductors but offered it special low-interest funding and tax benefits, such as exemption from import taxes when importing the necessary facilities for producing semiconductors.[57] Only with these government incentives did South Korean electronics companies begin to move toward producing semiconductors.

The South Korean government's initiative in the development of semiconductors continued in the early 1980s. The new military government, which seized power by a series of coups in 1979–1980, issued a Basic Plan for the Promotion of Electronics on January 14, 1981, in which the semiconductor industry was identified as "the core industry" to promote.[58] The government would offer special funding for the research, development, and manufacturing of selected items, such as digital integrated circuits and small computers: in the case of digital integrated circuits, the list indicated "I-chip Micro-computer, Microprocessor (design of circuit), and Memory (wafer process technology)" as the top three goals. The plan also proposed adding new courses on semiconductors at selected universities. With these detailed plans, the government began to press the big three—Samsung, GoldStar, and Hyundai—"to make serious commitments" to semiconductors.[59] And it worked. In his recent memoir, President Chun Doo-Hwan (1980–1988) proudly recalls his strong support of and critical role in semiconductor development during his tenure.[60] When Samsung, GoldStar, and Hyundai hesitated about developing 4Mb DRAM together in 1986 because of its astronomical cost, he summoned the heads of the three conglomerates to the presidential residence and pressed them to follow the government's policy.[61] They had no choice but to do so, along with the newly established Electronics and Telecommunication Research Institute (ETRI, the former KIET), and succeeded in developing 4Mb DRAM about fifteen days before Chun's retirement in February 1988.[62] By then, the initiative in semiconductor R&D had already moved from government to industry.

It is interesting to examine the contribution of government research institutes to the development of semiconductors. Ever since KIST had been established in 1966, it and its spin-off institutes had played very important roles in many fields, such as steel, automobiles, shipbuilding, electronics, nuclear power, and telecommunications.[63] The standard process during the 1960s, 1970s, and even 1980s was that a specific government research institute developed the necessary technology and then transferred it to the corresponding industry. However, KIST, KIET, and, later, ETRI had done relatively little in the development of semiconductors: their research on semiconductors was intermittent, and the few valuable experiences acquired were not accumulated there properly.[64] The development of 4Mb DRAM in the mid-1980s was therefore the first major—but also the last—achievement in semiconductors by the government research institutes. Especially after the mid-1980s, the major South Korean semiconductor companies wanted to develop the necessary technology by themselves, and they had enough manpower and capital to do so: as a result, they declined the government's offer to collaborate with them in developing 64Mb and 256Mb

DRAMs, as has been done for the 4Mb DRAM.[65] ETRI's own history indicates that its major contribution after its 1986 inception was not in semiconductors but in telecommunications, such as in the developments of time division exchange (TDX) technology and code division multiple access (CDMA) technology.[66]

In contrast to the South Korean government's early efforts to develop semiconductors, the country's electronics industry was not seriously interested in semiconductors until the mid-1970s. Instead, foreign companies—such as Fairchild, Signetics, Motorola, IBM, and Control Data—came and built their own production lines in South Korea to make transistors and various semiconductors for export.[67] These American companies built their production lines in South Korea in order to use its cheap labor. Few technology transfers were made by these companies during the 1960s and 1970s, and most of them moved to other countries in the 1980s. However, skilled laborers who had worked for them later transferred to South Korean semiconductor companies and contributed to the rapid development of semiconductor chips.

With pressure from the government and the rapidly increasing demand for semiconductors in the manufacture of radios, televisions, microwaves, and watches, the South Korean electronics industry began to develop semiconductor chips from the mid-1970s on. GoldStar, Anam Semiconductor, and Samsung Electronics began to produce transistors and a few integrated chips in these years, but all electronics companies still imported more advanced semiconductor devices from Japan and other countries. From the point of view of South Korean electronics companies, it was too risky to develop more advanced semiconductor chips by themselves because they possessed neither a high enough level of technology nor sufficient capital. The establishment of Hankuk Semiconductor and its immediate failure in 1974 indicated both the high demand for advanced semiconductor chips and the unprepared environment in South Korea in the mid-1970s. Established by Kang Ki-Dong, a Korean American semiconductor engineer, Hankuk Semiconductor succeeded in producing the complementary metal-oxide semiconductor (CMOS) chip for digital watches, the first large-scale integration (LSI) chip in South Korea, but faced financial difficulties within a year.[68] Lee Kun-Hee, the future head of the Samsung conglomerate, privately bought 50% of the company's stock, paving the way for its future success (Figure 1.2). In 1977 Samsung bought the remaining stock and changed the company's name to Samsung Semiconductor, though this new business produced few significant results until 1983.[69]

This mediocre situation changed dramatically in the early 1980s, when the new military government strongly pressed the electronics industry to develop and manufacture more advanced, VLSI-level semiconductors. Lee Byung-Chul (founder of the Samsung conglomerate) carefully studied the market and necessary technology for a year and finally made the decision to enter the memory business (DRAM) in February 1983.[70] With technical assistance from Micron Technology (US) and Sharp (Japan), Samsung's two research teams, one in the suburbs of Seoul and the other in Silicon Valley, each succeeded in developing 64Kb DRAM by November of that year. The Samsung conglomerate decided to invest about $133 million—an astronomical amount at that time, when South Korea's total annual exports were just over $10 billion—to build a production line of 64Kb DRAM. Samsung also succeeded in developing 256Kb DRAM by the end of 1984, 1Mb DRAM by July 1986, and

FIGURE 1.2   The development of the semiconductor industry in South Korea owed much to both Lee Byung-Chul and his son and successor Lee Kun-Hee of Samsung Conglomerate. Both Kim Choong-Ki and Morris Chang of TSMC indicated that the two Lees' decisive leadership was the secret of South Korea's success in semiconductors. In the picture, Lee Byung-Chul (fourth from the left in the front row, with a red tie) and Lee Kun-Hee (third from the left in the front row) inspect the building site for the first chip factory in 1983.

Source: *Hankuk Kyungje*, October 28, 2015.

4Mb DRAM (with GoldStar and Hyundai) by February 1988.[71] Steep drops in the price of DRAMs—for example, the price of a 64Kb DRAM dropped from $3.50 in mid-1984 to $0.30 in mid-1985—led the Samsung conglomerate to the edge of bankruptcy, but the personal computer (PC) boom revived the demand for 256Kb DRAM and erased the entire accumulated deficit by 1988. GoldStar and Hyundai, Samsung's domestic rivals, followed Samsung's model of developing and manufacturing DRAM and other VLSI semiconductors from the mid-1980s on. By the time Samsung Electronics announced the development of 64Mb DRAM for the first time in the world in September 1992, the South Korean semiconductor industry was no longer either a copycat or a "fast follower" but a major player in the world market.

What, then, did the academy contribute to the development of semiconductors? Many books and articles emphasize both government and industry's contributions but strangely neglect the role of the academy.[72] Nevertheless, it was the academy that provided both government research institutes and the semiconductor industry with the manpower required to carry out semiconductor research. To develop a series of DRAMs, for example, Samsung Electronics needed not only US-trained Chin

Dae-Je, Hwang Chang-Kyu, and Kwon Oh-Hyun as heads of each developing team but also dozens or hundreds of well trained rank-and-file engineers and scientists. Who supplied them? The academy, headed by KAIST!

South Korean industry and the government began to realize the importance of the academy only in the mid-1980s, when not only Samsung, GoldStar, and Hyundai but several medium-sized companies entered the semiconductor market. This immediately created a serious shortage of skilled manpower, and many universities hurriedly established programs to train semiconductor specialists. However, it required a huge amount of money to set up even a "minimum facility" tailored to semiconductor training, which most South Korean universities could not afford. The only solution was to receive support from government or industry. The government had sufficient funding only for KAIST or SNU, and the electronics industry preferred to support proven programs, such as Kim Choong-Ki's group, rather than new ones.[73] Therefore, industry money and pleas for cooperation concentrated on KAIST between 1985 and 1995, and KAIST carried out several joint projects and education programs for semiconductors during this period.

The strength of the academy in the triangular relationship increased in the early 1990s. It was during this period that SNU's semiconductor group presented a serious challenge to KAIST's dominance. Despite SNU's organizational shortcomings, preference for theory over practice, and lack of eminent leaders in semiconductor research, its unflinching fame, sheer size, and location in Seoul were enough to make both the South Korean government and industry invest great sums in its semiconductor research and education. SNU also began to aggressively recruit young semiconductor specialists as its faculty members.[74] Several KAIST professors of electrical engineering soon recognized this threat but had no choice except to work harder and do better than the competition.[75] The government's much-enlarged R&D budget and the electronics industry's aggressive investment in semiconductor education also enabled other universities, such as Pohang University of Science and Technology (POSTECH), Hanyang University, Yonsei University, and Korea University, to develop and expand their special programs on semiconductors during the 1990s. The establishment of the Integrated Circuit Design Education Center (IDEC) at KAIST in 1995 and its successful operation are another good example of the academy's contribution to the development of semiconductors: as a consortium of several universities, companies, and government research institutes, IDEC sought to "contribute to the Korean semiconductor industry by promoting collaborations between universities and industries," and was headed by Kim's former student Kyung Chong-Min (profiled in Chapter 3).[76] With these greatly enhanced educational bases, the academy could provide industry with enough manpower during the 1990s and thereafter.

In short, academia's primary contribution to the development of semiconductors in South Korea has been in education, not research. The exception to this occurred between 1975 and 1995, when Kim Choong-Ki's laboratory at KAIST actually led the country's semiconductor research because industry didn't yet have enough skilled manpower to conduct its own research. By the mid-1990s, however, industry possessed more semiconductor specialists than the academy did: in 1995, for example, Samsung Electronics' semiconductor R&D center had 100 engineers with doctorates and 400 with master's degrees on its payroll.[77] Nevertheless, industry has continued to generously support research in the universities, largely because the primary goal of industrial research was (and still is) not to make new discoveries but to solve immediate problems. As one retired Samsung Electronics director indicated, it's cheaper and safer for a company to give a risky project to a university and see what happens than to carry out the project itself.[78] The academy is also an important place where more fundamental, large-scale research, such as nano-technology, is carried out with government support: accordingly, the Korea National NanoFab Center (NNFC) opened at KAIST in March 2005.[79] This division of research labor between the electronics industry and the academy seems to have worked well to date.

NOTES

1   Young-Iob Chung, "The Impact of Chinese Culture on Korea's Economic Development," in Hung-Chao Tai (ed.), *Confucianism and Economic Development: An Oriental Alternative?* (Washington, D.C.: The Washington Institute Press, 1989), 149–165 on 155.

2   Sang-Woon Jeon, *A History of Korean Science and Technology* (Singapore: National University of Singapore Press, 2011).

3   Michael J. Seth, *A Concise History of Korea*, 2nd ed. (Lanham, Md.: Rowman & Littlefield, 2016), 148.

4   For the history of Korean ceramics, see Chang Yang-Mo et al., *Uri Doja Iyagi* (Ichon, Kyeongkido: Eksupo, 2004); Yoon Yong-I, *Uri yet Dojagiui Arumdaum* (Kyeongki-do, Paju-si: Dolbegae, 2007); Pang Pyoug-Son et al., *Hanbandoui Heukdojagiro taeonada* (Seoul: Kyeongjin Munhwasa, 2010); and Kang Kyung-Seok, *Hanguk Dojasa* (Seoul: Yegyeong, 2012).

5   Carter Eckert, "Korea's Transition to Modernity: A Will to Greatness," in Merle Goldman and Andrew Gordon (eds.), *Historical Perspectives on Contemporary East Asia* (Cambridge, Mass.: Harvard University Press, 2000), 119–154 on 123.

6   Yoon, *Uri yet Dojagiui Arumdaum*, 249–250.

7   For a brief history of Shim Soo-Kwan pottery, see https://chinjukanpottery.com, www.chin-jukan.co.jp/history.html and https://peopleofkagoshima.com/arts-and-crafts/satsuma-ware-chin-jukan-kiln/. For Yi Sam-Pyeong in the Sueyama Shrine, see http://arita-toso.net/#sp_section1 (searched on August 7, 2021).

8   Son Seung-Chul, "Imjinwaeran Piroin," *Encylopedia of Korean Culture*, http://encykorea.aks.ac.kr/Contents/Item/E0073566; Son Seung-Cheol, "Joseontongshinsa Piroin Swaehwankwa geuhanke," *Jeonbuk Sahak, 42* (2013), 167–200; Cho Yong-Joon, *Ilbon Dojagi Yeohaeng: Kyushu 7dae Chosun Gama* (Seoul: Do Do Publication, 2016), 146–147. Not only was the Chosun government unenthusiastic about bringing these captives back to Korea, but those who returned were despised because Confucianism never forgave those who became captives by an invading state.

9   For the positive influence of Confucianism on South Korean economic development, see Ezra F. Vogel, *The Four Little Dragons: The Spread of Industrialization in East Asia* (Cambridge, Mass.: Harvard University Press, 1991), 92–101, and Linsu Kim, *Imitation to Innovation: The Dynamics of Korea's Technological Learning* (Boston: Harvard Business School Press, 1997), 68–69, 72, 73, and 204.

10 Sungook Hong, "The Relationship between Science and Technology in Korea from the 1960s to the Present Day: A Historical and Reflective Perspective," *East Asian Science, Technology and Society, 6* (2012), 259–265 on 260.

11 Dong-Won Kim and Stuart W. Leslie, "Winning Markets or Winning Nobel Prizes? KAIST and the Challenge of Late Industrialization," *Osiris, 13* (1998), 154–185. The word "Technology" was only added in 1981, when renaming the institute KAIST.

12 Joong-Ang Ilbo, "Guknae Hakja 68%ga Gwahakagye Munjejeom 'Igongge Chabyeol' kkoba," *Joong-Ang Ilbo* (September 18, 2012), www.joongang.co.kr/article/9354007#home (searched on March 14, 2019). Discrimination against science and technology was chosen as the largest obstacle (68%). The second largest (36%) was the poor research environment.

13　Chosun Ilbo, "Gukmin boda Sujuni hwolssin najeon Saibi Bosujeongchi-ui Silpae," *Chosun Ilbo* (November 28, 2016), www.chosun.com/site/data/html_dir/2016/11/28/2016112800222.html (searched on March 14, 2019).

14　Park No-Chun, "Munchebanjeong," *Encyclopedia of Korean Culture*, http://encykorea.aks.ac.kr/Contents/Item/E0019697 (searched on March 16, 2022).

15　For the history of Keijo Imperial University, see Seoul National University, *Seouldeahakgyo 50nyeonsa [Fifty-Year History of Seoul National University]*, *Vol. 1* (Seoul: Seoul National University Press, 1998), 6.

16　The best analysis of the combined science and engineering faculty at Keijo Imperial University is found in Kim Geun-Bae, *Hankuk Geundae Gwahakgisulinryeok-eu Chulhyeon* (Seoul: Munhakgwa Jiseongsa, 2005), 450–480.

17　For detailed discussions of Korean scientists and engineers during the Japanese occupation period, see Park Seong Rae et al., *A Study of the Formation of Modern Scientists and Engineers in Korea* (in Korean) (Daejeon: Korea Science and Engineering Foundation Report, 1995), and idem, *A Study on the Foundation of Modern Scientists and Engineers in Korea II: Studying in America* (in Korean) (Daejeon: Korea Science and Engineering Foundation Report, 1998).

18　Seoul National University, *Seouldaehakgyo 50nyeonsa*, *Vol. 1*, 3–41.

19　Among those who left South Korea were Korea's two most important scientists of the first half of the twentieth century, the physical chemist Ree Tai-Kyu and the chemical engineer Li Seung-Ki. Both had returned to South Korea in 1945 from teaching posts at Japan's Kyoto Imperial University and had made important contributions to the establishment of SNU's science and engineering faculties. However, in 1948, Ree emigrated to the United States to continue his research with Henry Eyring at the University of Utah, whereas Li moved to North Korea to continue his research on synthetic fiber. For more details, see Dong-Won Kim, "Two Chemists in Two Koreas," *Ambix, 52:1* (2005), 67–84.

20　For more details about the Atomic Energy Research Institute, see Ko Dae-Seung, "The Establishment of the Office of Atomic Energy" (in Korean) (master's thesis, Seoul National University, 1991). The Minnesota Project brought many SNU professors in science, engineering, medicine, and agriculture to the University of Minnesota for advanced training, mostly to earn master's degrees. For the Minnesota Project, see Seoul National University, *Seouldaehakgyo 50nyeonsa, Vol. 1*, 92–98, and idem, "Minnesota Project, Seouldaehakgyo Jaegeoneul wihan Noryeok," www.snu.ac.kr/about/history/history_record?md=v&bbsidx=131605 (searched on September 7, 2022).

21　Korea Electronics Association, *Gijeokui Sigan, 50 (1959–2009): The Miraculous Time* (Seoul: Korea Electronics Association, 2009), 48.

22　Ibid., 46.

23　LG Electronics, *LGjeonja 50nyeonsa [LG Electronics 50-Year History]*, *Vol. 4: English Edition* (Seoul: LG Electronics, 2008), 26.

24　Ibid., 27.

25　For short summaries of the five-year economic plans, see Kim Il-Gyeong, "Gyeongjaegaebalgyehoek," *Encyclopedia of Korean Culture*, http://encykorea.aks.ac.kr/Contents/Item/E0002782; National Archives of Korea, "Girokeuro boneun Gyeongjaegaebal 5gaenyeon Gyehoek," https://theme.archives.go.kr//next/economicDevelopment/overview.do (searched on March 16, 2022).

26　For the record of South Korea's major economic growth, such as GNP and GDP, see Macrotrends, "South Korea (Economy)," www.macrotrends.net/countries/KOR/south-korea/gnp-gross-national-product (searched on March 16, 2022). For the export of tungsten in the 1950s and 1960s, see Weekly Chosun, "Sege Choebinguk Hankuk eotteoke Suchul Ogangi doeeotna?" *Weekly Chosun* (November 4, 2015), http://weekly.chosun.com/client/col/col_view.asp?Idx=244&Newsnumb=20151118668 (searched on March 18, 2022).

27　For the contribution of Kim Wan-Hee to the development of electronics in South Korea, see Korea Electronics Association, *Gijeokui Sigan, 50*, 106–108, and Myung Oh and James Larson, *Digital Development in Korea* (New York: Routledge, 2011), 36–37.

28　For the early history of the company, see Samsung Electronics, *Samsungjeonja 30nyeonsa [Thirty-Year History of Samsung Electronics]* (Seoul: Samsung Electronics, 1999), 108–152.

29　Ibid., 129.

30　Linsu Kim, *Imitation to Innovation*.

31　Ibid., chapter 7, especially 153–161.

32 Ibid., 14.
33 This section is based on Dong-Won Kim and Stuart W. Leslie, "Winning Markets or Winning Nobel Prizes?" Some parts are reproduced from this article.
34 For the early history of KIST, see KIST, *KIST 25nyeonsa* [*Twenty-Five Year History of KIST*] (Seoul: KIST, 1994).
35 Ministry of Science and Technology, "Policy and Strategy for Science and Technology" (1975), 40, in Frederick E. Terman Papers, Department of Special Collections, Stanford University, Stanford, CA, SC 160 Vi, 20/7.
36 Dong-Won Kim and Stuart W. Leslie, "Winning Markets or Winning Nobel Prizes?"
37 Frederick E. Terman et al., "Survey Report on the Establishment of the Korea Advanced Institute of Science," in Terman Papers, SC 160. X313, 9.
38 Ibid.,17 and 26.
39 Dong-Won Kim and Stuart W. Leslie, "Winning Markets or Winning Nobel Prizes?" 170.
40 These titles are selected from the list of master's and doctoral theses during the 1970s. See https://library.kaist.ac.kr/search/ctlgSearch/thesis.do (searched on July 8, 2020).
41 Dong-Won Kim and Stuart W. Leslie, "Winning Markets or Winning Nobel Prizes?" 172.
42 Ibid., 176–177. For a brief introduction to Park Song-Bai, see Department of Electrical Engineering, "Park Song-Bai," in Faculty, https://ee.kaist.ac.kr/en/professor_s6?language=en&combine=Park%2C%2BSong-Bae (searched on March 24, 2022).
43 These are some examples from the list of the most important works that the faculty members of the Department of Electrical Engineering produced between 1973 and 2008. Department of Electrical Engineering, KAIST, *2008/2009 Annual Report*, "Brief History," https://ee.kaist.ac.kr/en/node/21 (searched on March 24, 2022), 10–11.
44 For more details about KAIST's move to Daejeon, see Dong-Won Kim and Stuart W. Leslie, "Winning Markets or Winning Nobel Prizes?" 178–180.
45 Ibid., 173.
46 Kim Dong-Won et al., *Hankukgwahakgisulwon Sabansegi: Miraereul hyanghan kkeunimeopneun Dojeon* [*The First Quarter Century of KAIST: An Endless Challenge to the Future*] (Daejeon: KAIST, 1996), 123, 213, 232.
47 For the activities of these two SNU institutes during the 1980s and 1990s, see Seoul National University, *Seouldaehakgyo 50nyeonsa, Vol. 2*, 472–475 and 491–495.
48 South Korea has had very few "non-Korean" university presidents since its independence in 1945. When SNU was founded in 1946, its first president was an American, Captain Harry B. Ansted, but this happened when South Korea was under the control of the American Occupation Forces. Sogang University, established in 1960 as a Jesuit college, also appointed foreign Catholic priests as its first three presidents during the 1960s and 1970s.
49 Laughlin's final plan was announced in December 2004 with the title "KAIST Vision 2005." For more details of Laughlin's ambitious plans, see Kim Dong-Won (ed.), *Miraereul hyanghan kkeunimeopneun Dojeon: KAIST 35nyeon* (Daejeon: Somang, 2005), 213–222.
50 Hankuk Kyungje, "KAIST Jeongnyeonbojangsimsa daegeo Talak, Gyosusahoe Cheolbaptong kkaejina?" *Hanguk Kyungje* (October 12, 2007), https://sgsg.hankyung.com/article/2007101143171 (searched on April 4, 2022). About 40% of the applicants were denied tenure that year. Although many of those who failed had another chance to apply, this was an unprecedently high rate of failure by South Korean standards.
51 This section is a reprint, with some modifications, of the section "How Do We Position Kim Choong-Ki in the Triangular Relation of Government-Industry-Academy?" in Dong-Won Kim, "Transfer of 'Engineer's Mind': Kim Choong-Ki and the Semiconductor Industry in South Korea," *Engineering Studies, 11:2* (2019).
52 Daniel I. Okimoto, Takuo Sugano, and Franklin B. Weinstein (eds.), *Competitive Edge: The Semiconductor Industry in the U.S. and Japan* (Stanford: Stanford University Press, 1984), "Chapter 2: Background."
53 For American and Japanese efforts to develop VLSI semiconductors, see ibid.; and Kiyonori Sakakibara, *From Imitation to Innovation: The Very Large Scale Integrated (VLSI) Semiconductor Project in Japan* (Cambridge, Mass.: MIT Press, 1983).

54 For the development of the semiconductor industry in Taiwan, see John A. Matthews and Dong-Sung Cho, *Tiger Technology: The Creation of a Semiconductor Industry in East Asia* (Cambridge: Cambridge University Press, 2000), especially "Chapter 4: A Cat Can Look at a King: How Taiwan Did It"; and Chun-Yen Chang and Po-Lung Yu (eds.), *Made By Taiwan: Booming in the Information Technology Era* (Singapore: World Scientific, 2001).

55 Government of the Republic of Korea, *Je4cha Gyeongjegaebal 5gaenyeon-gyehoek* [*The Fourth Five-Year Economic Development Plan (1977–1981)*] (Seoul: Government of the Republic of Korea, 1976), 59 and 88.

56 Government of the Republic of Korea, *Jeonjagongeop Yukseong-gyehoek* [*Electronics Promotion Plan*] (Seoul: President Secretarial Office, September 25, 1976), 7. The report was prepared for the regular meeting of the ministers of economics, commerce, science and technology, etc.

57 Ibid., 13.

58 Ministry of Government Administration, *The Official Gazette, No. 8740* (in Korean) (Seoul: Ministry of Government Administration, January 14, 1981), 47–60 on 47.

59 Myung Oh and James F. Larson, *Digital Development in Korea*, 26. The government planned to invest about $400 million in the development of semiconductors, which was "ten times larger than anything attempted up to then."

60 Chun Doo-Hwan, *Chun Doo-Hwan Hoegorok, Vol. 2* (Seoul: Jajaknamusup, 2017), 218–225. While collaborating to develop 4Mb DRAM under ETRI's management, Samsung and GoldStar secretly carried out their own programs to develop 4Mb DRAM with different methods.

61 For more details about the development of 4Mb DRAM during the 1980s, see Sangwoon Yoo, "Innovation in Practice: The 'Technological Drive Policy' and the 4Mb DRAM R&D Consortium in South Korea in the 1980s," *Technology and Culture, 61:2* (2020), 385–415.

62 KIET was merged with other related institutes to become ETRI in March 1985. For the history of KIET and ETRI, see ETRI, *ETRI 30nyeonsa* [*Thirty-Year History of ETRI*] (Daejeon: ETRI, 2006), or idem, *ETRI 35nyeonsa* [*Thirty-Five Year History of ETRI*] (Daejeon: ETRI, 2012) (www.etri.re.kr/kor con/sub7/sub7_13.etri).

63 For KIST and its sister institutes in the 1970s, see Moon Man-Yong, "Hankukgwahakgisulyeonkuwon (KIST) ui Byeoncheon-gwa Yeonkuhwaldong," *Hankuk Gwahaksa Hakhoeji, 28:1* (2006), 81–115.

64 Kim Choong-Ki, "Guknae Bandoche Gongeop-eu Baljeon Hoego," *Jeonjagonghakhoeji, 13:5* (1986), 436–438.

65 Linsu Kim, *Imitation to Innovation*, 162–163.

66 ETRI, *ETRI 35nyeonsa*, 48–49, 147–153. ETRI claimed that it created about a USD 141 billion-worth economic effect between 1976 and 2011, among which the semiconductor share was about 12.5%, while telecommunication technology's was about 66%.

67 Korea Electronics Association, *Gijeokui Sigan*, 50, 123–127.

68 For the establishment of Hankuk Semiconductor, see Samsung Semiconductor and Telecommunications, *Samsung Bandochetongsin 10nyeonsa* [*Ten-Year History of Samsung Semiconductor and Tele-communications*] (Seoul: Samsung Bandochetongsin,1987), 88–92; Kim Choong-Ki, "Guknae Bandoche Gongeop eu Baljeon Hoego," 439. Kang Ki-Dong, the founder of Korea Semiconductor, wrote a private memoir (*Hankang-eu Gijeok, Hankuk Bandoche: Sijakeu Jinsil*) and uploaded it to the website (www.kdkelectronics.com/korea_semi_pdf/kdk-2014-0823mm.pdf). He claims that the Samsung conglomerate stole his company.

69 Kim Kwang-Ho, who was managing the semiconductor division from 1979 to the end of the 1980s, remembered the poor, almost devastating conditions, with little support in the first five years. See Korea Electronics Association, *Gijeokui Sigan*, 50, 179–180.

70 Lee Byung-Chul, *Hoamjajeon* [*Autobiograpny of Lee Byung-Chul*] (Seoul: Joong-Ang Ilbo, 1986), 233–244; Samsung Semiconductor and Telecommunications, *Samsung Bandochetongsin 10nyeonsa*, 187–189; Samsung Electronics, *Samsungjeonja 30nyeonsa*, 198–201.

71 For the development of semiconductors at Samsung Electronics in the 1980s and 1990s, see Samsung Semiconductor and Telecommunications, *Samsung Bandochetongsin 10nyeonsa*, 187–210, 256–278; Samsung Electronics, *Samsungjeonja 30nyeonsa*, 198–212; Han Sang-Bok, *Oebaljajeongeoneun Neomeojiji anneunda* (Seoul: Haneulchulpansa, 1995); Kang Jin-Ku, *Samsungjeonja Sinhwawa keu Bikyeol* (Seoul: Koryeowon, 1996); Song Sungsoo, "The Growth of Samsung's Semiconductor Sector

and the Development of Technological Capabilities" (in Korean), *Hangukgwahaksa Hakhoeji*, *20:2* (1998), 151–188.

72 For example, see Linsu Kim, *Imitation to Innovation*, 151–167; Youngil Lim, *Technology and Productivity: The Korean Way of Learning and Catching Up* (Cambridge, Mass.: The MIT Press, 1999), 102–117; and Myung Oh and James Larson, *Digital Development in Korea*, 35–39. See also Samsung Electronics, *Samsungjeonja 30nyeonsa*, 199–201, and Kang Jin-Ku, *Samsungjeonja Sinhwawa keu Bikyeol*, 203–219.

73 In fact, there was another way to train necessary manpower for semiconductor development. Samsung Electronics founded a "corporate college" within the company in the late 1980s to train or retrain various engineers, including semiconductor specialists, but it was not so successful.

74 Between 1988 and 1996, for example, SNU's Department of Electronic Engineering alone recruited six semiconductor specialists, who made up almost half of its new recruits during the period.

75 Lee Yong-Hoon remembers that in the mid-1990s SNU's Department of Electronic Engineering came to possess more professors and attract more research funds from government and industry than his own Department of Electrical Engineering at KAIST. See Interview with Lee Yong-Hoon (conducted by Jeon Chihyung, June 27, 2016), Department of Electrical Engineering (KAIST), 10.

76 Department of Electrical Engineering (KAIST), https://ee.kaist.ac.kr/en/node/10981 (searched on December 18, 2020).

77 Youngil Lim, *Technology and Productivity*, 100–101.

78 Interview with an anonymous director of Samsung Electronics by Dong-Won Kim (January 23, 2011, in Boston).

79 For more details about the Korea National NanoFab Center at KAIST, see www.nnfc.re.kr. The Semiconductor Institute at SNU entered into a cooperative contract with the NNFC in May 2005.

Part I

Mentor

2 Godfather of the South Korean Semiconductor Community
Kim Choong-Ki

On February 23, 1993,[1] the Samsung Welfare Foundation announced that its prestigious Ho-Am Prize for Science and Engineering would be awarded to Kim Choong-Ki of Korea Advanced Institute of Science and Technology (KAIST) for his great contributions to "the research of semiconductor devices and integrated circuits for over two decades" and for "building a solid foundation for Korea's semiconductor industry."[2] Among the four prize categories, the science and engineering prize has been the most celebrated from the beginning, and its prize money is twice that of the other three prizes.[3] Its first recipient, in 1991, was not an individual but the Electronics and Telecommunications Research Institute (ETRI). The second recipient, in 1992, was Kim Jihn-E, a physicist at Seoul National University (SNU). As the third recipient, Kim Choong-Ki was therefore the first South Korean engineer to receive this coveted prize (Figure 2.1).

Kim's award attracted public attention because of Samsung Electronics' miraculous success in dynamic random-access memory (DRAM) over the preceding ten years. In the 1990s and 2000s, people in the South Korean semiconductor community called Kim's former students "Kim's Mafia" because they took up many leading roles and positions in industry, government research institutes, and academia.[4] Following suit, the South Korean media began to call Kim the "godfather" of the South Korean semiconductor community. Yet even though many of Kim's former students appeared in the mass media along with their success stories, especially during the 1990s and 2000s, Kim himself had rarely been in the public eye. How, then, did this lesser known engineering professor become the "godfather" of the South Korean semiconductor community? In other words, who is Kim Choong-Ki?

EARLY YEARS

Kim Choong-Ki was born in Seoul on October 1, 1942, with a textile engineer as his father and an elementary school teacher as his mother. His paternal ancestors were not typical Korean intellectuals (scholar-bureaucrats) but lower-ranking military men and merchants in the Seoul area. After an epidemic in the late nineteenth century had killed all the other male members of his paternal family, Kim's grandfather moved from Seoul to Kaeseong to join his maternal uncle and became a rice dealer.

DOI: 10.1201/9781003353911-4

FIGURE 2.1   Kim Choong-Ki at the award ceremony for the Ho-Am Prize for Science and Engineering in 1993. In the lower photo, Lee Kun-Hee, chairman of the Samsung Conglomerate, the sponsor of the prize, is second from the left, and Kim is second from the right.

Source: Courtesy of Chang Hea-Ja.

At almost the same time, he converted to Christianity (the Methodist Church), and he later became a full-time missionary. He created a very unusual milieu in his family, in which traditional Confucian values and customs had little influence. None of his children recalled studying the Confucian classics; instead, they enjoyed reading various subjects and playing with mathematics, two distinctive traditions that were passed down to Kim Choong-Ki's generation. If Kim Choong-Ki has surprised many South Korean students and engineers with his nonauthoritarian style, a major cause must be his exceptional family background, dating back to the early twentieth century.

It was Choong-Ki's father, Kim Byung-Woon, who most influenced Choong-Ki's future career as an engineer. Born in 1912, two years after the Japanese annexation of Korea, he was the first engineer in his family. He studied textile engineering at Kyungseong (Seoul) Technical High School, one of the two higher-education institutions that the Japanese colonial government had established to train engineers.[5] Immediately after graduating in 1935, Kim Byung-Woon was hired by Kyungseong Bangjik (better known as Kyungbang), the largest Korean textile company during the Japanese colonial period, and sent to Japan to learn spinning technology. In 1936, he and other young Korean engineers succeeded in building the first spinning factory at Kyungbang.[6] Kim Byung-Woon soon emerged as the most valuable and respected engineer in the company. When Kyungbang decided to establish an affiliate company in Manchuria, it sent him there as chief engineer.[7] After Korea was liberated from Japanese rule on August 15, 1945, Kim Byung-Woon was promoted to chief manager of the main factory in Seoul, at the relatively young age of 33. During and after the Korean War, he was responsible for rebuilding the factory and assembly lines from the ashes and ruins.[8] He frequently visited the United States to import the necessary technology and machines and became renowned among his American counterparts as an extremely meticulous engineer. In the early 1960s, Lone Star, a textile company in Texas, invited him to teach their engineers.[9]

Kim Byung-Woon possessed a true "engineer's mind" from the day he entered Kyungseong Technical High School in April 1932 until his death in December 1972. Unlike other Korean engineers of his generation, he did not crave higher management positions or academic jobs. He was, however, promoted to various managerial posts because Kyungbang needed his technical competence whenever it faced a crisis. He also served as a lecturer in the Department of Textile Engineering at Seoul National University (SNU) in the late 1940s and published the first Korean textbook on textile engineering (in Korean) in 1949.[10] Yet he did not pursue a professorship at SNU but remained an engineer at Kyungbang for 37 years, emphasizing practice. Kim Choong-Ki often told his cousins and students the following story: "When my father was the chief manager of the factory in Seoul, he made a daily tour of the factory. He told me that he could detect which machines were in trouble, and why, just by listening to them during his tour. Now, after some decades, I fully understand what that meant."[11] Many anecdotes illustrate Kim Byung-Woon's punctuality and dexterity as an engineer. Kim Choong-Ki and his brother Joon-Ki would become their father's most successful disciples with their contributions to the semiconductor and computer industry in the last quarter of the twentieth century.[12]

When Kim Choong-Ki was asked to give a short speech summarizing his life and work at a ceremony in his honor on February 25, 2008, he spent almost half the time recalling how deeply his father had influenced his life and career.[13] After many anecdotes describing how Kim Byung-Woon had taught him to have an "engineer's mind," he concluded:

> When Korea was liberated from Japanese rule in 1945, my father was 33 years old. Between then and 1972, when he died, he made a significant contribution to the development of the textile industry as an engineer. … He might have become a rich man if he had started his own textile business with his knowledge and experience, but he didn't. Instead, he liked to share his knowledge and know-how on textiles with other engineers. … When I returned to Korea in 1975, I was 33 years old. Between then and now, I have done my best to make some contribution to the semiconductor industry. If I have achieved any success, it's because I have done my best to keep abreast of my father.

Kim Choong-Ki was a typical model student at Deoksu Elementary School, Kyunggi Middle School and High School, and SNU. He was always one of the top students in his class and was obedient and silent at school, concealing his intellectual curiosity perfectly in front of teachers and friends. At home, his sister, who was one year his elder, was often his intellectual sounding board. Although his relatives pressed him to study textile engineering, he chose to enter the Department of Electrical Engineering at SNU. He does not seem to have been very satisfied with the education there, which emphasized "the designs and theories of the transformer, generator, power transmission, motor, etc."[14] He instead attended many courses offered by the Department of Applied Physics.

Nonetheless, when Professor Kim Wan-Hee of Columbia University asked SNU to recommend their brightest student, the Department of Electrical Engineering selected Kim Choong-Ki, who then went to Columbia for advanced study in 1965. After passing the qualifying examinations, he chose Edward S. Yang, a specialist in transistor theory, as his advisor. Yang also came to Columbia in 1965, so Kim became his second Ph.D. candidate.[15] Kim's doctoral dissertation, "Current Conduction in Junction-Gate Field-Effect," suggests a plausible theory to explain complicated experimental data: "The theory presented in this dissertation explains the important features of the experimentally observed steady-state drain characteristics by clarifying the contribution of the various physical mechanisms to the current conduction."[16] After receiving his doctorate from Columbia (Figure 2.2), Kim Choong-Ki was hired by the Research and Development Laboratory of Fairchild Camera and Instrument Corporation in Palo Alto, California, in the summer of 1970. He married Chang Hea-Ja in December of that year, and the new couple settled down in Mountain View, not far from his office.

FIGURE 2.2 Kim Choong-Ki's stay in the United States between 1965 and early 1975 converted him into a top-notch semiconductor engineer. **Left:** At Columbia University in New York, he learned semiconductor theory. **Right:** At Fairchild Camera and Instrument in Palo Alto, California, he became a semiconductor specialist with a true "engineer's mind."

Source: Courtesy of Chang Hea-Ja.

FAIRCHILD CAMERA AND INSTRUMENT CORPORATION IN PALO ALTO, CALIFORNIA

Starting as the Fairchild Aerial Camera Corporation in 1920, Fairchild had been the world's leading developer of imaging equipment, including radar cameras, radio compasses, and X-ray machines.[17] In 1957, the company launched a new division, Fairchild Semiconductor, to fabricate transistors and integrated circuits from silicon—an innovative move because most semiconductor devices at the time used germanium. The venture spawned dozens of products, including the first silicon integrated circuit, thus fueling the rise of Silicon Valley. By the late 1960s, more than two dozen startup companies, often dubbed "Fairchildren," had been founded by engineers who had once worked at Fairchild Semiconductor.[18]

It was during his four and a half years at Fairchild that Kim Choong-Ki was transformed from a shy model student into a self-assured and communicative semiconductor engineer. His engineering potential, which had lain dormant during his schooling in South Korea, began to bud at Columbia and blossomed at Fairchild. He had learned theories on semiconductors at Columbia, but it was at Fairchild that he learned the practices needed to make theories actually work.

Two things at Fairchild influenced Kim deeply. The first was that he encountered state-of-the-art charge-coupled device (CCD) technology there. Just the year before, in 1969, Willard S. Boyle and George E. Smith of Bell Laboratories had proposed the new idea of the CCD, which eventually brought them Nobel Prize in Physics in 2009.[19] A CCD is

> an integrated circuit that captures and stores light and displays it by turning it into an electrical charge. Each CCD chip is composed of an array of Metal-Oxide-Semiconductor (MOS) capacitors, and each capacitor is a pixel. When electrical voltages are applied to the CCD's top plates, charges can be stored in silicon under the top plate. Then, digital pulses applied to the top plates can shift these charges among the pixels, creating a picture representing charged pixels.[20]

CCD image sensors would soon be widely used in consumer, medical, and scientific applications, such as digital photography, digital radiography, and observational astronomy.

Kim Choong-Ki's years at Fairchild were devoted to this emerging field of technology. Fairchild was a pioneer in CCD technology along with RCA, Texas Instruments, and Kodak, and the company asked Kim to work on this field when he arrived at the Palo Alto research center.[21] Kim and his colleagues at Fairchild developed a new kind of CCD—buried-channel charge-coupled devices—which "offer many advantages over [original] surface channel devices at the expense of only one additional step in the fabrication process" (Figure 2.3).[22] This new method of making CCDs led them to develop a high-performance CCD area image sensor that greatly improved detecting the image in low light and to create the world's first CCD linear image sensor that demonstrates "the ease of use and high quality of image reproduction."[23] Later, Kim remembered that "about a half hour before closing the office, people came to my office to ask something about CCD. Soon they began to call me 'Professor CCD.' "[24] In short, Kim was one of few South Koreans

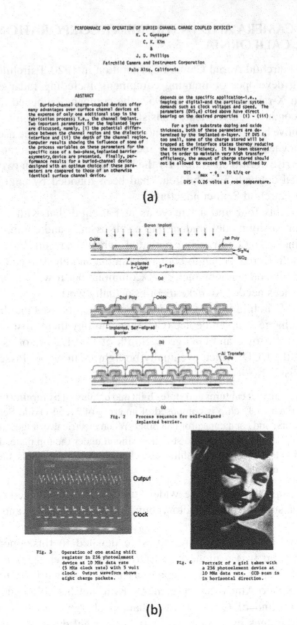

PERFORMANCE AND OPERATION OF BURIED CHANNEL CHARGE COUPLED DEVICES*

K. C. Gunsagar
C. K. Kim
&
J. D. Phillips

Fairchild Camera and Instrument Corporation
Palo Alto, California

ABSTRACT

Buried-channel charge-coupled devices offer many advantages over surface channel devices at the expense of only one additional step in the fabrication process; i.e., the channel implant. Two important parameters for the implanted layer are discussed, namely, (i) the potential difference between the channel region and the dielectric interface and (ii) the depth of the channel region. Computer results showing the influence of some of the process variables on these parameters for the specific case of a two-phase, implanted barrier asymmetry, device are presented. Finally, performance results for a buried-channel device designed with an optimum choice of these parameters are compared to those of an otherwise identical surface channel device.

depends on the specific application—i.e., imaging or digital—and the particular system demands such as clock voltages and speed. The parameters (DVS,ε) cited above have direct bearing on the desired properties (i) – (iii).

For a given substrate doping and oxide thickness, both of these parameters are determined by the implanted n-layer. If DVS is too small, some of the charge stored will be trapped at the interface states thereby reducing the transfer efficiency. It has been observed that in order to maintain very high transfer efficiency, the amount of charge stored should not be allowed to exceed the limit defined by

DVS = ϕ_{max} - ϕ_s = 10 kT/q or

DVS = 0.26 volts at room temperature.

(a)

Fig. 2   Process sequence for self-aligned implanted barrier.

Fig. 3   Operation of one analog shift register in 256 photoelement device at 10 MHz data rate (5 MHz clock rate) with 5 volt clock. Output waveform shows eight charge packets.

Fig. 4   Portrait of a girl taken with a 256 photoelement device at 10 MHz data rate. CCD scan is in horizontal direction.

(b)

FIGURE 2.3   Kim Choong-Ki's two monumental papers on CCD published during his years at Fairchild. (a) The 1973 paper presents a new kind of CCD—buried-channel CCD. (b) Kim's 1974 paper submits the world's first linear image sensor.

Sources: K. C. Gunsagar, C. K. Kim, and J. D. Phillips, "Performance and Operation of Buried Channel Charge Coupled Devices," 21; and Choong-Ki Kim, "Two-Phase Charge-Coupled Linear Imaging Devices with Self-Aligned Implanted Barrier," 57.

who had substantial hands-on experience and state-of-the-art theoretical knowledge in semiconductors in the early 1970s.

Edward Yang, who directed Kim's doctoral thesis at Columbia, nicely summarizes Kim's achievement at Fairchild as follows:

> James Early, one of the transistor pioneers at Bell Labs who had moved to Fairchild in 1970, and who recruited Choong-Ki, later asked if Columbia had any more graduates like Kim for Fairchild. At that time Bell Labs had a large program on CCD, probably over 50 PhDs. They had the basic charge transfer idea [that eventually earned] Boyle and Smith the Nobel Prize, yet they did not know how to build a functionally useful product. Within a year after joining Fairchild, Choong-Ki made use of the silicon-gate technology and fabricated the first self-aligned buried-channel line and area imagers, the first marketable CCD devices. Fairchild's—or, better call them Choong-Ki's—CCDs, made possible the wide applications in high-resolution cameras. Without Choong-Ki's inventions, there would have been no Nobel [Prize] for CCD. To me, Choong-Ki deserved to be recognized with Boyle and Smith. That is the true level of achievement.[25]

The second thing at Fairchild that fundamentally changed Kim's attitude toward engineering was the company culture. This can be divided into two aspects. First, it was at Fairchild that Kim realized that book learning alone was not sufficient for practicing engineering in the real world. When he was forced to attend a lecture course at the company campus, he quickly realized that it was quite different from the subject matter he had studied at Columbia,

> though the course used the exactly same textbook that I had studied at Columbia. Each chapter was taught by a different lecturer who had worked on that specific field for more than ten years. Lecturers taught us what they had actually experienced rather than what was written in the textbook. I really learned a lot from this course.[26]

Kim also began to read internal technical reports and memos that he could find at the company library and "really learned something new from these informal reports and memos that I had never learned at the university." He later brought many ideas from these documents to South Korea and taught them at KAIST.

The other lesson was how to communicate with, manage, and lead other engineers in the company. Kim had always been quiet and introverted, as was expected of a South Korean model student. However, Fairchild's company culture demanded that he adopt very different ways to survive and thrive there:

> After two and a half years passed, the company ordered me to attend a training course to become a manager. … The instructor asked us to appraise each member's pros and cons to become a good manager. When it became my turn, the instructor told me the collected opinions: "unmask." The other members of my group had first considered me a strange Asian who could not speak English since I was shy and silent. However, as the days went on, they found that I could speak English well and had many good ideas but didn't want to reveal them to others. So, my colleagues' advice was to "unmask" myself and express my ideas to others for better communication. Their analysis was quite correct. I had been very, very introverted. I had just commuted between school and home when I was in Korea. In my leisure time, I just read books at home. This tendency doubled after I went to Columbia and also to Fairchild because of the unfamiliar

environments and English being a second language. However, after receiving such advice at the management course, I determined to change myself. From then on, I intentionally became much more conversational, telling jokes or talking about my interests to anyone around me. After one or two months passed, I found that my popularity began to rise. People came to my office about a half hour before the closing to chat or to ask me about CCDs.[27]

The newly voluble Kim Choong-Ki would become the "loudest-speaking" professor at KAIST, one whose absence made the whole campus seemed quiet.[28]

Although Kim Choong-Ki rose quickly and successfully within the company hierarchy, in early 1974 he began to seriously consider returning to South Korea. There were three reasons for this. First of all, his beloved father had died in December 1972. His two elder sisters and his younger brother were all in the United States and seemed to have settled down there. As the eldest son, he felt a heavy responsibility to care for his widowed mother in South Korea.[29] Second, the racial discrimination in the United States hurt his pride. In the early 1970s, racial discrimination was not an accident but a norm. Kim remembers that some Caucasian colleagues at Fairchild openly discriminated against him.[30] Third, he found an ideal place in South Korea where he could train true engineers—KAIST.

THE CRADLE OF FUTURE SEMICONDUCTOR SPECIALISTS

Kim Choong-Ki became an associate professor in the Department of Electrical Engineering at KAIST in the spring of 1975.[31] KAIST, with its special emphasis on practice, was the ideal place for him to begin training the next generation of semiconductor engineers. Kim was only the fourth professor hired in the department, following Na Jung-Woong (1971), Park Song-Bai (1973), and Kim Jae-Kyoon (1973). Park Song-Bai, head of the department, gave Kim 100 million won (roughly USD 200,000 at that time), which had been reserved for a semiconductor specialist, to build a clean room—the essential facility for semiconductor teaching and research.

During the 1970s, the Department of Electrical Engineering was one of the three most influential and popular departments at KAIST (along with the departments of Chemistry and Mechanical Engineering). It was also the most ardent supporter of Frederick Terman's original goal for KAIST—namely, that of its graduates prioritizing "winning markets" (see Chapter 1). Kim, of course, faithfully supported the department's policy from the beginning. He remembers:

> When I became professor at KAIST in 1975, I thought that the proper job of an engineering professor was training engineering students for the industry, not publishing papers in academic journals. So, I published almost nothing in the beginning. One day, the president of KAIST summoned me and scolded me for neglecting "the duty of publishing papers." He said that he could not promote me to full professor without papers. So, I reluctantly began to write and publish papers for academic journals.[32]

Kim would soon become the icon of the "KAIST spirit" and would remain so over the following decades.

Kim's semiconductor laboratory immediately attracted many talented and ambitious master's and doctoral candidates. The primary reason for the popularity of his laboratory was quite obvious: the semiconductor. The South Korean government and industry had recognized the importance of semiconductors since the beginning of the 1970s. Companies such as Samsung Electronics, GoldStar (the future LG Electronics), and Daewoo Electronics needed a large number of semiconductors to manufacture their radios, TV sets, desk calculators, and electronic watches, but they didn't have the technology to produce sophisticated ones. They therefore depended heavily on Japanese semiconductor manufacturers such as Sony, Toshiba, Nippon Electric Company (NEC), and Sharp. Until the end of the 1970s, the South Korean government and industry's efforts to develop semiconductors produced no significant results, largely because of a lack of specialists who could carry out research and development on semiconductors.[33]

Unfortunately, few South Korean universities paid any serious attention to training semiconductor engineers during the 1970s. The Department of Electronic Engineering at SNU, for example, was still dominated by old subjects, such as communication or circuit theory.[34] Min Hong-Sik, who came to SNU in 1976, was the first professor to teach semiconductors there. However, he did not receive the proper support to install the facilities needed for research and education, so he had no choice but to teach only theories of semiconductors for many years. Nonetheless, his lectures on semiconductors fascinated many undergraduate students, and some of them, such as Kim Ki-Nam, went on to KAIST after graduation for practical training. Whereas KAIST possessed at least six professors of semiconductor engineering by 1987, including Kim Choong-Ki's two former students, Kyung Chong-Min (profiled in Chapter 3) and Han Chul-Hi, Min Hong-Sik remained the only teacher on semiconductors at SNU until 1988, when Park Young-Joon was hired. SNU began to erect a new building dedicated to semiconductor research in 1985 and started recruiting more semiconductor specialists from the late 1980s on. It was only in the mid-1990s that SNU emerged as a serious challenger to KAIST on semiconductors.

Kim Choong-Ki's laboratory at KAIST was therefore virtually the only true semiconductor training ground in South Korea from 1975 to 1995. Between 1991 and the early 2000s, more universities began to train semiconductor engineers, and even at KAIST, other professors, including Kim's two former students, trained their own students. Kim's laboratory, however, continued to be one of the best places to produce the most able semiconductor specialists.

The research at Kim's semiconductor laboratory can be divided into three periods, based on the 78 master's theses he supervised: 1975–1981, 1982–1991, and 1992–2007.[35] In the first period, between 1975 and 1981, Kim and his students were still building their laboratory, and its facilities and staff were not yet complete. This first group of master's candidates worked primarily on the design and fabrication of various semiconductor devices. Examples of their theses include "Design of MOS Shift Register" (Chung Jin-Yong, 1976), "Design and Fabrication of a Charge-Coupled Device" (Ju Dong-Hyuk, 1976), "Characterization of Diodes for Integrated Circuits" (Kwon Oh-Hyun, 1977), "A Study on the High Voltage Planar Diodes" (Park Zoo-Seong, 1978), "Design and Fabrication of a Seven Segment Decoder/Driver with

PMOS Technology" (Lim Hyung-Kyu, 1978), "A Study on the Oxidation Process of Silicon" (Choi Yearn-Ik, 1978), "Fabrication of the Integrated Injection Logic by New Process" (Han Chul-Hi, 1979), "Design and Fabrication of a 2-Bit ALU (Arithmetic Logic Unit) Using N-MOS Technology" (Chung Bong-Young, 1979), "Experimental Study on the Aluminum Anodization Using Ammonium Tartrate and Its Application to I. C. Fabrication" (Shin Yun-Seung, 1980), "A 512-Bit Mask Programmable ROM Using PMOS Technology" (Shin Hyun-Jong, 1980), "The Design and Fabrication of 64-Bit Silicon-Gate N-MOS Static RAM" (Suh Kang-Deog, 1981), and "An Experimental Study on the Threshold Voltage and Punch-Through Voltage Reduction in Short-Channel NMOS Transistors" (Lee Won-Shik, 1981). Most of these topics were not cutting edge by global standards, but they were very advanced in South Korean terms. In other words, Kim was training his students to meet the future needs of the South Korean market rather than its present needs in the late 1970s. Lim Hyung-Kyu (profiled in Chapter 4), who later led the development of memories and nonmemories, especially NAND flash memory, during the 1990s and 2000s at Samsung Electronics, remembered, "When I returned to the company after receiving my master's degree, I found myself the top specialist in semiconductors within the company."[36]

During the second period, between 1982 and 1991, Kim's laboratory began to work on cutting-edge subjects as well as more conventional ones but still focused on design and fabrication. Some examples are "Design and Fabrication of the Area Image Sensor Using Charge-Coupled Device with Double Polycrystalline Silicon Gates" (Oh Choon-Sik, 1982), "Platinum Silicide Barrier IR Detector" (Park Bo-Young, 1983), "Experimental Analysis of CMOS Latch-Up Phenomena" (Koh Yo-Hwan, 1985), "Fabrication of 32×32 Flat Panel Display Using Liquid Crystal" (Kim Jeong- Gyoo, 1985), "Fabrication of Hydrogenated Amorphous Silicon Thin-Film Transistors for Flat Panel Display" (Kim Nam-Deog, 1986), "Schottky Barrier Lowering by Ion Implantation for High Performance PtSi Schottky Barrier Infrared Detector" (Eo Ik-Soo, 1987), "A Three-Dimensional Dynamic RAM Cell with Polysilicon Transistor on a Storage Capacitor" (Lee Jung-Yeal, 1987), "Design Study of a Rapid Thermal Annealing System and Temperature Control" (Cho Byung-Jin, 1987), "Fabrication of $Hg_{1-x}Cd_xTe$ Photovoltaic Infrared Detector" (Han Pyong-Hee, 1989), "Design of Metal Gate CMOS Processing Using Rapid Thermal Process" (Park Sung-Kye, 1990), "Design and Implementation of Lamp-Heated LPCVD System and Poly-Si Deposition" (Ha Yong-Min, 1990), and "Thin Film Evaporation of ZnS for the Anti-Reflection Layer of $Hg_{1-x}Cd_xTe$ IR Photodetector" (Park Gun-Woo, 1991).

The third period, between 1992 and 2007, was a transition period when KAIST moved from the Seoul campus to the newly constructed Daejeon campus and also annexed the Korea Institute of Technology (KIT). The relocation was complete by the end of 1992, but it required a few more years to settle down. Moreover, from 1994 to 1997, Kim accepted several administrative positions, including that of vice president of KAIST, so he could not train as many students as before. The number of master's theses during the period therefore decreased to 16, but their high quality remained intact. Some examples are "Fabrication of Poly-Si TFT and Shift Register for LCD Driver by Low-Temperature Process" (Kim Tae-Sung, 1992), "Effect of Implanted Oxygen on the Crystallization of Silicon Films and Polysilicon TFT Characteristics" (Lee Seong-Hoon, 1993), "Design and Implementation of Active Addressing Controller for STN [Super-Twisted Nematic] LCD" (Park Sung-Hoon, 1994), "Characterization of NMOSFET's Fabricated on Misfit Dislocation-Free p/p+ Silicon Substrate Using Protection Ring" (Yoon Nan-Young, 1995), "Single Crystal Silicon Substrates for High-Resolution Transmissive Active-Matrix LCD" (Yoon Jun-Bo, 1995), "Novel TFT with Active Layer Buried in Glass Substrate" (Han Ki-Ho, 1996), "Noise Analysis of Readout Integrated Circuit for IR Detector" (Kim Dae-Ki, 1999), "CMOS Process Using Impurity Diffusion" (Kyung Ki-Myung, 2000), "High Q Variable Capacitor for RF Application Fabricated by 3-Dimensional MEMS [Micro-Electro-Mechanical Systems] Technology" (Chang Jae-Hong, 2000), and "A New Digital Signal Processing Method for Optical Fiber Gyroscope" (Oh Hyoung-Seok, 2002).

Between 1975 and 2006, Kim Choong-Ki also trained 39 doctors of electrical engineering, who can be divided into two groups—those who received their doctorates in 1975–1991 and those in 1992–2007. Most of Kim's doctoral candidates had also received their master's degrees under him, but three students had transferred from Han Chul-Hi's laboratory to Kim's after Han's premature death in 2001. Many of Kim's doctoral candidates continued the research that they had started in their master's years, but about one-third changed subjects when they began their doctoral work.

Kim produced only 13 doctors of engineering during the first period (1975–1991). The first was Kyung Chong-Min, who received his doctorate in 1981 for his thesis entitled "A New Charge-Coupled Analogue-to-Digital Converter." Others soon followed: "Graded Etching of Silicon Dioxide Layer and Its Application to Schottky Diodes" (Choi Yearn-Ik, 1981), "Type Conversion of Polycrystalline Silicon Heavily Doped with both Boron and Phosphorus Due to High Current Conduction" (Kim O-Hyun, 1983), "Substrate Fed Threshold Logic" (Han Chul-Hi, 1983), "The Effect of Si-SiO$_2$ Interface on the Excess Point Defect Distribution in Silicon" (Shin Yun-Seung, 1984), "MOSFET Source and Drain Structures for High-Density CMOS Integrated Circuit" (Oh Choon Sik, 1986), "Rapid Thermal Diffusion of Boron and Phosphorus into Silicon Using Solid Diffusion Sources" (Kim Kyeong-Tae, 1988), "Latch-Free Self-Aligned Power MOSFET and IGBT Structure Utilizing Silicide Contact Technology" (Koh Yo-Hwan, 1989), "Two-Step Rapid Thermal Diffusion of Phosphorus and Boron into Silicon from Solid Diffusion Sources" (Kim Jeong-Gyoo, 1989), "Hot-Carrier Effects in BC [Buried-Channel]-MOSFET and New CMOS Logic Circuits for High Reliability" (Park Heung-Joon, 1989), "Hydrogenated Amorphous Silicon Thin-Film Transistor with Two-Layer Amorphous Silicon Nitride Gate Insulator" (Kim Nam-Deog, 1990), "A New Analytical Design Method for Optimum Field-Limiting Ring Systems in Semiconductor Power Devices" (Suh Kang-Deog, 1991), and "Modeling of Rapid Thermal Diffusion of Phosphorus into Silicon and Its Application to VLSI Fabrication" (Cho Byung-Jin, 1991).

From the late 1980s on, Kim accepted and trained more doctoral candidates, producing 26 doctors in the second period (1991–2007). Some examples of their theses are "Characteristics of Low-Resistive Plasma Enhanced Chemical Vapor Deposited Tungsten (PECVD-W) Thin Film" (Kim Yong-Tae, 1992), "Parametric Investigation of Zone Melting Recrystallization of Polysilicon and Threshold Voltage Model for Thin SOI MOSFET" (Choi Jin-Ho, 1992), "The Effect of Input Power Conditions on the Quality of the Zone-Melting Recrystallized Silicon Film Using Radiative Heat Source" (Yoon Byoung-Jin, 1993), "Device Design for Suppression of Floating Body Effect in Fully-Depleted SOI MOSFETs" (Park Sung-Kye, 1994), "Device Structure of Fabrication Process for High-Performance Polysilicon Thin Film Transistors" (Ha Yong-Min, 1994), "Reliability Study and Effect of NH_3 Annealing on the Electrical Characteristics of Polysilicon Thin Film Transistors" (Choi Deuk-Sung, 1995), "Suppression of Misfit Dislocations in Heavily Boron-Doped Silicon Layer for Micro-Machining (Lee Ho-Jun, 1996), "Indium Rapid Thermal Diffusion into p-HgCdTe and Its Application to Infrared Detector Fabrication" (Park Seung-Man, 1997), "Surface Treatment Effects on the Electrical Properties of Insulator/ HgCdTe Interface" (Lee Seong-Hoon, 1998), "Three-Dimensional Microstructure Technology for Microfluidic Systems and Integrated Inductors" (Yoon Jun-Bo, 1999), "Improved CMOS Integrated Circuit Design and Implementation for the Readout of Infrared Focal Plane Arrays" (Yoon Nan-Young, 2000), "Dual-Gate Poly-Si TFT with Intermediate Lightly Doped Region and Analog Buffers for Integrated Data Drivers of Poly-Si TFT-LCD's" (Chung Hoon-Ju, 2002), "New Pixel Circuits for Active-Matrix Organic Light-Emitting Diode Displays" (Goh Joon-Chul, 2004), "A Single-Chip 5 GHz CMOS Receiver for 802.11a" (Chang Jae-Hong, 2005), and "A Wide Dynamic Range CMOS Image Sensor with In-Pixel Analogue Memory for Pixel Level Integration Time Control" (Han Sang-Wook, 2006).

Although the primary goal and function of Kim Choong-Ki's laboratory was to train the semiconductor specialists that South Korean industry required, his lab actually led South Korean semiconductor research between 1975 and 1995, when most universities and companies didn't have proper research facilities. During these twenty years, Kim and his students did research on the design of metal-oxide-semiconductor (MOS) integrated circuits, the oxidation process of silicon, the design of integrated injection logic, the charge-coupled device (CCD), random-only memory (ROM), and the rapid thermal process.[37] They published their results in both international and domestic journals, including fourteen papers in the prestigious Institute of Electrical and Electronics Engineers (IEEE) journals. One of Kim's students, Suh Kwang-Seok, became the first Korean to publish his research carried out in South Korea in a foreign journal, *IEEE Transactions on Electron Devices*, in 1980, and another, Kyung Chong-Min, was the first Korean to read results from his doctoral dissertation at an international conference, the 1981 Custom Integrated Circuit Conference.[38] In 1984, Oh Choon-Sik and Koh Yo-Hwan became the first Koreans to present the results of research carried out in South Korea at the International Electron Devices Meeting (IEDM), the most prestigious meeting on semiconductor devices.[39] For the South Korean semiconductor industry, Kim and his students developed various processing techniques for the rapid thermal process, which "constitutes an essential part of high density memory products beyond 16Mb DRAM" (Figure 2.4).[40]

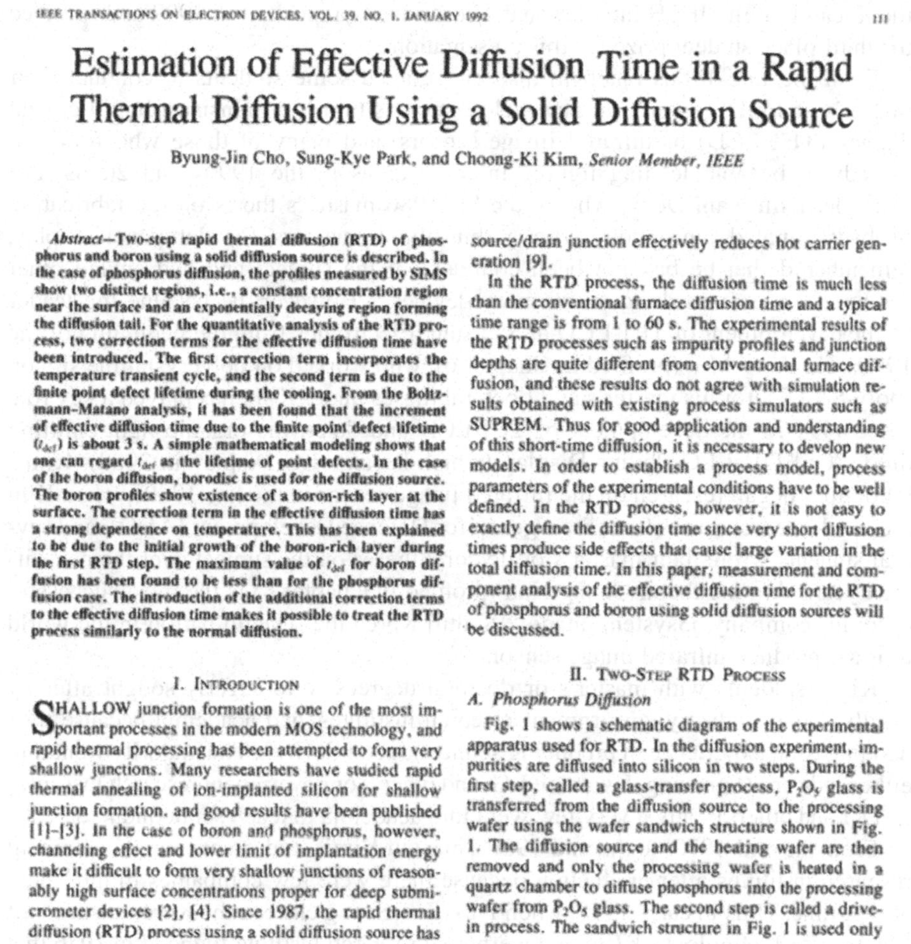

IEEE TRANSACTIONS ON ELECTRON DEVICES, VOL. 39, NO. 1, JANUARY 1992
111

Estimation of Effective Diffusion Time in a Rapid Thermal Diffusion Using a Solid Diffusion Source

Byung-Jin Cho, Sung-Kye Park, and Choong-Ki Kim, *Senior Member, IEEE*

Abstract—Two-step rapid thermal diffusion (RTD) of phosphorus and boron using a solid diffusion source is described. In the case of phosphorus diffusion, the profiles measured by SIMS show two distinct regions, i.e., a constant concentration region near the surface and an exponentially decaying region forming the diffusion tail. For the quantitative analysis of the RTD process, two correction terms for the effective diffusion time have been introduced. The first correction term incorporates the temperature transient cycle, and the second term is due to the finite point defect lifetime during the cooling. From the Boltzmann-Matano analysis, it has been found that the increment of effective diffusion time due to the finite point defect lifetime (t_{def}) is about 3 s. A simple mathematical modeling shows that one can regard t_{def} as the lifetime of point defects. In the case of the boron diffusion, borodisc is used for the diffusion source. The boron profiles show existence of a boron-rich layer at the surface. The correction term in the effective diffusion time has a strong dependence on temperature. This has been explained to be due to the initial growth of the boron-rich layer during the first RTD step. The maximum value of t_{def} for boron diffusion has been found to be less than for the phosphorus diffusion case. The introduction of the additional correction terms to the effective diffusion time makes it possible to treat the RTD process similarly to the normal diffusion.

I. INTRODUCTION

SHALLOW junction formation is one of the most important processes in the modern MOS technology, and rapid thermal processing has been attempted to form very shallow junctions. Many researchers have studied rapid thermal annealing of ion-implanted silicon for shallow junction formation, and good results have been published [1]-[3]. In the case of boron and phosphorus, however, channeling effect and lower limit of implantation energy make it difficult to form very shallow junctions of reasonably high surface concentrations proper for deep submicrometer devices [2], [4]. Since 1987, the rapid thermal diffusion (RTD) process using a solid diffusion source has

source/drain junction effectively reduces hot carrier generation [9].

In the RTD process, the diffusion time is much less than the conventional furnace diffusion time and a typical time range is from 1 to 60 s. The experimental results of the RTD processes such as impurity profiles and junction depths are quite different from conventional furnace diffusion, and these results do not agree with simulation results obtained with existing process simulators such as SUPREM. Thus for good application and understanding of this short-time diffusion, it is necessary to develop new models. In order to establish a process model, process parameters of the experimental conditions have to be well defined. In the RTD process, however, it is not easy to exactly define the diffusion time since very short diffusion times produce side effects that cause large variation in the total diffusion time. In this paper, measurement and component analysis of the effective diffusion time for the RTD of phosphorus and boron using solid diffusion sources will be discussed.

II. TWO-STEP RTD PROCESS

A. Phosphorus Diffusion

Fig. 1 shows a schematic diagram of the experimental apparatus used for RTD. In the diffusion experiment, impurities are diffused into silicon in two steps. During the first step, called a glass-transfer process, P_2O_5 glass is transferred from the diffusion source to the processing wafer using the wafer sandwich structure shown in Fig. 1. The diffusion source and the heating wafer are then removed and only the processing wafer is heated in a quartz chamber to diffuse phosphorus into the processing wafer from P_2O_5 glass. The second step is called a drive-in process. The sandwich structure in Fig. 1 is used only

---

FIGURE 2.4 Kim Choong-Ki and his students had worked on the rapid thermal process since the mid-1980s and produced many high-quality papers for IEEE journals and South Korean journals. The paper shown was written with two of his students, Cho Byung-Jin and Park Sung-Kye (profiled in Chapters 5 and 7, respectively).

Source: Byung-Jin Cho, Sung-Kye Park, and Choong-Ki Kim, "Estimation of Effect Diffusion Time and Rapid Thermal Diffusion Using a Solid Diffusion Source," 111.

All these research activities, "despite poor conditions for experiments in South Korea," were quite impressive in the 1980s.[41] High-quality research continued at Kim's laboratory during the 1990s, despite his departure to hold high administrative positions at KAIST in 1994–1997. For example, with help from Kim and from Han Chul-Hi, Yoon Jun-Bo succeeded in fabricating various three-dimensional (3D) on-chip inductors by micromachining technology, which was expected to reduce the weight, electricity consumption, and cost of mobile phones.[42] Yoon read the results of the research at the IEEE International Microwave Symposium in 1999 and received the third-place student prize for this presentation.[43]

From the mid-1980s on, Kim also encouraged some students to conduct their master's or Ph.D. research in new fields, such as thin-film-transistor liquid-crystal display (TFT-LCD) or infrared image sensors, and many of those who followed his advice became leading figures in these areas in the 1990s and 2000s. For example, Kim Nam-Deog, who wrote his 1986 master's thesis on the fabrication of hydrogenated amorphous silicon thin-film transistors for flat panel display, remembered that he had not been confident of the subject and had envied other graduate students working on memory devices.[44] However, he became the leader of the development of TFT-LCD at Samsung Electronics from the beginning of the 1990s. Ha Yong-Min (profiled in Chapter 6), who worked on polycrystalline silicon (poly-Si) TFT for his master's and doctoral theses under Kim, has played a key role since 1994 in the development of TFT-LCD and, later, the organic light-emitting diode (OLED) at LG Philips Display (which became LG Display in 2008). Kim's laboratory began research on the infrared image sensor in the late 1980s, when Kim received research funds from the Agency for Defense Development (ADD), and several students wrote their master's or doctoral theses on the subject over the next fifteen years. Among them, Chung Han (profiled in Chapter 8) is the most successful case: his company, i3system, made "[South] Korea the seventh nation in the world to mass-produce infrared image sensors."[45]

Kim's students with master's or doctoral degrees were eagerly sought after by South Korean industry, government research institutes, and academia because they had hands-on experience in producing semiconductor devices. The majority went into either industry (Samsung, Hyundai, LG, and others) or government research institutes (ETRI and others); only a very few went into academia to teach semiconductors.[46]

From the mid-1970s to the mid-1980s, most of Kim's students went to government research institutes after graduation because there were few companies to hire them at that time. Thus in early 1978, when Park Zoo-Seong entered the Korea Institute of Electronics Technology (KIET, a government research institute founded in 1976 that later became ETRI), Kim's former students Yoo Young-Ok and Kwon Oh-Hyun were already there, and several others joined them over the next few years.[47] As the popularity of government research institutes declined rapidly by the mid-1980s, many of those who had worked there moved to industry or academia, although some—such as Jung Hee-Bum and Eo Ik-Soo at ETRI and Park Seung-Man at ADD—remained for several decades.

Two cases illustrate the very different effects such decisions could have on a young engineer's subsequent career. Kwon Oh-Hyun is a good example of a semiconductor engineer who moved from government to industry employment. He began work at KIET after earning his master's degree in early 1977. In 1980, he went to Stanford University for advanced study, and, upon receiving his Ph.D. there in 1985, Samsung Electronics scouted him. Kwon contributed greatly to developing 64Mb DRAM in 1994, as well as to advancing both the memory and the nonmemory business in the 2000s. He eventually rose to be vice chairman and CEO of Samsung Electronics (2012–2018) and then chairman (2018–2019). In contrast, Jung Hee-Bum remained at government research institutes. He entered KIET (later ETRI) after receiving his master's in 1983 and contributed greatly to developing system-on-a chip (SoC) for information and communication technology until the mid-2000s. From the mid-2000s on, his major job was not research and development of specific technologies but the management of several of ETRI's ongoing projects.[48] It is clear that the role of the government research institute for semiconductor development was almost over by the early 2000s.

The popularity of Kim's students within South Korean industry increased dramatically from the mid-1980s on, when not only Samsung Electronics but also GoldStar Electronics (which became LG Electronics in 1995) and Hyundai Electronics (which became Hynix in 2001) entered the semiconductor field, mostly to produce DRAM. This demanded large numbers of well trained semiconductor specialists, but the supply was still very limited. Many of Kim's former master's and doctoral candidates went to Samsung Electronics, especially during the 1980s: Kwon Oh-Hyun, Lim Hyung-Kyu, Kim Oh-Hyun, Chung Bong-Young, Shin Yun-Seung, Suh Kang-Deog, Lee Won-Sik, Kim Kyung-Tae, Lee Yun-Tae, Kim Nam-Deog, Lee Won-Woo, Sim Sang-Pil, Chang Sung-Jin, and Kim Tae-Sung went to either Samsung Electronics or its sister electronics companies to develop various semiconductor devices. Kim Choong-Ki was careful not to send his talented students only to Samsung, however. From the mid-1980s on, he encouraged his students to go to less developed companies (from the point of view of semiconductors) such as Hyundai Electronics or GoldStar. His students obediently followed their mentor's advice without complaint. As a result, Oh Choon-Sik, Koh Yo-Hwan, Yoon Byung-Jin, Cho Byung-Jin, Choi Deug-Sung, Jeong Jae-Hong, Chung Han, Lee Ho-Joon, Park Gun-Woo, Lee Seong-Hoon, Yoon Nan-Young, and Kyung Ki-Myung went to Hyundai Electronics (later Hynix) after graduation, where they contributed greatly to making the company number two in memory production. Other graduates, such as Han Chul-Hi, Ha Yong-Min, Park Sung-Kye, Jang Seong-Jin, and Chung Hoon-Ju, went to GoldStar Semiconductor (later Goldstar Electronics and then LG Semiconductor) or to LG Philips Display (later LG Display): some of them, such as Park Sung-Kye and Jang Seong-Jin, moved to Hyundai when it bought out LG Semiconductor in 1999. Many of those who went into industry later rose to be vice presidents or to hold even higher positions in their companies. In short, it is not an exaggeration to say that Kim's former students were "everywhere" in the South Korean semiconductor industry from the mid-1980s on.

Academic positions were therefore not the coveted choices among Kim's former students, at least in the beginning. Kim himself did not encourage them to remain in the academy but repeatedly emphasized that an engineer's proper position is in industry, not the ivory tower. It is therefore ironic that, as the years passed, more and more of his former students found jobs in the academy, although they remain only about one-fifth of the whole. Unlike most other South Korean professors, most of Kim's former students who became professors did so only after they had spent several years either in industry or at government research institutes. Many South Korean universities, which hurriedly established departments or courses in semiconductor engineering from the mid-1980s on, desperately needed such experienced specialists to train students who would work in the industry after graduation. Park Zoo-Seong, who received his master's degree in 1978, for example, worked at KIET for more than five years and then moved to Pusan National University to teach semiconductors. Kim Oh-Hyun, who earned his doctorate under Kim in 1983, is another good example. He developed DRAMs at Samsung Semiconductor and Telecommunications (which merged with Samsung Electronics in 1988) for three years after graduation. When he finished his obligatory period at Samsung in 1986 to compensate for his exemption from military service, he read by chance that the newly established Pohang University of Science and Technology (POSTECH) was seeking a semiconductor specialist to teach the subject.[49] He sent an inquiry without expecting much but was soon approached by none other than the president of the university, who asked that he come to teach at the university right away. Kim Oh-Hyun established a small but a strong program at POSTECH and trained many able semiconductor specialists over the next three decades.

Three of Kim Choong-Ki's doctoral students who became professors at KAIST followed a similar trajectory: Kyung Chong-Min spent two years as a postdoctoral fellow at Bell Labs in Murray Hill, New Jersey, and then returned to KAIST to teach from 1983 on; Han Chul-Hi worked for four years at GoldStar Semiconductor before joining KAIST in 1987; and Cho Byung-Jin (profiled in Chapter 5) worked for three years at Hyundai Electronics after graduation, became a star engineering professor at the National University of Singapore (NUS) for ten years, and then returned to KAIST in 2007.[50]

By 2010, Kim Choong-Ki's former students were located in many parts of South Korea to train the next generations of semiconductor engineers. Some, such as Kyung Chong-Min and Kim Oh-Hyun, went so far as to produce an equal or greater number of master's and doctoral students than their mentor; moreover, in 2011, one of Kyung's former doctoral candidates was hired as a professor at KAIST. Seeds thrown on good soil truly produce a lot of grain.

UNIQUE MENTORING

It was not simply the popular subject of semiconductors, however, or KAIST's virtual monopoly on training in the field, that made Kim Choong-Ki a mentor without peer in South Korean semiconductor circles. Rather, it was his unique training, his unusual attitude and down-to-earth words, and his idea of engineering. These at first puzzled his students, then impressed and influenced them profoundly, and finally changed their view of engineering permanently. To become a competent semiconductor engineer, Kim demanded that each student properly use his or her hands as well as "something malleable within the hard nut on your neck" (i.e., "your brain within the skull on your neck").[51] He made his students learn and understand these lessons through practice, as the following excerpts from the collective memoir of his former students, *Uri Kim Choong-Ki Seonsaengnim* (*Our Teacher, Kim Choong-Ki*), clearly indicate:

- I still clearly remember those years between 1975 and 1981 when I worked with my fellow students to install necessary equipment at the laboratory. Since

Professor Kim emphasized the importance of practice, and the money was not enough, we had to make and repair all the equipment by ourselves. Through these unprecedented lessons, he provided South Korean industry and higher education with a proper model of engineering education. (Kyung Chong-Min)

- When I first entered his laboratory, I had no clear idea of engineering. He repeatedly told us that "Fast decisions, though wrong, are better than slow ones" and "Be prepared to encounter the worst." These lessons still influence me as basic principles. He is the man who has influenced my life most profoundly. (Kwon Oh-Hyun)

- When I first came to Professor Kim's laboratory, I expected to learn something about the famous CCD. However, he did not say anything about it but ordered me to clean the empty room every day, emphasizing that the semiconductor laboratory must be clean. ... To upgrade the pipelines and DI water system in the laboratory, I often went to Euljiro market to purchase necessary parts. My colleagues then gave me the nickname "Pipe Park." (Park Zoo-Seong)

- My laboratory decided to purchase an air conditioner, and the discussion started. Professor Kim led the discussion to calculate the proper capacity of the air conditioner based on the size of the room and the quantities of input and output air. People usually tell the seller the size of the room when they order it, but he didn't. That taught me a lot. ... Thanks to this lesson, I often try modeling to deduce the results whenever I encounter similar issues. (Lee Yoon-Tae)

- Two most important lessons that I have learned from Prof. Kim are change of the mindset and an experimenting spirit. ... His famous buttocks theory is that "If you just repeat others' ideas, you will never overcome them but just follow their buttocks." Once I was struggling with the results of simulations. He said to me, "Did you make it? Make it by yourself! Then we will start discussions." So, I designed and constructed the circuit and carried out measurements. I then realized why he had asked me to make it. First of all, I loved the circuit that I had constructed. I paid much more attention to whether or not that circuit worked properly: when it didn't work, I began to analyze the cause of its failure. I reported the results of my experiments in the laboratory meeting, and he gave me a lot of advice on how to improve it further. If I had received that advice before I carried out the experiment, I would not have understood much of it. I learned that what is so obvious in theory in a book or an article may not be so obvious in reality, that something ideal is in fact only "ideal," that I must consider every contingency, and that I must be open to every possibility. (Oh Hyung-Seok)[52]

Many episodes recounted in *Uri Kim Choong-Ki Seonsaengnim* indicate how Kim's students learned problem-solving and improvising abilities that no textbooks had taught them in the classroom. Some of them recalled that they became specialists at repairing pipes, which surprised their parents when they asked their offspring, "What did you study at KAIST?" They also happily confessed that they repeated Kim's famous dicta in their own laboratories, workplaces, and classrooms: "Don't work on the subjects that others have thrown into the trashcan," "Scientists consider *why* first, but engineers must think *how* first," "Be prepared for the worst case," "If you work harder, everyone in your laboratory will be more comfortable. If you are comfortable, everyone in your laboratory will have to work harder."

The "engineer's mind" that Kim wanted to cultivate in his students was a balanced attitude between theory and practice. Since most South Korean engineering students in the 1970s and 1980s viewed engineering as another book learning discipline, Kim's equal emphasis on practice was quite a shock to them: many anecdotes about Kim are therefore related to hands-on practice. This emphasis on practice was not confined to Kim's laboratory but was also widely employed by other professors at KAIST during these years, especially those in the Department of Electrical Engineering. However, it was Kim Choong-Ki who epitomized this approach for more than three decades, through his inimitable words and unflagging persistence.

Although most of Kim's former students remember him as kind, humorous, nonauthoritarian, meticulous, and hard-working, they also recall that he was strict, hot-tempered, and often "terrifying" if students didn't fulfill their duties or showed any hint of laziness or inaccuracy. For example, Kim severely rebuked several students who neglected to clean the areas of the laboratory for which they were responsible. Also, if a student made a vague report to him, Kim immediately fired back, asking, "Many? How many? One hundred? One thousand? Or ten thousand?" or "10 millimeters? 10 meters? or 10 kilometers?"[53] Those who approached their lab jobs casually could be met with a burst of anger too. There is a legend among Kim's students that shows how they were afraid of him: according to the story, when KAIST was still in Seoul, some of his students bypassed his office by using a ladder to go up to the roof and then down into the lab.[54]

In short, the discipline of Kim's laboratory was very strict. This was absolutely necessary because his students were dealing with toxic chemicals, high-voltage machines, and very delicate instruments in a Class 1 clean room. Any negligence or small mistake could have evolved into a big or even fatal accident. All Kim's former students have therefore been severely criticized by him at least once or twice. That was a part of his teaching method, and no serious accidents took place.

As the laboratory settled down, Kim began to share his responsibilities and his "role of villain" with his senior students. From the beginning of the 1980s, senior students began to maintain discipline in the lab as well as to teach their juniors basic skills, and some made a lasting impression for their harsh treatment. As a result, Kim Choong-Ki's status was elevated along the lines of his being, for example, just "a little more dreadful than Kyung Chong-Min (who became assistant professor in 1983)."[55] For instance, Yoon Jun-Bo, who was a master's and doctoral candidate during the 1990s, remembers: "Once I arrived late at the regular students' meeting in the laboratory. My senior ordered me to come to the front and stand there during the meeting [just like an elementary school wrongdoer]. Professor Kim was the person whom these fearful seniors were truly afraid of" (Figure 2.5).[56] It seems that Kim's former students fully understand why he had to be so critical because to this day they happily recall his strict standards.

Another important factor was that Kim was always willing to cooperate with anyone and that his laboratory was open to outsiders from the very beginning. Kim liked to work with younger professors in the Department of Electrical Engineering,

FIGURE 2.5   Kim Choong-Ki's laboratory at KAIST was famous for its strict discipline and physical hardship. Regular lab meetings were often the occasions when Kim severely scolded some students' laziness, negligence, or selfishness. This photograph, taken in November 1992 at KAIST's new campus in Daejeon, seems to be a rare one in which Kim and some students are smiling.

Source: Courtesy of Park Sung-Kye.

and happily shared his lab space with them. He also liked to visit the labs of other departments and talk with professors and students there to get new ideas and perspectives. To encourage his own students to do the same thing, he organized a monthly dinner meeting and invited professors and advanced doctoral candidates from other laboratories and departments to meet his students. Kim Jun-Ho's recollections emphasize this point:

> My recent research subject is Bio-MEMS (Biomedical Micro-Electro-Mechanical Systems), which requires interdisciplinary research with biology or chemistry groups. Although I have only studied electronics, I can carry out this research without much difficulty, perhaps owing to Professor Kim's training. If I had not received his training that emphasized interdisciplinary cooperation, I would have a lot of trouble.[57]

As already noted, between 1975 and 2007, Kim produced 78 master's and 39 doctoral graduates.[58] At the same time, he influenced scores of others, directly or indirectly, through his openness. A good example is Chin Dae-Je, the first star semiconductor engineer in South Korea, who successfully developed 16Mb DRAM at Samsung Electronics in 1989, rose to become president of the company, and then served as

Minister of Information and Technology (2003–2006). Although Chin was not Kim's student at KAIST, he acknowledges his debt to Kim in his autobiography as follows:

> In the early 1970s, South Korea was just a beginner in the semiconductor field. There was no systematic education on the subject so that we had to study it by ourselves. In late 1974 [sic], a South Korean who had studied semiconductors in the US returned, and the situation began to change. That person was Professor Kim Choong-Ki of KAIS, who is often called the "godfather" of semiconductors in South Korea. For those like me who wanted to study semiconductors, he was a savior. … So, my colleagues and I at SNU actively used this opportunity, inviting him to organize seminars or to have discussions. There was an intense spirit of competition between SNU and KAIS in those years. Since I often went to KAIS to study the semiconductor with Professor Kim, I soon became a "problem" student at SNU.[59]

Chin later reiterated his gratitude to Kim as follows: "Professor Kim Choong-Ki of KAIST and the late Chairman Lee Byung-Chul (of Samsung Conglomerate) are two true teachers who made today's Chin Dae-Je."[60]

Chin was not the only problem student at SNU in the mid-1970s. Park Young-Jun, who became an influential professor of semiconductors at SNU from 1988, was another. In the preface of his 1995 book *Theory of VLSI Device* (in Korean), he writes:

> This book is similar to A. S. Grove's *Physics and Technology of Semiconductor Devices*, published in 1967, which is still widely read. I first learned of this book under Professor Kim Choong-Ki of KAIST, along with Chin Dae-Je of Samsung Electronics and Min Sung-Ki of IDT in the US. I still vividly remember the thrill when I first opened the book, which clearly explains every theory of semiconductor devices without using difficult jargon. That was the moment that I made up my mind to major in semiconductors. I still happily remember the days when I attended the lectures [by Kim] with that book.[61]

Kim not only helped those who studied semiconductors in the academy but enthusiastically assisted those who worked in industry and government research laboratories. For example, in the early 1980s, he often brought his students to KIET at Gumi during the summer vacation in order to give them some experience. He even served as director of research at KIET in 1982–1983, when, under his guidance, his former students there, along with other researchers, successfully developed 32Kb and 64Kb ROM.[62]

Kim was also eager to spread recent knowledge and know-how on semiconductors through KAIST's industry–academy cooperation workshop programs during the 1980s. These workshops were so popular and successful that Samsung, LG, and Hyundai decided to set up their own special, tailored educational programs at KAIST to train the semiconductor manpower that South Korea would need in the 1990s. During the 1980s and 1990s, this kind of close cooperation with the semiconductor industry helped Kim Choong-Ki and, later, his former student Kyung Chong-Min build and operate on the KAIST campus the Integrated Circuit Laboratory, Semiconductor Workshop for Young Engineers, Center for Electro-Optics, Center for High-Performance Integrated Systems (CHiPS), and Integrated-Circuit Design Education Center (IDEC), most of which were funded by either government or industry (Figure 2.6).

FIGURE 2.6 Kim Choong-Ki actively shared his knowledge and experience with young semiconductor engineers outside KAIST, through a series of workshops. This photo was taken in February 1992, at a workshop that Kim organized.

Source: Courtesy of Chang Hea-Ja.

Perhaps Kim's unique style of mentoring, along with his distinctive personality in doing so, can be summarized as "charismatic authority," a term that the famous German sociologist Max Weber coined to describe strong and enduring leadership in politics, sociology, religion, or academics.[63] In Kim's case, his long-lasting concern for his students, emphasis on earnestness, and unconcern for his own personal interest must have strengthened his charismatic authority.

BRIEF INTERRUPTION AND RESUMPTION OF RESEARCH

In the spring of 1994, Kim Choong-Ki was appointed the director of KAIST's newly created Office of Overall Planning and then served as vice president between 1995 and 1997. During these years, he therefore paid more attention to the management and future planning of KAIST than to training semiconductor specialists in his laboratory. He actually tried not to accept any new master's or doctoral candidates during this period but eventually accepted a few: for example, both Yoon Nan-Young and Yoon Jun-Bo visited him several times to persuade him to take them on, and, after becoming Kim's students, they often visited his office in the central administrative building for instructions.[64] Some younger professors in the department lamented the fact that Kim, then in his early fifties, was not mentoring as many students as before but was instead carrying out several administrative jobs. At the time, Kim thought that his mission as a pioneering teacher in semiconductors was almost complete, since by the early 1990s there were many young semiconductor professors not only at KAIST but also at other universities, and their numbers were rapidly increasing.

Kim's sudden departure to administrative positions in early 1994 sprang in part from his sense of crisis about the future of KAIST. KAIST's move from Seoul to Daejeon and its merger with KIT at the turn of the 1990s had elicited some anxiety and discontent among KAIST faculty members. Many feared that KAIST could not continue to attract the best students if it left Seoul, and this soon proved true: applications to KAIST from SNU and other top private universities dropped dramatically after KAIST moved to Daejeon.[65] The merger with KIT (an undergraduate-only college) might have compensated for this setback. However, most KAIST professors, who had taught only graduate students for the prior twenty years, didn't wish to teach undergraduates, and many even considered their new responsibilities to be a waste of time; moreover, the "chemical fusion" between KAIST and KIT faculty members was problematic. The changing political milieu of the early 1990s, the emergence of a stronger staff union, and the professors' desire to directly elect their president only added to these difficulties. In the spring of 1994, when for the first time the professors' union sent a list of candidates for the new KAIST president to the board of trustees, Kim's name was on it.[66] The board instead elected a senior renowned chemist, Shim Sang-Chul, and Shim appointed Kim director of the newly created Office of Overall Planning (the equivalent of vice president). When Shim resigned the next year, his successor, Yoon Duk-Yong, appointed Kim vice president. As the number two man at KAIST for about four years, Kim was determined to solve the problems that KAIST had encountered after relocating to Daejeon and merging with KIT. The result was a partial success. Kim contributed greatly to strengthening the undergraduate program and to improving relations between former KAIST and KIT faculty members, but he was not so successful in other areas.

Strengthening the undergraduate program was Kim's most signification achievement during his four-year administrative period. Unlike some of his colleagues who had also moved from the Seoul campus to Daejeon, he taught an introductory course on electrical engineering with much enthusiasm. His undergraduate courses became popular and created many fans among students, some of whom later became his doctoral candidates. Kim Joo-Ho, for example, remembers Kim's class as the most interesting one he took during his undergraduate years and went on to earn his master's and doctoral degrees from him.[67] And when Yoon Jun-Bo didn't explain an idea clearly in Kim's undergraduate class, Kim criticized Yoon severely, which made Yoon decide to work harder to impress his teacher.[68] Kim continued teaching undergraduates even after he became director of the Office of Overall Planning (and later vice president), even though he was technically exempted from doing so. His graduate students during this period often complained that Kim paid more attention to teaching undergraduates than to directing their research.[69] As a respected senior professor, Kim's approach must have influenced other professors' attitude toward undergraduate education. In fact, KAIST had no choice but to pay more attention to its undergraduates in order to secure a steady supply of able students for its graduate program. Kim also paid attention to strengthening humanities and social sciences courses for undergraduates.

A by-product of Kim's interest in undergraduate education during this period was that he began to pay serious attention to proper science and engineering education from a wider perspective. In a 1997 interview, he outlined his future goals:

> The economic development of Korea was dependent on reverse engineering and following advanced countries in engineering. In ten or twenty years we cannot do that any longer because we will be at the leading edge. … Education in science and engineering in Korea was like teaching [students] how to read maps, to find out where you want to go. We have been teaching how to read maps. And who made the maps? Advanced countries. We now have to change our educational policy and teach our students how to draw the maps.[70]

Kim would soon realize that it was much more difficult to change the educational policy and environment so as to teach students "how to draw maps" than it was to train a hundred semiconductor specialists. His ambitious dream of producing creative-minded scientists and engineers only began to become slowly realized in the first two decades of the twenty-first century.

Apart from strengthening undergraduate education and advancing harmony between former KAIST and KIT faculty members, Kim was not successful as an administrator. The primary reason was that he was not political enough. In fact, he had never been political, a serious defect for a high-ranking administrator. One of his former students, who first met Kim as an undergraduate student and then became his master's and doctoral student during the 1990s, remembers: "When he was the vice president, it seemed that he was disliked by many, perhaps because of his pure mind and strong stubbornness. However, those who know him certainly understand him and love him greatly."[71] Neither a "pure mind" nor "strong stubbornness" is required for a high-ranking administrator such as a vice president, who is supposed to be political instead. Kim was certainly aware of this and seemed to suffer from the situation that surrounded him. Like his father, Kim had been coping with diabetes since his early forties, and his health deteriorated during these years, slowly improving only after he quit his administrative job.

In late 1997, Kim finally finished his four-year administrative stint and returned to his laboratory. He accepted a very limited number of new students and concentrated on supervising his current students. Between 1998 and 2007, he produced six master's and eleven doctoral graduates, including five students who transferred from Han Chul-Hi's laboratory because of Han's illness and premature death. Kim was preparing for his compulsory retirement in 2008. While reducing his duties supervising degree candidates, he also focused on directing the newly established Center for Electro-Optics (Figure 2.7). Opened in December 1994, the center was funded by the Ministry of Defense (through its Agency for Defense Development) and consisted of three laboratories for thermal imaging, optical fiber, and laser research. Kim described the major goals of the center as "accumulating basic technologies on these three subjects, training specialists in them through cooperation among industry, the academy, and research institutes, and contributing to the self-reliance national defense capability by reducing dependency on foreign technologies."[72] He served as director of the center from its inception in 1994 to the end in

FIGURE 2.7 Besides closely working with the South Korean semiconductor industry, Kim Choong-Ki also cooperated with government research institutes. This picture was taken at a workshop organized by the Center for Electro-Optics in July 1995. The center was established in December 1994 with financial support from the Agency for Defense Development. It became the cradle for many state-of-the-art defense-related semiconductor devices, including the infrared sensor.

Source: Courtesy of Chang Hea-Ja.

2003. Technologies developed at the center were often transferred to industry for commercial production.

In the spring of 2007, less than a year before his compulsory retirement at the age of 65, Kim was elected one of KAIST's first three "Distinguished Professors," a position that virtually abolishes the compulsory retirement age and provides the elected professors with a special research fund.[73] KAIST still needed his continuing service as a mentor.

COMPARISON OF KIM CHOONG-KI WITH TWO TAIWANESE SEMICONDUCTOR SPECIALISTS

Kim Choong-Ki was also something of an exception on the international stage. It is important to remember that the South Korean government was not alone in making microelectronics and semiconductor manufacturing an important part of its strategy for rapid industrialization. Particularly in East and Southeast Asia, this industry was prioritized. Taiwan offers the closest parallel to South Korea's path, so it is helpful to compare Kim with two semiconductor specialists, Morris Chang and Simon Min Sze, who played similarly facilitating roles in the growth of Taiwan's microelectronics and semiconductor industry.

Morris Chang, the "Godfather of High Technology in Taiwan," is often described as the man who "created the semiconductor industry here [in Taiwan]."[74] Born in Ningbo, China, and raised in Hong Kong and Shanghai, Chang studied mechanical engineering at MIT and joined Texas Instruments as an engineering manager in 1958.[75] He soon turned his attention to electrical engineering, and the company sponsored his doctoral work on the subject at Stanford University. For a quarter century, he worked hard at Texas Instruments and rose to be the senior vice president responsible for the global semiconductor business. It was Chang who first recognized the importance of the foundry in the semiconductor industry. He quit Texas Instruments in 1983, after being refused the top position and sidelined within the company. The Taiwanese government then asked him to come to help its burgeoning semiconductor industry. In 1985, he moved to Taiwan and became the head of the Industrial Technology Research Institute. Two years later, in 1987, Chang established the Taiwan Semiconductor Manufacturing Company (TSMC) with the initial investment supplied by the Taiwanese government. The government gave Chang free rein to manage the company and has been TSMC's largest stockholder ever since the company was privatized in the early 1990s. TSMC was a new kind of business model, "focusing solely on manufacturing customers' products. By choosing not to design, manufacture, or market any semiconductor products under its own name, the company ensures that it never competes directly with its customers."[76] TSMC eventually became the "world's largest semiconductor foundry" in the twenty-first century, and Chang became the fifteenth richest man in Taiwan.[77]

Simon Sze, who was born in Nanjing, China, and raised in Taiwan, was also thoroughly trained in the United States. After receiving his bachelor's degree in electrical engineering at National Taiwan University (NTU), he received his master's degree from the University of Washington in Seattle and a Ph.D. from Stanford University. He then worked at the Bell Labs for 27 years.[78] At Bell Labs, he made several important discoveries, including the floating-gate transistor with Dawon Kahng in 1967, which became the foundation of nonvolatile memory (NVM) such as the flash memory. He also wrote and edited more than twenty textbooks on the field: his 1969 *Physics of Semiconductor Devices*, for example, became "the most cited work in contemporary engineering and applied science publications" and "a 'Must Study' [that] has been constantly used and referenced by worldwide semiconductor and integrated circuit researchers, graduate school faculty/students, and engineers across the entire electronic and photonic industry."[79] Sze returned to Taiwan in 1990 to become an influential professor at the National Chiao Tung University, where he trained many semiconductor specialists for Taiwanese industry.

Kim Choong-Ki, Morris Chang, and Simon Sze share an American education and company experience but differ in *how* they have influenced and contributed to the development of the South Korean and Taiwanese semiconductor industries, respectively. Kim and Sze's principal role has been that of a teacher who trains semiconductor specialists at universities in their home countries. They are quite dissimilar, however, in terms of *when* they returned home and *what* they taught the next generations. Kim returned to South Korea in 1975, at the age of 33, when the South Korean semiconductor industry had barely started, and therefore had to teach his students how to think with an engineer's mind so that they could go on to develop semiconductors and survive in the industry. Sze, in contrast, returned to Taiwan in 1990, at the age of 54, when the Taiwanese semiconductor industry was already growing; as a world-renowned specialist in semiconductors, his role was the more conventional one of writing textbooks, doing research, and consulting, which was enough to stimulate the Taiwanese semiconductor community. Chang's role has been, in turn, quite different from those of Sze and Kim. He suggested the future direction of the Taiwanese semiconductor industry—foundry work—and became an entrepreneur himself to realize his idea.[80] As an engineer-turned-entrepreneur, Chang may seem to be the most likely of the three to have become an influential leader in a high-tech industry like semiconductors: Gordon E. Moore and Robert Noyce's establishment of Intel in 1968 is a good earlier example of such a trajectory. The situation that Kim Choong-Ki encountered in South Korea in the spring of 1975, however, was quite different from the one that Chang and Sze found in Taiwan in 1985 and 1990. And Kim responded properly to the request of the time, perhaps inadvertently positioning himself to become the future "godfather" of the South Korean semiconductor community.

HOW DO WE APPRAISE KIM CHOONG-KI?

How can we summarize Kim Choong-Ki's contributions to the South Korean semiconductor community over the last 50 years? As described in Chapter 1, the success of the South Korean semiconductor industry is due largely to the close and unique cooperation among government, industry, and academia during this period. Kim Choong-Ki was both a major beneficiary and a major promoter of this unique triangular cooperation. On the one hand, he was a major beneficiary of this three-way relationship. The Park government's creation of a new graduate-only institution, KAIS, enabled Kim to return to South Korea and train the first two generations of South Korean semiconductor specialists in a laboratory equipped with all the necessary facilities. It was also the South Korean government that continuously emphasized the importance of semiconductors in the 1970s and early 1980s and that prepared detailed plans to push the reluctant electronics industry into the semiconductor market. At the same time, Kim was a major beneficiary of the country's industrial sector, which welcomed his students and supported his research. He often told his students, "Whatever the government prepared for the development of semiconductors, and however much engineers led semiconductor technology, the memory business could not have grown so successfully if [Samsung conglomerate founder] Lee Byung-Chul had not decided to invest in the memory business [in 1983]"—a very rare and frank recognition by a South Koran academician.[81]

On the other hand, Kim Choong-Ki was a significant promoter of this unique South Koran triangular relationship. Even though he was neither the first nor the only person to introduce semiconductor technology into South Korea, he was the first who systematically taught it at the graduate level in South Korea as early as 1975. Moreover, Kim was the model teacher whom both the South Korean government and the electronics industry had sought for a long time: he taught his students both theory and practice, along with how to develop and use a true engineer's mind. Most of his research papers have dealt with practical subjects that can be applied to real problems in the semiconductor industry. This has been especially evident when he and his students published their papers in Korean.[82] The South Korean government and the electronics industry had finally found someone who could and would provide the industry with practice-minded semiconductor engineers. The academy also discovered a model to emulate so as to produce semiconductor specialists for the industry. Since South Korea's human resources were limited and its companies very insular until the end of the twentieth century, training a sufficient number of Korean semiconductor engineers was critical to the success of the semiconductor industry.

As virtually the sole teacher and researcher in semiconductors in South Korea between 1975 and 1995, Kim Choong-Ki's contributions were significant enough to be recognized by the country's government, industry, and academy: the South Korean government awarded him the Moran Medal (Order of Civic Merit) in 1997 and selected him "Person of Distinguished Service to Science and Technology" in 2019 (Figure 2.8); the Samsung conglomerate awarded him its prestigious Ho-Am Prize in 1993; and SNU's College of Engineering honored him with a Distinguished Alumni Award in 1997; and KAIST gave him a Distinguished Professorship in 2007.[83] On August 27, 2022, *IEEE Spectrum* uploaded a biographical essay on Kim Choong-Ki entitled "The Godfather of South Korea's Chip Industry: How Kim Choong-Ki

FIGURE 2.8 Kim Choong-Ki's contribution to the development of semiconductor industry in South Korea has been recognized by the South Korean government. The photo was taken at the ceremony that appointed him a 2019 "Person of Distinguished Service to Science and Technology" in the fall of 2021 (delayed due to the COVID-19 pandemic). Kim is the first semiconductor engineer to receive this honor.

Source: Courtesy of Chang Hea-Ja.

FIGURE 2.9   Kim Choong-Ki has maintained a close relationship with his former students and their families even some decades after their graduation. This hiking group was organized by his former students to celebrate Kim's 60th birthday in 2002. They brought their families to Mount Deokyu in the Muju region of South Korea. Some of Kim's students have already reached middle age and have become important figures in the South Korean semiconductor community.

Source: Courtesy of Chang Hea-Ja.

Helped the Nation Become a Semiconductor Superpower," which made him the first Korean electrical engineer whose life, work, and contributions are detailed in this prestigious journal.[84]

For the new generations of South Korean semiconductor engineers in the twenty-first century, Kim Choong-Ki is just a legend, the great figure of the past who trained their professors or their bosses. For those who were trained by him, however, Kim remains a definite model to emulate (Figure 2.9). Kim's former students often wish they could borrow a page from his exacting approach, saying, "Professor Kim did so and so to us in his forties. Why can't I do so to my juniors (or subordinates) in my forties?" The respect and affection that Kim's students feel for their mentor is well reflected in the 2002 collective memoir *Uri Kim Choong-Ki Seonsaengnim*, in which 67 contributors happily recall how severely they were scolded and how thoroughly they were trained by Kim during their KAIST years, as well as how much and in what ways Kim's training and his famous dicta influenced their personal and professional lives since graduation. Whether or not they were especially close to their mentor during their years at KAIST, many still visit Kim to chat or to have lunch together.[85] One of Kim's former students, who has become a very successful semiconductor engineer, told the author, "Perhaps not 100% but at least 90% of Kim Choong-Ki's former students must be truly grateful for his teaching at KAIST because it has shaped their future careers."[86]

In conclusion, Kim Choong-Ki is a truly successful teacher and mentor, some of whose former students have certainly surpassed him in their contributions to the development of semiconductor devices in South Korea over the last four decades. As the Chinese sage Xunzi (third century BCE) put it: "Learning must never stop. Blue dye derives from the indigo plant, and yet it is bluer than the plant. Ice comes from water, and yet it is colder than water."[87] The relationship between Kim Choong-Ki and his students is a valuable example of how this energetic and fruitful dynamics can shape an entire industry—and the modern world.

NOTES

1 This chapter is an expanded version of Dong-Won Kim's "Transfer of 'Engineer's Mind': Kim Choong-Ki and the Semiconductor Industry in South Korea," *Engineering Studies, 11:2* (2019), 83–108.

2 Joong-Ang Ilbo, "3hoe Ho-Amsang Susangja Seonjeong," *Joong-Ang Ilbo* (February 23, 1993), www.joongang.co.kr/article/2788676 (searched on October 25, 2018). For more details on Kim's award, see The Ho-Am Foundation, http://hoamprize.samsungfoundation.org/eng/award/part_view.asp?idx=10.

3 The prize was established in 1990 by Samsung chairman Lee Kun-Hee to commemorate his late father, Lee Byung-Chul, founder of the Samsung conglomerate, and is awarded each year to "individuals who have contributed to academics, the arts, and social development, or who have furthered the welfare of humanity through distinguished accomplishments in their respective professional field." It was originally divided into four areas—science/engineering, medicine, mass media, and community service, but in 1995 science and engineering became separate areas. The awarding organization was also changed from the Samsung Welfare Foundation to the new Ho-Am Foundation in 1997. For more information about the Ho-Am Prize, see The Ho-Am Foundation, http://hoamprize.samsungfoundat ion.org/eng/foundation/intro.asp.

4 Park Sang-In et al., *Uri Kim Choong-Ki Seonsaengnim [Our Teacher, Kim Choong-Ki]* (Daejeon: Privately printed, 2002), 44.

5 Seo Moon-Seok, "Iljeha Gogeupseomyugisuljadeuleu Yanseong-gwa Sahoejinchule gwanhan Yeongu," *Kyungjesahak, 34* (2003), 83–116.

6 Hyun Won-Bok, *Uri Gwahak, keu Baeknyeon eul Bitnaen Saramdeul: Hankukeu Gwahak Gisulin Baeknyeon, Vol. 1* (Seoul: Gwahak Sarang, 2009), 267.

7 Kyungseong Bangjik, *Kyungseong Bangjik 50nyeon (1919–1969)* (Seoul: Kyungseong Bangjik, 1969), 120.

8 Ibid., 138–139.

9 Hyun Won-Bok, *Uri Gwahak*, 268.

10 Kim Byung-Woon, *Myeonbangjeok [Cotton Spinning]* (Seoul: Eulyumunhwasa, 1949).

11 Interview with Kim Choong-Ki by Kim Dong-Won (June 5, 2014).

12 Kim Joon-Ki, Choong-Ki's younger brother, was a computer scientist/engineer. He worked at IBM's Thomas J. Watson Research Center in Yorktown Heights, New York, for more than two decades. In 1997, the Samsung conglomerate invited him to join the company as a "S[uper] class" scientist/engineer. Between 1997 and 2007, he worked at Samsung Advanced Institute of Technology and other related Samsung companies.

13 KAIST, "A Ceremony in Honor of Professor Kim Choong-Ki," KAIST Archive, video footage 2–10237. According to South Korean law, all professors must retire at the age of 65. Kim's colleagues therefore prepared a special ceremony when he reached this age. As "Distinguished Professor," however, he continued working in his laboratory at LG Innovation Hall (formerly LG Semicon Hall), which was constructed in 1997 with a donation from LG Semiconductor to train semiconductor designers and to accelerate the cooperation between KAIST and the semiconductor industry.

14 Interview with Kim Choong-Ki by Jeon Chihyung (March 7, 2016), 1. Jeon conducted a series of interviews of former chairmen of the KAIST Department of Electrical Engineering to write the history of the department. The recordings were transcribed and deposited in the department, which Kim Choong-Ki chaired in the 1980s.

15 Yang inherited a PhD candidate when he moved to Columbia in 1965, and Kim was Yang's first PhD candidate whom he trained from the beginning.

16 Choong-Ki Kim, "Current Conduction in Junction-Gate Field-Effect" (Ph.D. thesis, Columbia University, 1970), 2.

17 Unfortunately, in the 2020s it is very difficult to trace the history of the Fairchild Camera and Instrument Corporation. The company's last website (www.fairchildimaging.com/our-history) was permanently deleted in the late 2010s. Wikipedia therefore provides the best information about the company. https://en.wikipedia.org/wiki/Fairchild_Camera_and_Instrument (searched on July 31, 2022).

18 Computer History Museum, "Fairchildren," https://computerhistory.org/fairchildren/ (searched on August 31, 2022). See also Leslie Berlin, *Troublemakers: Silicon Valley's Coming of Age* (New York: Simon & Schuster, 2017).

19 Willard S. Boyle and George E. Smith, "Charge Coupled Semiconductor Devices," *The Bell System Technical Journal, 49:4* (April 1970), 587–593.

20 Engineering and Technology History Wiki, "Charge-Coupled Device," http://ethw.org/Charge-Coupled_Device (searched on March 12, 2019), with some modifications.

21 Interview with Kim Choong-Ki by Jeon Chihyung (March 7, 2016), 1. Fairchild was the first company to manufacture the commercial CCD in 1973.

22 K. C. Gunsagar, C. K. Kim, and J. D. Phillips, "Performance and Operation of Buried Channel Charge Coupled Devices," *1973 International Electron Devices Meeting (IEDM)*, Washington, D.C. (December 3–5, 1973), 21–23 on 21. See also Choong-Ki Kim, "Carrier Transport in Charge-Coupled Devices," *1971 IEEE International Solid-State Circuits Conference, University of Pennsylvania*, Philadelphia (February 19, 1971), 158–159; Choong-Ki Kim, J. M. Early, and G. F. Amelio, "Buried-Channel Charge-Coupled Devices," presented at Northeast Electronics Research and Engineering Meeting (NEREM), Boston (November 1–3, 1972); Choong-Ki Kim and E. H. Snow, "P-Channel Charge-Coupled Devices with Resistive Gate Structure," *Applied Physics Letters, 20* (1972), 514; and Choong-Ki Kim, "Design and Operation of Buried Channel Charge-Coupled Devices," presented at CCD Applications Conference sponsored by the Naval Electronics Laboratory Center, San Diego (September 18–20, 1973).

23 Choong-Ki Kim, "Two-Phase Charge-Coupled Linear Imaging Devices with Self-Aligned Implanted Barrier," *1974 International Electron Devices Meeting (IEDM)*, Washington D.C. (December 9–11, 1974), 55–58 on 56.

24 Interview with Kim Choong-Ki by Jeon Chihyung (March 7, 2016), 5.

25 Email from Edward S. Yang to Kim Dong-Won (February 12, 2022).

26 Interview with Kim Choong-Ki by Jeon Chihyung (March 7, 2016), 10–11.

27 Ibid., 4–5.

28 Ibid., 5.

29 One of Kim Choong-Ki's uncles reminded him of the sacred responsibility of Korean eldest sons.

30 Interview with Kim Choong-Ki by Jeon Chihyung (March 7, 2016), 4.

31 The name of the department has changed several times since its inception, in accord with organizational changes over the years. For convenience, I adopt the name "Department of Electrical Engineering," which is the wording that most KAIST professors and students commonly use.

32 Interview with Kim Choong-Ki by Kim Dong-Won (September 21, 2020).

33 Kim Choong-Ki, "Guknae Bandoche Gongeop-eu Baljeon Hoego," *Jeonjagonghakhoeji, 13:5* (1986), 435–444.

34 See Seoul National University, *Seouldaehakgyo 50nyeonsa* [*Fifty-Year History of Seoul National University*], *Vol. 2* (Seoul: Seoul National University), 147–155. By the end of the 1970s, most professors in the SNU Department of Electronics had been trained at Japanese universities, and few had received their doctoral degrees from US universities. It was quite the opposite at KAIST, where most professors had received their doctorates from US or other Western universities.

35 This total of 78 master's theses includes six co-advised theses.

36 Park Sang-In et al., *Uri Kim Choong-Ki Seonsaengnim*, 29.

37 Kim Choong-Ki, "Guknae Bandoche Gongeop eu Baljeon Hoego," 440.

38 Kwang-Seok Seo and Choong-Ki Kim, "On the Geometrical Factor of Lateral p-n-p Transistors," *IEEE Transactions on Electron Devices, ED-27* (January 1980), 295–297; and Chong-Min Kyung and Choong-Ki Kim, "Charge-Coupled A/D Converter," *1981 Custom Integrated Circuit Conference*, Rochester, New York (May 1981), 621–626.

39 Choon-Sik Oh, Yo-Hwan Koh, and Choong-Ki Kim, "A New P-Channel MOSFET Structure with Schottky-Clamped Source and Drain," *International Electron Devices Meeting (IEDM)*, San Francisco (December 1984), 609–613.

40 From http://hoamprize.samsungfoundation.org/eng/award/part_view.asp?idx=10. For his research on the rapid thermal process, see Kim Choong-Ki, "Rapid Thermal Processing for Submicron Devices," *Teukjeongyeonku Gyeolkwa Balpyohyoe Nonmunjip 1* (1988), 155–159; Kim Choong-Ki and Kim Jeong-Kyu, "Two-Step Rapid Thermal Diffusion of Phosphorous and Boron into Silicon from Solid Diffusion Sources," *Teukjeongyeonku Gyeolkwa Balpyohyoe Nonmunjip 1* (1989), 192–196; Kim Choong-Ki, Lee Dong-Yup, and Jo [Cho] Byung-Jin, "Metal Alloy and Implantation Annealing by Rapid Thermal Processing," *Teukjeongyeonku Gyeolkwa Balpyohyoe Nonmunjip 1* (1989), 202–205; Jo [Cho] Byung-Jin, Choi Jin-Ho, and Kim Choong-Ki, "Slip Elimination in Rapid Thermal Processing," *Teukjeongyeonku Gyeolkwa Balpyohyoe Nonmunjip 1* (1989): 206–209.

41 The Ho-Am Foundation's website states in Korean that "Professor Kim made excellent research achievements in semiconductor devices and integrated circuits through experiments despite very poor conditions for experiments in South Korea." In the English version on the website, the phrase "despite poor conditions for experiments in South Korea" is omitted. See http://hoamprize.samsungfoundation. org/eng/award/part_view.asp?idx=10.

42 Jun-Bo Yoon, Chul-Hi Han, Euisik Yoon, and Choong-Ki Kim, "High-Performance Three-Dimensional On-Chip Inductors Fabricated by Novel Micromachining Technology for RF MMIC," *IEEE MTT-S International Microwave Symposium Digest, 4* (1999), 1523–1526.

43 Jeonja Shinmun, "KAIST Kim Choong-Ki Gyosutim, 3chawon jipjeop-inductor chut Gaebal," *Jeonja Shinmun* (July 15, 1999), www.etnews.com/199907150074 (searched on October 11, 2021).

44 Park Sang-In et al., *Uri Kim Choong-Ki Seonsaengnim*, 66.

45 Dong-A Ilbo, "10nyeon dwi Hankukeul bitnael 100in: 'Hankyeneun eupda' ... Silpaedo jeulgimyeo dalyeo-on 100gaeui kkum," *Dong-A Ilbo* (April 2, 2013), www.donga.com/news/People/article/all/ 20130402/54131148/1 (searched on September 1, 2021); POSCO TJ Park Foundation, "2021 POSCO TJ Park Prize Awardees," www.postf.org/en/page/award/history.do (searched on September 1, 2021).

46 The list of Kim Choong-Ki's former students at KAIST can be found in Park Sang-In et al., *Uri Kim Choong-Ki Seonsaengnim*, 134–135.

47 Ibid., 31.

48 Interview with Jung Hee-Bum by Kim Dong-Won (May 23, 2021).

49 Interview with Kim Oh-Hyun by Kim Dong-Won (December 3, 2021).

50 Yoon Jun-Bo, who joined KAIST's Department of Electrical Engineering as a professor in 2000, is the only exception to this trend. With no experience in either industry or a large research institute and with just one year's experience as a postdoctoral fellow at the University of Michigan, he was hired largely due to his excellent research record.

51 Park Sang-In et al., *Uri Kim Choong-Ki Seonsaengnim*, 86–87.

52 Ibid., 24–25, 27, 30–31, 62–63, and 131–132, respectively.

53 Ibid., 61.

54 Ibid., 114.

55 Ibid., 49.

56 Ibid., 117.

57 Ibid., 106–107.

58 As previously noted, six of these 72 master's theses were cosupervised with other professors in the KAIST Department of Electrical Engineering.

59 Chin Dae-Je, *Yeoljeong-eul Gyeongyeong hara* (Paju, Kyeongkido: Kimyeongsa, 2006), 175–176.

60 Daejeon Ilbo, "Insaeng-ui Vision boindamyeon miri Junbihaneun Jase Pilyo: Kim Choong-Ki KAST Teukhun-gyosu," *Daejeon Ilbo* (March 13, 2007), www.daejonilbo.com/news/articleView. html?idxno=675390 (searched on November 1, 2020).

61 Park Young-Jun, *VLSI Soja Iron* [*Theory of VLSI Device*] (in Korean) (Seoul: Kyohaksa, 1995), i.

62 ETRI, *ETRI 35nyeonsa* [*Thirty-Five Year History of ETRI*] (Daejeon: ETRI, 2006), 130–131.

63 Max Weber, "The Principal Characteristics of Charismatic Authority and Its Relation to Forms of Communal Organization," in idem, *The Theory of Social and Economic Organization*, translated by A. M. Henderson and Talcott Parsons, edited with an introduction by Talcott Parsons (London: The Free Press of Glancoe, Collier-Macmillan Ltd., 1947), 364–369 on 367.

64 Park Sang-In et al., *Uri Kim Choong-Ki Seonsaengnim*, 114.

65 Kim Dong-Won et al., *Hankukgwahakgisulwon Sabansegi: Miraereul hyanghan kkeunimeopneun Dojeon* [*The First Quarter Century of KAIST: An Endless Challenge to the Future*] (Daejeon: KAIST, 1996), 103–111 and 176–177.

66 Kim Choong-Ki silently declined the nomination.

67 Park Sang-In et al., *Uri Kim Choong-Ki Seonsaengnim*, 106.

68 Ibid., 116.

69 Ibid., 105.

70 Dong-Won Kim and Stuart W. Leslie, "Winning Markets or Winning Nobel Prizes: KAIST and the Challenges of Last Industrialization," *Osiris, 13* (1998), 182. Dong-Won Kim and Stuart Leslie interviewed Kim Choong-Ki on October 6, 1997.

71 Park Sang-In et al., *Uri Kim Choong-Ki Seonsaengnim*, 102.

72 Kim Choong-Ki, "Jeonjagwanghak Yeonkuso" [Center for Electro-Optics]," *Gukbang-gwa Gisul, 263* (January 2001), 36–41 on 36.

73 See the official announcement by KAIST, posted by the Department of Electrical Engineering at https://ee.kaist.ac.kr/node/10278?language=ko.

74 Mark Landler, "The Silicon Godfather; The Man Behind Taiwan's Rise in the Chip Industry," *The New York Times* (February 1, 2000), www.nytimes.com/2000/02/01/business/the-silicon-godfather-the-man-behind-taiwan-s-rise-in-the-chip-industry.html (searched on February 7, 2019).

75 For Chang's life, see Sahil Bloom, "The Amazing Story of Morris Chang," *The Curiosity Chronicle* (January 25, 2021), https://sahilbloom.substack.com/p/the-amazing-story-of-morris-chang; and "Morris Chang: Chinese-Born Entrepreneur," *Britannica*, www.britannica.com/biography/Morris-Chang (searched on February 17, 2022).

76 Taiwan Semiconductor Manufacturing Company, "Dedicated IC Foundry," www.tsmc.com/english/dedicatedFoundry (searched on February 17, 2022). For more about TSMC as well as other Taiwanese foundry companies, see Muh-Cherng Wu, "Chapter 5: IC Foundries: A Booming Industry," in Chun-Yen Chang and Po-Lung Yu (eds.), *Made by Taiwan: Booming in the Information Technology Era* (Singapore: World Scientific, 2001), 133–152.

77 Forbes, "Morris Chang," www.forbes.com/profile/morris-chang/?sh=1680cbbe5fc4 (searched on February 17, 2022).

78 For a brief biography of Simon Min Sze, see an official profile at the National Chiao Tung University, "Simon M. Sze," https://eenctu.nctu.edu.tw/en/teacher/p1.php?num=127&page=1; Computer History Museum, "Oral History of Sze, Simon," www.computerhistory.org/collections/catalog/102746858; Semicon China, "Dr. Simon M. Sze," www.semiconchina.org/en/788; and IEEE Xplore, "Simon M. Sze," https://ieeexplore.ieee.org/author/37294788800 (searched on February 17, 2022).

79 IEEE Xplore, "Simon M. Sze," and Future Science Prize, "2021 The Mathematics and Computer Science Prize Laureate: Simon Sze," www.futureprize.org/en/laureates/detail/56.html (searched on February 17, 2022).

80 For more information about Morris Chang's view and work on foundries at TSMC, see Kevin Xu, "Morris Chang's Last Speech," *Interconnected* (September 12, 2021), https://interconnected.blog/morris-changs-last-speech/ (searched on February 17, 2022).

81 Park Sang-In et al., *Uri Kim Choong-Ki Seonsaengnim*, 14.

82 For example, Kim Choong-Ki's group published 43 papers in Korean during the 1980s, and most of them were about the "design," "fabrication," or "construction" of semiconductors.

83 The Moran Medal is the second highest medal in the Order of Civic Merit, which is awarded to those who contribute significantly to the progress of politics, economics, society, education, or scholarship in South Korea. Kim Choong-Ki was one of twelve scientists and engineers who received

the 2019 "Person of Distinguished Service to Science and Technology" honor. See Person of Distinguished Service to Science and Technology, "Godfather of Semiconductor Industry Who Led the Technological Development by Problem-Solving Research and Education," www.koreascienti sts.kr/eng/merit/merit-list/?boardId=bbs_0000000000000051&mode=view&cntId=58&category= 2019&pageIdx=. The College of Engineering at SNU began to present the Distinguished Alumni Award in 1993, and Kim Choong-Ki was its first recipient who taught at a South Korean university (see http://eng.snu.ac.kr/node/13).

84 Dong-Won Kim, "The Godfather of South Korea's Chip Industry: How Kim Choong-Ki Helped the Nation Become a Semiconductor Superpower," *IEEE Spectrum* (August 28, 2022), https://spectrum. ieee.org/kim-choong-ki (searched on August 29, 2022), and *IEEE Spectrum* (October 2022), 32–39.

85 In December 2021, when I invited Kim Choong-Ki to lunch at an Italian restaurant near his home, he instead suggested a Japanese restaurant far from his house since he visited that Italian restaurant so often to have lunch with his former students.

86 The interviewee asked to remain anonymous.

87 Xunzi, *Xunzi: The Complete Text*, translated and with an introduction by Eric L. Hutton (Princeton, N.J.: Princeton University Press, 2014), 1.

Part II

Mentees

3 Teacher, Chip Designer, Patron of Startups, and Spokesperson
Kyung Chong-Min

Kyung Chong-Min is the one of the best known semiconductor professors in South Korea. He is also a famous chip designer, a prominent supporter of startups, and a leading spokesperson for the South Korean semiconductor community. His life and career have been filled with challenges to achieve seemingly unattainable goals: in the 1990s, for example, when the South Korean semiconductor industry was still in its infancy, he challenged himself to design microprocessor chips that would be fully compatible with Intel's 80386 and 80486 chips and succeeded. Kyung launched two startup companies in the late 1990s and 2000s, both of which failed. He nonetheless continued preaching that South Korean academia should support startups more aggressively and launched another startup company when he retired. He has also been an effective spokesperson for the South Korean semiconductor community through his many newspaper articles on various subjects. By filling these diverse roles simultaneously over the course of 40 years, Kyung has left clear footprints in the history of semiconductors in South Korea. The title of his contribution to the 1994 book that collects twelve essays by KAIST's professors and graduates is "Youth without Challenge Can't Be Called Youth," which nicely sums up his view of life.[1] As Kim Choong-Ki's first doctoral candidate at Korea Advanced Institute of Science (KAIS, later KAIST), Kyung's relationship with Kim has been special and close, despite their very different characters and different contributions to the South Korean semiconductor community.

EARLY YEARS

Kyung Chong-Min was born on June 21, 1953, in Seoul. His parents paid special attention to their first son's education and expected him to have a successful career. Kyung excelled in school and entered Kyunggi Middle and High Schools, the most prestigious secondary schools in South Korea at that time. However, he went through a hard time during puberty and was far from a model student: he began to smoke and drink, became an expert at billiards and the game go, joined the climbing club, and played soccer.[2] It was therefore a surprise that in early 1971 he was admitted to the Department of Electronic Engineering at Seoul National University (SNU), one of

DOI: 10.1201/9781003353911-6

the most difficult departments to enter in those years. These events reveal his brilliant mind but also his troubled and defiant mind in his youth. As the eldest son, Kyung didn't want to disappoint his parents, but he also resisted the way South Korean society trained members of its selected elites. This conflicting attitude continued during his freshman and sophomore years at the university. He participated in some of the political demonstrations that swept through South Korean universities in the early 1970s, although he was not seriously enthusiastic about them. Instead, when the university was closed in the fall of 1971 due to a presidential garrison decree, Kyung climbed several mountains in South Korea (Figure 3.1) and became enamored of some Western philosophies, especially existentialism.[3] He was not quite sure of his future as an engineer but couldn't throw it away either. Kyung's wandering period ended only in his junior year, when he had to consider his future after graduation. He began to study hard to catch up with his classmates and soon became one of the top students in his class. Since he was a healthy young man, admission to KAIS was the only way to continue his career as an engineer without the three-year interruption of military service. He took the entrance examination and passed it. A very different life was awaiting him at KAIS.

In the fall of 1975, his second semester at KAIS, Kyung chose Kim Choong-Ki as his advisor for his master's thesis, thus beginning his lifelong connection with Kim and with semiconductors. Kim had begun teaching at KAIS the preceding spring and was busy building a small clean room for semiconductor research. He had inherited seven master's candidates from other professors and also selected four new ones that year, including Kyung and Kwon Oh-Hyun (who later served as vice chairman and CEO of Samsung Electronics between 2012 and 2017). Kyung recalls that when he

FIGURE 3.1 Kyung Chong-Min loves mountain climbing, which he took up seriously during his troubled college years. He remembers that he was once stranded on a cliff and had to wait several hours for the arrival of a rescue team. **Left**: A photo taken when he climbed Mount Halla during college. **Right**: Kyung climbed Mount Seorak in 1994.

Source: Courtesy of Kyung Chong-Min.

and his classmates first arrived at Kim's laboratory, it was almost "empty."[4] Since Kim believed that engineers learn more by constructing their own necessary facilities and instruments, he required his students to do so. Kim often visited the semiconductor laboratories at Stanford University and described the details he observed to his students when he returned. Kyung's task was one of the most dangerous—to set up the darkroom, where many poisonous chemicals were stored. He remembers that he was once shocked by high voltage during the setup but fortunately was not seriously hurt.[5]

Kyung's master's thesis was on the "Low Level Currents in Buried Channel MOS Transistor," whose experimental data he found were in "a good agreement with the result of the numerical calculation despite many approximations and assumptions involved in modelling the device."[6] Kim carefully observed Kyung's progress and suggested that he become his first doctoral candidate. Kyung was not sure about doing so in the beginning: not only was he seriously considering taking a job to support his family after receiving his master's, but South Korean society did not regard holders of domestic doctorates highly at the time, preferring those who had received their doctorates from foreign universities. Kyung's family, however, strongly supported his pursuing a doctoral degree at KAIS.[7]

Kyung chose a new type of analog-to-digital converter, using a charge-coupled device (CCD), as the subject of his doctoral thesis. Since Kim's laboratory was still under construction and Kyung was one of the main workers building it, Kyung could not always concentrate on his own project. He was also a student supervisor who managed juniors at the laboratory: many students in Kim's laboratory in the late 1970s and early 1980s remember Kyung as the second most intimidating figure in the laboratory, surpassed only by Kim himself. Kim helped Kyung design the device, but it was solely Kyung's job to purchase and install the necessary equipment for making the target device. He even went to GoldStar Semiconductor in the city of Gumi, about a five-hour bus trip from Seoul, to make the necessary photomasks. He also received some assistance from engineers at Samsung Semiconductor and Motorola Korea while he was constructing the semiconductor device. His efforts finally began to pay off in early 1980, and he published the outcomes in the *IEEE Journal of Solid-State Circuits*.[8] In February 1981, he became the first doctor of engineering at the Department of Electrical Engineering at KAIST (KAIS had been renamed KAIST in January 1981). In his relatively short acknowledgments, Kyung expressed his deep gratitude to many people, including his fiancée, who had helped him organize the final draft of his dissertation (Figure 3.2).

After a short telephone interview with George E. Smith, the coinventor of CCD with Willard Boyle, Kyung was hired by Bell Labs in Murray Hill, New Jersey, as a postdoctoral fellow. He arrived there in April 1981, just two months after his graduation, with his bride. He experienced "three troubles"—his first social life after school, marriage, and his first foreign experience—simultaneously for the next two years.[9] His pride at being KAIST's first doctor of electrical engineering was often badly bruised by other scientists and engineers at Bell Labs, and sometimes by his own misstep or inexperience. Nevertheless, he learned a valuable lesson from his two-year stay at Bell Labs. He found that the CCD had few things to cultivate further and began to research new areas for the subject of his future career. He became interested

FIGURE 3.2   Kyung Chong-Min was trained thoroughly by Kim Choong-Ki, first as a master's candidate and then as Kim's first doctoral candidate. **Left**: Kyung (far right) poses with his colleagues Choi Sung-Hyun (second from right) and Kwon Oh-Hyun (third from right), and his mentor Kim (far left). **Right**: In the spring of 1981, Kyung was hired as a postdoctoral fellow by Bell Labs in Murray Hill, New Jersey, where he brought his bride.

Source: Courtesy of Kyung Chong-Min.

in design technology, especially that of microprocessor chips with computer-aided design (CAD): he was slowly becoming a designer of microprocessors.

In the late fall of 1982, Kyung received a letter from Kim Choong-Ki inviting him to return to KAIST as an assistant professor in the Department of Electrical Engineering. Kim later recalled that semiconductor research and teaching grew rapidly at KAIST, yet there were few reliable candidates except Kyung to fill the new post.[10] Kyung accepted the offer without hesitation and returned to KAIST in January 1993. Familiar but very different surroundings awaited him.

PROFESSOR AT KAIST

Kyung's position as an assistant professor at KAIST, at the tender age of 30, was a little bit strange and even uncomfortable at the outset. Senior professors in the department, who had been his teachers just two years before, were not used to treating him as a colleague and an equal. Graduate students still considered him their senior rather than their teacher. The Korean custom of demonstrating respect by using honorific speech, which dictates changing the form of almost every word according to one's senior or junior position relative to the person one is addressing, worsened the situation. Moreover, the other professors in the department had all published more and better research results in the prestigious international journals than Kyung had at the time. As a result, few talented students wanted to join his laboratory in the beginning.[11] To become a respected and competent professor in the department, Kyung first had to become a renowned researcher. He recalls his early years as an assistant professor as follows:

> I was the first graduate of the department to become a professor there, and naturally very nervous in the beginning. All senior professors in the department were notorious for pushing themselves and their students to work hard day and night. Since I was just a beginner as a researcher compared with them, I devoted all my time and energy to research and teaching preparation. Sunday church service and the weekly Bible study on Friday night were the only two exceptions that freed me from those burdens. Until KAIST moved to the new campus in the Daeduk area [in the early 1990s], I worked at my laboratory from 8 a.m. to 11 p.m., skipping most holidays and vacations.[12]

Kyung was so busy at this early stage that, to save commuting time, he refused to move from campus housing to a condominium complex in which units were being offered to KAIST professors at a special discount.[13] Since the price of that complex skyrocketed within a few years, he lamented this decision for a long time and regrets it for his wife's sake to this day.

Kyung's first priority as a junior professor was to become a competent researcher and prolific writer. He succeeded in doing this within a decade, publishing 27 research papers in international journals and 49 in South Korean journals between 1983 and 1993. His most frequent subject during this period was algorithms and hardware for image analysis using CAD. Titles of his articles include "New Hardware Architecture for Fast Raster Image Generator," "Hardware Accelerator for Outline Font Generation," "Fast Image Generation Method for Animation," "FAMOS: An Efficient Scheduling Algorithm for High-Level Synthesis," "A One-Pass Standard Cell Placement Algorithm Using Multi-Stage Graphic Model," "Graphic Workstation" (in Korean), "Functional-Level Design and Simulation of a Graphic Processor" (in Korean), "Development of Three-Dimensional Solid Modeling and Rendering" (in Korean), and "A Graphic Accelerator for Hidden Surface Removal and Color Shading" (in Korean).[14] He also published on various other subjects, such as the channel routing system using a complementary metal-oxide semiconductor (CMOS) standard cell, the impurity profile in semiconductors, adaptive cluster growth, hierarchical alternating linear ordering (HALO), and parallel processing by very large-scale integration (VLSI) processor.

Kyung's contribution to the introduction and spread of CAD in the South Korean semiconductor community deserves special attention. In a very early article in 1984, he emphasized the importance of CAD as the design of the semiconductor chip became more and more complicated:

> The recent changes of design technology in semiconductor chips resulted from more use of the computer when chip designers manage vast amounts of data and use various simulation programs. ... The design of VLSI is changing into the design of CAD software that designs VLSI. ... Advanced countries like the USA and Japan strictly restrict the export of the CAD software for chip design. Although they allow it, South Korea does not have enough manpower to use these CAD tools.[15]

In another article that year on the manufacturing process of the integrated circuit, Kyung showed the relationship among the various CAD tools in predicting the circuit operation.[16] His efforts were soon recognized in the semiconductor community. The Ministry of Science and Technology awarded a special fund to his research team in 1986 for developing CAD software for South Korean engineers. Kyung's team published the summary of its report the next year, in which it presented various CAD software as well as the integrated system to hold it together.[17]

It was, however, the development of microprocessor chips, not the advancement of CAD for the design of chips, that made Kyung one of the most celebrated semiconductor professors in South Korea. He began to apply his algorithm and CAD design technology to the development of various chips in 1988, when he converted his CAD laboratory into a VLSI system laboratory. He was ambitious enough to develop microprocessors that would be compatible with Intel's 80386 and 80486 chips. With the support from the Ministry of Industry, he made a contract with Hyundai Electronics in 1991 to codevelop them. This seemed an audacious or even ridiculous dream in the beginning, considering the technology level of South Korea in the early 1990s: even Samsung Electronics, South Korea's front-runner in the semiconductor industry, did not try it. Kyung's team carried out this project so enthusiastically that many of its members did not have time to publish their usual research papers for four or five years.[18]

In July 1994, Kyung's team announced the development of the HK387 chip that was completely compatible with Intel 80387 (a coprocessor to accelerate the speed of calculation of the main microprocessor).[19] By the end of 1994, his team also completed the design of a chip that was compatible with Intel 80386.[20] When the first test chips arrived, Kyung's students examined them to see whether they really were compatible with Intel's chips: they pulled Intel's chips (80386 and 80387) from the computer board and inserted their chips instead. The new chips worked smoothly with the software![21] On September 2, 1995, Hyundai Electronics and KAIST jointly announced that they had succeeded in developing the HK386, which was fully compatible with Intel 80386 microprocessor.[22] Although the 80486 and the Pentium were two mainstream CPUs by then, it was quite sensational news in South Korea to have developed a fully working microprocessor for the first time that was much more complicated than memories and that had a wider market. Almost all major newspapers in South Korea reported extensive stories on this accomplishment, often including interviews with Kyung.[23] The full description of the HK386 was presented at the 1997 Asia and South Pacific Design Automation Conference (ASP-DAC) (Figure 3.3).[24] Hyundai's support to develop a microprocessor compatible with Intel 486 had been halted in 1996, largely because Hyundai did not want to provoke powerful Intel. Kyung's team, however, continued their efforts and finished designing the HK486 and also another chip compatible with Intel Pentium by the summer of 1997.

The success of Kyung's development of the microprocessor greatly influenced the establishment of two important centers at KAIST—the Center for High-Performance Integrated Systems (CHiPS) in 1994, and the Integrated Circuit Design Education Center (IDEC) in 1995 (Figure 3.4). As the successor of the Center for High-Performance Integrated Circuits that had been founded in 1991 with the support from

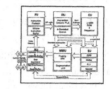

HK386 : An x86-compatible 32bit CISC Microprocessor

C.M.Kyung, I.C.Park, S.K.Hong, K.S.Seong, B.S,Kong, S.J.Lee, H.Choi
S.R.Maeng, D.T.Kim, J.S.Kim, S.H.Park and Y.J.Kang
KAIST
email : kyung@dalnara.kaist.ac.kr

Abstract— In this paper, we describe the implementation and design methodology of a microprocessor, called HK386. The microprocessor is compatible with Intel 80386 with respect to the behavior of each instruction set. As the extraction of the exact behavior of each instruction set is the single most important step in compatible chip design, we focused our effort on establishing the reliable verification strategy ensuring the complete instruction level compatibility. The HK386 was successfully designed and fabricated using 0.8 um CMOS technology.

I. INTRODUCTION

We have developed 32-bit microprocessor called HK386 which is fully instruction-level and pin-to-pin compatible with Intel 80386. The HK386 is simply plugged into the real PC instead of Intel 80386 and run all application softwares. Maintaining the compatibility with previous generation processors saves huge effort in developing application software which forms huge market. Since the behavior of the Intel 80386 is very complex and veiled, the most time-consuming part in the development of HK386 was to verify the compatibility. We focused our effort to main-

II. ARCHITECTURE

As indicated in the block diagram of Fig.1, HK386 consists of 6 blocks; an instruction prefetch unit(FU), a decode unit(DU), a control unit(CU), an execution unit(EU), a memory management unit(MMU), and a bus interface unit(BU).

Fig. 1. Block diagram of HK386

FIGURE 3.3 The development of the microprocessor chip compatible with Intel 80386 in 1995 made Kyung Chong-Min one of the most celebrated semiconductor engineers in South Korea. **Left**: South Korean media reported extensively on the success of the development of the chip. **Right**: Kyung's team presented its HK386 to the 1997 Asia and South Pacific Design Automation Conference.

Sources: *Dong-A Ilbo*, September 3, 1995, 12; and Chong-Min Kyung et al., "HK386: an X86-compatible 32-bit CISC Microprocessor," 661.

Korea Science Foundation, CHiPS's new goal was to research and develop nonmemory chips such as microprocessors and application-specific integrated circuits (ASIC).[25] The new building for CHiPS was a gift from Chung Mong-Hun, chairman of Hyundai Electronics, who had been so impressed by Kyung's successful development of the microprocessor. In the 2000s, CHiPS extended its target areas to the development of chips for wireless communications, such as chips for next-generation wireless communication, routers, and multimedia.[26] By the spring of 2011, when Kyung quit its directorship, the center had faithfully accomplished its original mission, and its major functions were dispersed into other laboratories in IDEC.

IDEC was more successful and enduring than CHiPS. Kim Oh-Hyun, who earned his doctoral degree under Kim Choong-Ki in 1983 and became an influential professor at Pohang University of Science and Technology (POSTECH), considers IDEC to be Kyung's most important contribution to the South Korean semiconductor community.[27] IDEC was created to train design specialists in nonmemory chips, with financial support from both the Ministry of Information and Communication and South Korea's three major semiconductor companies. Its primary goals were to (1) distribute CAD tools and workstations; (2) educate semiconductor designers in CAD; (3) connect design teams with semiconductor companies so as to actually manufacture chips; (4) exchange information, data, and opinions among members; and (5) support small and medium-sized semiconductor companies in designing and manufacturing chips.[28] For these purposes, IDEC prepared textbooks and videotapes

FIGURE 3.4 The establishment and operation of the Center for High-Performance Integrated Systems (CHiPS) in 1994 and of the Integrated Circuit Design Education Center (IDEC) in 1995 to train chip designers were Kyung Chong-Min's most important achievements. The groundbreaking and opening ceremonies for the CHiPS building were attended by many VIPs. **Left**: The groundbreaking ceremony, including Chung Mong-Hun, the chairman of Hyundai Electronics (second from left). **Right**: The opening ceremony, including Chung Kun-Mo, the minister of Science and Technology (sixth from left).

Source: Courtesy of Kyung Chong-Min.

that recorded the lectures held at the center.[29] Although it was located on KAIST's campus, IDEC was a kind of consortium of several universities, companies, and government research institutes that sought to "contribute to the Korean semiconductor industry by promoting collaborations between universities and industries."[30]

The CHiPS building was used to accommodate IDEC, especially after Hyundai Electronics gave up developing microprocessors. Kyung was chosen as the first director of IDEC, and selected 30 teams from various universities and several medium-sized companies as the first partners of the consortium.[31] When the promised government funding arrived, Kyung purchased 150 workstations and distributed them to the member institutes, along with the necessary CAD tools. He served as director of IDEC for seventeen years, until the end of 2011, which indicates his affection for the center and commitment to training young engineers. He later recalled that he was almost "obsessed" with the development of IDEC for its first ten years.[32] His successor was Park In-Cheol, Kyung's former doctoral candidate who had worked on the development of the microprocessor compatible with the Intel 80386 chip in the early 1990s.[33]

During his 35-year tenure as a KAIST professor, Kyung published more than 180 high-quality research papers (about 120 in international journals and 60 in domestic journals) with his students. Some examples from IEEE journals are "Two Complementary Approaches for Microcode Bit Optimization" (1994), "Design Verification of Complex Microprocessors" (1997), "MetaCore: An Application-Specific Programmable DSP Development System" (2000), "Conforming Block Inversion for Low Power Memory" (2002), "CeRA: A Router for Symmetrical FPGAs Based on Exact Routing Density Evaluation" (2004), "PrePack: Predictive Scheme for Reducing Channel Traffic in Transaction-Level Hardware/Software

Co-Emulation" (2006), "Temperature-Aware Integrated DVFS and Power Gating for Executing Tasks with Runtime Distribution" (2010), "Energy-Aware Video Encoding for Image Quality Improvement in Battery-Operated Surveillance Camera" (2012), "Runtime Thermal Management for 3-D Chip-Multiprocessors with Hybrid SRAM/ MRAM L2 Cache" (2015), "A Low-Complexity Pedestrian Detection Framework for Smart Video Surveillance Systems (2017)," "In-Pixel Aperture CMOS Image Sensor for 2-D and 3-D Imaging" (2018), "A Memory- and Accuracy-Aware Gaussian Parameter-Based Stereo Matching Using Confidence Measure" (2019), and "On-Chip Depth and Image Sensing System with Offset Pixel Apertures" (2020). He also wrote or edited several textbooks for semiconductors, such as *The Structure and Design of the High-Performance Microprocessor* (2000, in Korean), *Smart Sensors and Systems* (2015), *Theory and Applications of Smart Cameras* (2016), and *Nano Devices and Circuit Techniques for Low-Energy Applications and Energy Harvesting* (2016).[34]

Kyung and his students presented their research results at many international conferences and even won several Best Paper Awards: "Reducing Cross-Coupling among Interconnected Wires in Deep-Submicron Datapath Design" (36th Design Automation Conference, 1999), "FLOVA: A Four-Issue Media Processor with 3D Geometry" (International Conference on Signal Processing Applications and Technology, 1999), "A Regular Layout Structured Multiplier Based on Weighted Carry-Save Address" (International Conference on Computer Design, 1999), "SDRAM-Stacked Multimedia Application Core (MAC) System-in-Package Design" (International SoC Design Conference, 2009), and "Runtime 3-D Stacked Cache Data Management for Energy Minimization of 3-D Chip Multiprocessors" (International Symposium on Quality Electronic Design, 2014).

Kyung was also responsible for the introduction and spread of "concurrent engineering" (or simultaneous engineering) in the South Korean semiconductor community. To reduce the time required to develop a new product, this method aims to carry out design, manufacturing, and other processes simultaneously and has been adopted most notably by the aerospace industry.[35] Kyung recalls that he had a hard time persuading his students to adopt this simultaneous method at first because they were only familiar with the sequential engineering method. He insisted, however, and achieved success:

> My research team carried out the project to develop the 32-bit RISC microprocessor, compiler, assembler, and its application board from 1991 on. I decided to apply the concurrent engineering method to this project. The design of the chip, software development, and the design of the board were all carried out simultaneously and ended almost at the same time. The designs of the chip and the board were sent to the manufacturers almost simultaneously. The board arrived about one and a half months later, when the software was almost ready. When the chip finally arrived at my laboratory about two and a half months later, the board and software were fully ready. Within three days we could observe the picture on the screen using the software. It was a victory of the concurrent engineering method. There was another merit to this new method. There were some errors in the designs of the circuit board and the chip. But, thanks to the concurrent engineering method, it took only a half day for students to find and correct the errors.[36]

Between 1983 and 2018, Kyung produced 72 master's and 55 doctoral degree holders, a slightly higher number than his mentor Kim Choong-Ki. He always emphasized to his students that they must work on practical subjects for the semiconductor industry. The majority therefore went on to work for large companies such as Samsung Electronics, SK Hynix, SK Telecom, Qualcomm, and Apple: three of his former students, for example, climbed to the presidency of Samsung Electronics. Others began startup companies such as SPINA Systems, The Magen, 3A Logics, Coreriver, and Emcrafts. Eleven of his former students became professors training the next generation of semiconductor specialists: Park In-Cheol became Kyung's colleague at the Department of Electrical Engineering and inherited the directorship of IDEC in 2011, and one of his foreign students, Khan Asim from Pakistan (master's in 2011), became a professor in the Department of Electrical Engineering at the University of Faisalabad.

What kind of teacher and mentor was Kyung? He must have been among the most eccentric professors at KAIST because in his first class he required students to write essays on the two questions, "Why do you live?" and "Why do you study?"[37] Although Kyung's personality is quite different from that of his mentor Kim Choong-Ki, they were both unique teachers who had a profound influence on their students not only professionally but in terms of lifestyle. Although there is no collective memoir by Kyung's students to offer a vivid picture of him as teacher and mentor, the acknowledgments in his students' theses provide hints as to how he inspired them. Kyung's students all thank him for his advice and criticism and express admiration for his deep and wide knowledge of their chosen subjects. However, this was not all they learned from him. There was something else too:

- "I would like to thank Prof. Kyung, who provided me with the direction in my life as an engineer and also as a human." (Kim Hyung-Won, 1993)
- "I am deeply grateful to Prof. Kyung, who taught me how to live and work as an engineer with pride and a sense of duty." (Yang Woo-Seung, 1998)
- "I am deeply thankful to Prof. Kyung, who has become my role model as an engineer and a man." (Lee Seung-Jong, 2002)
- "I was transformed from a reckless child into a true adult under Prof. Kyung's guidance. He always taught me to aim higher than just becoming an excellent engineer." (Kim Ah-Chan, 2009)
- "His openness and availability are praiseworthy, considering his outstanding academic stature. His sound advice has helped me achieve many academic and personal goals. His passion, dedication and intensity are inspirational. His work ethic has set a very high standard." (Muhammad Umar Karim Khan, 2014)
- "I particularly learned two important lessons from him as doctoral candidate: one is the cooperation and communication with other researchers, and the other is the systematic approach to both selection and solution of the problem." (Lee Seung-Han, 2014)

Although these are only partial descriptions of Kyung, and perhaps slightly biased, they nonetheless attest to what kind of teacher and mentor he was. Just like his own mentor Kim Choong-Ki, Kyung taught his students not only semiconductor knowledge and skills but also how to become a responsible engineer.

PATRON OF STARTUPS

KAIST has been the foremost incubator of startup companies in South Korea since the early 1980s, when its first doctor of computer science, Lee Buhm Cheon, established the first startup company in South Korea.[38] Lee quit his professorship at KAIST and founded the company Qnix with four KAIST graduates in 1981, producing personal computers, word processors, and other peripherals. Although Qnix closed in 1997 when the economic crisis hit South Korea, Lee has been always remembered as the "first" legend in the history of South Korean startup companies. The Department of Computer Science at KAIST subsequently produced many other legendary figures in South Korean startup companies, including Kim Jung-Ju (Nexn) and Lee Hae Jin (Naver). The Department of Electrical Engineering also produced many stars of startup companies, most notably Lee Min Hwa, who established a medical equipment company, Medison, in 1985. Kyung had been familiar with startup companies since 1983, when he returned to KAIST as an assistant professor, but paid little attention to them during the 1980s. The design and production of chips did not seem ideal business models for startup companies in South Korea at that time.

In the mid-1990s, however, when Kyung's ambitious project of the Intel-compatible chips was suddenly halted, the idea of founding a startup company came to mind. Hyundai Electronics, which had supported Kyung's project from the beginning, didn't want to damage its relationship with Intel and had therefore stopped the development of the compatible chips completely. Business representatives from Taiwan and Australia then suggested that Kyung begin startup companies with them to produce Intel-compatible chips using his knowledge and experience, but he declined both offers. He later regretted doing so:

> Opportunities come with a smile, but they never look back once they pass. I have regretted the decision to decline the offer of commercializing the Intel-compatible chips ever since. It resulted from my inexperience and lack of insight. My students' brilliance and efforts during the preceding five years were nullified by my fault. A young inexperienced professor threw away a truly golden opportunity.[39]

Kyung had a second chance to begin a startup in the late 1990s, when the startup boom took place in South Korea. He had originally carried out the project to develop advanced communication chips with a large company, but the company soon gave it up. Some venture capitalists then suggested that they provide him with necessary capital to begin his own startup company. In 1999, Kyung set up Paion to produce advanced communication chips for large communication equipment companies such as Cisco. Kyung became CTO of the company and recruited professionals as CEO and CFO. The company soon confronted many difficulties, however. The most serious problem was that Kyung's R&D team could not produce any visible results within the targeted schedule: he overestimated his team's ability and failed to maintain a balance between his high aims and the market's needs. Major investors soon left the company, and Kyung took up the burden to maintain it for some years. Chun Jung-Bum, one of his master's students who participated in Paion from the beginning, later specialized in CMOS image sensor technology and established his own startup company, Emcrafts, in 2010, which specializes in the scanning electron microscope (SEM).[40]

In 2000, Kyung and his graduate students established another startup, Dynalith System, to offer the design and testing tools for microprocessor chips (Figure 3.5). Once again, he became CTO of the company, in charge of research and development. His team had already developed a new product, iSAVE, which uses C-language to design chips. At that time, chip designers used either VHDL or Verilog, hardware description languages that were very difficult to learn and use. Kyung was confident that his program would allow more engineers to enter the chip-designing process. When he attended the Design Automation Conference in 2000 and 2001, many showed interest in the new technology, but few wanted to buy it because of its high price tag.[41] Once again, Kyung had misread the market: few chip designers were ready to abandon VHDL or Verilog, and only some educators were willing to buy the product, though at a much cheaper price. He left the company in mid-2000s. Nevertheless, Dynalith survived despite its early failures and continued to provide the market with high-quality electronic design automation (EDA) tools. In 2014, the company was chosen as one of seven model startups that KAIST's Technology Business Incubation Center had produced.[42] It finally closed in 2018.

FIGURE 3.5   Kyung Chong-Min became interested in startups and launched his own in the late 1990s. Dynalith System, his second startup, aimed to offer design and testing tools for chips. **Left**: An article on Dynalith's rosy beginnings, under the headline "Open the New Era of Chip Design." **Right**: Dynalith's first product, iSAVE, used common C-language to design and test chips.

Sources: *Chosun Ilbo*, July 1, 2000, 31; and Kyung Chong-Min, *Changeopga*, 52.

In his 2020 book *Changeopga* (*Startup Entrepreneur*), Kyung analyzes the causes of his early startup failures (Figure 3.6).[43] First, he did not examine the market properly before deciding to establish the company and didn't meet with actual customers but only with investors or market analyzers. Second, he did not discuss fully, with potential customers and with specialists, whether or not his products were superior to other similar ones or would better meet customers' needs. In other words, he confessed, he did not know everything he needed to know about his own product or technology. Third, as a professor, he had somewhere to return to if the company failed—which he sees as a "curse" for a startup entrepreneur. He adds that venture capitalists often prefer those professors who quit their professorships to those who keep them because the latter usually don't do their best. These are truly severe criticisms of Kyung's own failures, yet they are valuable lessons not only for him but for others. The shock was so immense that he needed about a decade to recover from it. Nevertheless, his interest in and enthusiasm for startup companies have never dwindled.

In the spring of 2011, the Ministry of Education, Science and Technology announced that it would support four new Global Frontier Projects for nine years with ten million dollars per project per year.[44] Encouraged by his colleagues, Kyung

FIGURE 3.6 Despite his personal failures in launching startup companies, Kyung Chong-Min continued preaching the importance of startups in the South Korean industry and economy, and became a chief patron of them. **Left**: A poster advertising his special lecture for future startup entrepreneurs at a university. **Right**: Kyung also wrote a book in which, based on his own failures, he details how to become a successful startup entrepreneur.

Sources: Courtesy of Kyung Chong-Min; and Kyung Chong-Min, *Changeopga*, cover.

applied for one of them, "Multi-Dimensional Smart Integrated System," whose primary goals were (1) to develop revolutionary smart sensors whose overall capacities would be a thousand times better than those available in 2011 and (2) to apply these sensors to the bio/health industry, handheld devices, robots, vehicles, and ubiquitous computer systems. Kyung won the competition and slightly changed its goals to focus on the development of smart sensors. The Center for Integrated Smart Sensors (CISS) opened at KAIST on May 30, 2012. At the outset, three hundred specialists on sensors, systems, and circuits from KAIST, SNU, POSTECH, the National NanoFab Center, and the Electronics and Telecommunications Research Institute (ETRI) participated in the project.[45] The number of cooperating institutions and companies increased sharply within a few years: not only many South Korean universities but several foreign universities, such as Stanford University, UC Berkeley, Cornell University, and National Tsing Hua University (Taiwan), worked with the center, and companies such as Samsung Electronics, SK Hynix, and LG Innotek, as well as Intel and Hewlett Packard, supported the center's activities from the beginning.[46]

Kyung planned to make CISS not just another university research center for big companies but a major springboard for startups. This goal was not included in the 2011 government document, but he inserted it silently in the center's major activities. His interview with the media in May 2013, about a year after the opening of the center, hinted at this goal:

> The Center aims to develop outstanding technologies in order to publish the results in the prestigious journals or to file patents. *The Center also intends to start business by using these new technologies.* Presently, we are developing some items, such as an electronic nose that can detect spoiled food or air/water pollution, a sensor to monitor the bridge's vibration, a radioactive sensor, an advanced glucose monitor, or a highly sensitive anemia monitor, which can be presented in the market within one or two years. We are also working on an advanced gastrofiberscope and an angioscope, a three-dimensional endoscope, a smart black-box, and a networkable camera. (emphasis added)[47]

He revealed his true intention nine years later:

> In 2011, I gave up other projects and directorships, and concentrated on building CISS. What would become the vision of CISS? I decided to challenge the startup again. … CISS aimed not to just publish research papers *but to develop marketable technologies and to establish startups with those technologies.* It is very challenging to begin startups in the university. Published research papers are like cash for researchers at the university, while startups are like bonds whose true value can be evaluated only in the future. Nevertheless, I strongly encouraged professors and researchers in CISS to develop marketable technologies and to establish startups with them. Over the last nine years there were many difficulties, but we established more than thirty startups. About 30 million dollars have been invested, and the present estimated value of those startups is about 200 million dollars. Some of them will grow into larger companies. (emphasis added)[48]

Among those 30 startups at CISS, Kyung selected the following four companies as the most successful examples:[49] (1) WARP Solution developed its own radio frequency

(RF) power amplifier package and module, "which is the core component for long-range wireless charging and 5Gb network";[50] (2) OBELAB manufactures a series of real-time portable brain image devices, "NIRSIT," which are applicable to concussion, depression, neuro-rehabilitation and education;[51] (3) Point2 Tech provides "low-power and high-speed point-to-point interconnect solutions designed to meet the bandwidth requirements of 5Gb and cloud-based data centers";[52] and (4) Pico Foundry (renamed Pico SERS) developed a new technology that "can conveniently arrange tens of nanometers of nanowires in two-dimensional and three-dimensional structures with nanotransfer printing technology."[53] These four companies attracted investors from around the world, began to produce their products, and "graduated" from CISS. In short, CISS has played the role of an incubator for startups, with Kyung as mentor. Due in part to analyzing his own shortcomings as a startup CTO in the 1990s and 2000s, Kyung became a very successful mentor of startups in the 2010s.

In the fall of 2020, the government's support of CISS finally ended. Freed from teaching after his retirement in 2018 and now also free of the directorship of CISS, Kyung decided to establish a new startup of his own again. The new company, Dexelion, aims to develop the software algorithm that can analyze not only images but also the distances between the points within the image (depth) and the 3-D structure of the image in real time, using the CMOS image sensor (CIS).[54] This new algorithm works with a single-lens CIS camera and offers a much better depth map and 3-D solution than other current distance-sensors. If the company goes as well as he plans, Kyung has the ambition to enter augmented reality and metaverse markets in the future.

SPOKESPERSON FOR THE SEMICONDUCTOR COMMUNITY AND BETTER ENGINEERING EDUCATION

Unlike his mentor Kim Choong-Ki, Kyung has written many popular articles for newspapers and popular journals since the mid-1980s. Many of these have focused on the importance of nonmemory areas and the development of the semiconductor industry in South Korea. His 1986 article "Future of the South Korean Semiconductor Industry and the Problems of Manpower," for example, emphasizes the close cooperation among industry, research institutes, and the academy, as well as a long-term approach to the country's manpower training plan.[55] About half of his popular articles, however, are about more general topics, such as more active promotion of startups, better engineering education, science and engineering policy, and the responsibilities of engineering professors. His 1993 article "Making a High-Tech Society" urges South Korean engineers to solve practical rather than purely academic problems and to pay more attention to the real world.[56] In a 1994 article, he criticized South Korean society for not treating professionals fairly.[57] In the 2000s, Kyung became even more outspoken, criticizing populism in South Korean society and suggesting that South Korea needs "leaders who respect common sense."[58] His 2004 essay "Secrets of Training Manpower" urged his fellow citizens to discard old prejudices and regulations in South Korean society and to build a new, efficient, unbiased system for the new century.[59]

However, the most interesting of his popular writings may be his "Seven Commandments for Engineering Professors," which first appeared in a 1993 newspaper article that was then expanded into a 1997 book.[60] They can be summarized as follows:

1. Draw the line between public and private interests: Engineering professors must be cautious of costs but also be less interested in their own profits.
2. Recognize your position in time and space: Engineering professors must not only recognize the cycle of engineering trends and prepare for the future accordingly, but also be familiar with technical trends and their distribution in the world.
3. Maintain a balance between theory and practice: Engineering professors must respect both theory and practice and harmonize between them.
4. Keep in mind the whole: Engineering professors must consider the interests of the much wider community and society.
5. Maintain a balance between education and research: Engineering professors must reach the level at which education and research help each other.
6. Have experience as a manager: Engineering professors must have at least five years of managerial experience.
7. Possess your own strong professional background: Engineering professors must not only be proficient in their own field but also be knowledgeable in related fields. They must widen their experience and perspective as they get older.

These are truly golden rules for "ideal" engineering professors, even though few, including Kyung himself, can satisfy these conditions.

Kyung's unceasing concern for better engineering education in South Korea was one of the major reasons he ran for election as president of KAIST in 2013, 2017, and 2021, and was one of three finalists in both 2017 and 2021. His 2021 campaign statement not only reveals his vision for new directions of KAIST but also represents the opinions of KAIST hardliners on engineering education:

> For the last fifty years, the graduates of KAIST have contributed to the development of science and engineering communities as well as that of various industries in both South Korea and the world. However, I believe, KAIST now must pay more attention to breeding startups in connect with its education and research. ... As Stanford University became the origin and core of the Silicon Valley in the last century, KAIST must encourage its students to become pioneers of startups rather than to become just good scholars and researchers. ...

> Some departments at KAIST, in response to students' demands, have recently reduced student's loads and even dropped difficult (and unpopular) experiment courses from their curricula. Those departments took the easier way and became popular. However, education is not to teach something that is easier or faster to deliver, but to teach something basic or actually useful, which often requires money and effort. Only after one achieves a strong basis can he or she communicate or cooperate with other specialists in order to challenge higher goals. Research is a process of challenge and [taking the]

trouble to search out important subjects and confront them. The startup mind (communication, cooperation, and challenging spirit) is absolutely necessary not only for startup entrepreneurs but also for all basic and applied researchers as well as educators. No one without strong basic training can make any contributions to new discoveries or developments in science and technology.[61]

Unfortunately, Kyung failed to win the election and didn't have an opportunity to realize his ambitious educational dreams.

Kyung's series of articles on the Taiwanese semiconductor community were not simply acute but prophetic (Figure 3.7). In the 1994 article "Government Research Institutes in Taiwan," he outlines how the Industry Technology Research Institute (ITRI) and its daughter institute, Electronics Research and Service Organization (ERSO), contributed to the rapid growth of the semiconductor and computer industry in Taiwan.[62] He focuses on their roles in promoting many successful spin-off companies, such as Taiwan Semiconductor Manufacturing Company (TSMC) and United Microelectronics Corporation (UMC). Sixteen years later, when Taiwan's semiconductor foundry and design capacity rivaled and sometimes exceeded those of its South Korean counterparts, Kyung wrote another newspaper article suggesting how South Korea could outmaneuver Taiwan.[63] His suggestion

FIGURE 3.7 Kyung Chong-Min wrote many newspaper articles and played a role of spokesperson for the South Korean semiconductor community. **Left**: He was a very rare figure in paying attention to the Taiwanese semiconductor community as early as 1994, when South Korea was mesmerized by Samsung's success in developing and producing DRAM. **Right**: Kyung was also the most critical and outspoken semiconductor professor to oppose the sale of Hynix to Micron and other foreign companies in 2002 and thereafter.

Sources: *Jeonja Shinmun*, June 7, 1994, 3, and April 3, 2003, 30.

was to establish more foundry and fabless (without a microchip fabrication plant) companies simultaneously. For that purpose, he wrote, the South Korean government should pay more attention to training the necessary manpower, especially chip designers and engineers who could connect the design with the production at a foundry. Unfortunately, Kyung's two articles attracted little attention when they were published in 1994 and 2010. Bolstered by Samsung Electronics' continuous success in dynamic random-access memory (DRAM), the South Korean people, government, and even industry simply ignored the rapidly growing Taiwanese semiconductor industry. Few newspapers had mentioned it, and even the name "TSMC" was practically unknown to most South Koreans until Kyung's second article was published in 2010. By the 2010s, when South Korea began to pay attention, the Taiwanese semiconductor industry, headed by TSMC, was already more than big enough to challenge South Korea's hegemony in the world semiconductor market: in the foundry market, for example, TSMC has maintained its absolute lead since the early 2000s and enjoyed a 53% share (in revenues) in 2021, while Samsung Electronics had 16%.[64]

In the 2020s, South Korea experienced a serious shortage of semiconductor specialists, largely due to the government's neglect of Kyung and other specialists' warnings over the preceding two decades. The Korean Broadcasting System (KBS), for example, aired a special news report on June 19, 2022, in which a reporter said, "According to the semiconductor industry, it lacks about 3,000 semiconductor specialists per year. The industry has requested increasing the number of semiconductor students at the university ... but there are few professors and facilities to teach them."[65] Kim Kwang-Kyo, a former head of the semiconductor research center at Samsung Electronics, even argued that "the present situation of the semiconductor community in South Korea is less favorable than that forty years ago. ... While the South Korean government has paid little attention to developing semiconductors ... and done almost nothing, ... the Taiwanese government has set up a plan to train 10,000 semiconductor specialists per year."[66]

Kyung's most influential topic in the popular media, however, was his strong objection to the South Korean government's decision to sell Hynix—the second largest semiconductor manufacturer in the country—to the US company Micron Technology in the early 2000s (see Figure 3.7). Hynix was in deep financial trouble, owing both to its chronic mismanagement and to the sharp drop in the price of DRAMs in the early 2000s. Micron was a major producer of DRAMs and flash memories in the world market and had tried hard to purchase Hynix ever since the fall of 2001, so as to compete with Samsung Electronics.[67] Despite the strong objections of Hynix employees and the South Korean semiconductor community, the government wanted to sell this loss center to Micron, with the excuse that no South Korean company was big enough to buy it except Samsung. Kyung was one of the leading voices, and perhaps the most critical one, objecting to this plan. He severely criticized the government's intention and negotiation style and stressed that Samsung Electronics could not maintain South Korea's continuous development of semiconductors by itself but needed a healthy rival within the country.[68] Micron finally gave up its effort to buy Hynix on May 2, 2002, but the South Korean government did not abandon the plan to sell Hynix to a foreign buyer. Kyung contacted semiconductor professors at other universities, and they decided to form a professors' consultative group for persuading the government to change its mind. Kyung even participated in a TV discussion on which he exchanged sharp words with government representatives.[69] On June 7, 2002, the Professors' Consultative Group to Consider the Nation's [Semiconductor] Industry published a manifesto in South Korea's four major newspapers, in which they clearly objected to selling Hynix to Micron, adding that "the sale of Hynix [to a foreign firm] will weaken the competitiveness of the South Korean semiconductor industry."[70] The South Korean government finally gave up the plan to sell Hynix, and the company was placed under the control of a government-controlled creditor consortium for the next ten years. In 2003 and 2004, Kyung published two more newspaper articles to review the "Hynix Incident" and thus help ensure that this kind of mistake not be repeated.[71]

In the spring of 2012, the SK conglomerate officially absorbed Hynix. Renamed SK Hynix, the company soon became one of the largest manufacturers of DRAM and flash memory in the world market, contributing greatly to the South Korean economy. In 2014, an editorial entitled "Who Revived Hynix?" spotlighted Kyung as one of the major contributors to the company's resuscitation:

> Professor Kyung Chong-Min of KAIST was the representative who objected to selling Hynix to foreign buyers. He explained the reasons as follows: "Imagine that South Korea sells Hynix to a foreign company and receives the money. What kind of business does it do with that money? Manufacturing shoes? Or, movies or the leisure industry? South Korea needs a workplace where young scientists and engineers can work. For the industrial strategy in the future, what business is more suitable than semiconductors?"[72]

The Hynix case must be Kyung's proudest achievement as a spokesperson of the South Korean semiconductor community.

ENDLESS CHALLENGE

Kwon Oh-Hyun, who shared space with Kyung for two years in Kim Choong-Ki's laboratory (see Figure 3.2) and later became CEO and vice chairman of Samsung Electronics, has described his classmate's character and contributions as follows:

> I once wondered whether Kyung could fulfill his role of professor properly because I had the strong prejudice that a professor must do research silently. I was certainly wrong. He has maintained his typical dynamism of his student years, by extending his research areas according to the changing environments, launching startup companies, and establishing and operating IDEC to train [the] necessary chip designers for the South Korean semiconductor community. I believe that Kyung has become the professor who best realizes Frederick Terman's original idea of KAIST.[73]

This is a very nice and critical summary of Kyung's entire life and work.

When asked how he would encapsulate his entire career as an engineering professor and whether he had any regrets, Kyung replied:

I have done almost everything that I wished to do as a professor. As an engineer, however, I may not have been so successful. I have a quick temper and am straightforward. If I find an interesting subject or an important task, I usually jump into it right away. Perhaps that's why I did not concentrate on a few carefully selected subjects during my career. My tenure as professor at KAIST can be described as "unique" due to the dynamic situation that the South Korean semiconductor community encountered during the late twentieth and early twenty-first centuries. I may have been a reformer or innovator rather than a traditional professor. "Try what you believe is really valuable" has been my motto.[74]

Indeed, Kyung's contributions to the development of the South Korean semiconductor community differ from those of his mentor Kim Choong-Ki. Whereas Kim concentrated on training the first and second generations of South Korean semiconductor engineers on the KAIST campus, Kyung extended the knowledge and experience that KAIST had accumulated to the whole country through IDEC and, later, CISS. His popular writings often awakened the public to paying more attention to the development of the semiconductor industry in South Korea. His very frank confessions and sharp analyses of his failures in launching startup companies have become an invaluable text for future entrepreneurs. As he himself says, he was a very unconventional professor even among many eccentric KAIST professors. His direct communication style and challenging attitudes have often invited some misunderstanding among or even antagonism from his colleagues. None, however, would question his contributions to the development of semiconductors in South Korea and his loyalty to KAIST. And, as his new startup began in 2020, the next chapter of Kyung's challenges has just begun.

NOTES

1 Kyung Chong-Min, "Dojeonhajianneun Jeoleumeun eopda," in Park Yoon-Chang et al., *Nobelsangeul Gaseume pumgo* (Seoul: Dong-A Ilbo, 1994), 131–145.
2 Ibid., 132–133.
3 Kyung Chong-Min, *Keun Namuga jaraneun Ttang* (Seoul: Sigma Press, 1999), 267–268.
4 Interview with Kyung Chong-Min by Kim Dong-Won (September 23, 2020).
5 Ibid.
6 Chong-Min Kyung, "Low Level Currents in Buried Channel MOS Transistor" (master's thesis, Seoul: KAIS, 1977), 4. A slightly revised version was published the next year: see Kyung Chong-Min and Kim Choong-Ki, "A Two-Dimensional Analysis of the Low-Level Currents in Buried Channel MOS Transistors," *Jeonjagonghakhoeji*, 15:6 (1978), 35–38.
7 Kyung Chong-Min, "Dojeonhajianneun Jeoleumeun eopda," 137–138. In the acknowledgments of his doctoral thesis, Kyung expresses special thanks to his eldest sister, who had supported his entrance to the doctoral program.
8 Chong-Min Kyung and Choong-Ki Kim, "Pipeline Analog-to-Digital Conversion with Charge-Coupled Devices (correspondence)," *IEEE Journal of Solid-State Circuits*, 15 (April 1980), 255–257.
9 Kyung Chong-Min, "Dojeonhajianneun Jeoleumeun eopda," 140–142.
10 Interview with Kim Choong-Ki by Kim Dong-Won (May 22, 2021).
11 Interview with Kyung Chong-Min by Kim Dong-Won (September 23, 2020).
12 Kyung Chong-Min, *Keun Namuga jaraneun Ttang*, 278–279.
13 Interview with Kyung Chong-Min by Kim Dong-Won (May 20, 2021).
14 The list of Kyung Chong-Min's research papers in international journals can be found at https://ieeexplore.ieee.org/search/searchresult.jsp?newsearch=true&queryText=Kyung%20Chong-Min (searched on January 28, 2021).

15 Kyung Chong-Min, "Bandoche Seolgyegisului Hyeonhwangkwa Jeonmang," *Jeonjajinheung*, *4:2* (1984), 12–15 on 14–15.

16 Kyung Chong-Min, "Jikjeophoero Jejogongjeongui Model," *Jeonjagonghakhoejapji*, *11:5* (1984), 1–4 on 1.

17 Park Song-Bae, Kyung Chong-Min, Im In-Chil, Cha Gyun-Hyun, and Kim Hyung-Gon, "Research on the Development of CAD Software," *Teukjeongyeonku Gyeolgwa Balpyohoe Nonmunjip* (1987), 69–72.

18 Kyung Chong-Min, *Changeopga* [*Startup Entrepreneur*] (Seoul: Yulgok, 2020), 45.

19 Jeonja Shinmun, "Intel 387 Hohwanchip Gaebal," *Jeonja Shinmun* (July 2, 1994), www.etnews.com/199407020050 (searched on February 2, 2021).

20 Dong-A Ilbo, "Microprocessor-gisul, 10nyeonane Segejeongsang," *Dong-A Ilbo* (January 10, 1995), 17.

21 Kyung Chong-Min, *Changeopga*, 45–47.

22 Dong-A Ilbo, "386hohwanchip Guknae cheot Gaebal," *Dong-A Ilbo* (September 3, 1995), 12.

23 For example, Chosun Ilbo, "Guknaechoecho Intel 386hohwan CPU Gaebal: Hankukgwahakgisulwon Kyung Chong-Min-gyosu," *Chosun Ilbo* (September 7, 1995), https://biz.chosun.com/site/data/html_dir/1995/09/07/1995090772304.html; Hankuk Kyungje, "Hitech geu Juyeokdeul: KAIST Kyung Chong-Min-gyosutim," *Hankuk Kyungje* (September 18, 1995), www.hankyung.com/news/article/1995091800521 (searched on February 2, 2021).

24 Chong-Min Kyung et al., "HK386: an X86-compatible 32-bit CISC Microprocessor," *Proceedings of the ASP-DAC 1997* (January 1997), 661–662.

25 Maeil Kyungje, "Hyundaijeonja Daeduke Goseongneugjikjeopsistem Yeonkucenter Gigong," *Maeil Kyungje* (November 23, 1994), www.mk.co.kr/news/home/view/1994/11/60172/ (searched on February 12, 2021).

26 Jeonja Shinmun, "KAIST Goseongneungjikjeopsistem Yeonkucenteo," *Jeonja Shinmun* (October 17, 2002), https://m.etnews.com/200210160056 (searched on February 12, 2021).

27 Interview with Kim Oh-Hyun by Kim Dong-Won (December 3, 2021).

28 Kyung Chong-Min, "Bandoche Seolgyegyoyukcenteo (IDEC)ui Saeopgyehoek," *Jeonjagonghakhoeji*, *22:10* (1995), 1122–1130 on 1123.

29 Kyung Chong-Min, "Bandoche Seolgyegyoyukcenteo (IDEC) Saeopeul tonghan Bimemory mit Systemseolgye Hwalseonghwa Bangan," *Hankuktongshinhakhoeji*, *13:11* (1996), 1232–1240 on 1235–1237.

30 Department of Electrical Engineering, KAIST, https://ee.kaist.ac.kr/en/node/10981 (searched on December 18, 2020).

31 Jeonja Shinmun, "Kyung Chong-Min–Naedal Munyeoneun 'Bandoche Seolgyegyoyukcenteo' Ssenteojang," *Jeonja Shinmun* (June 28, 1995), www.etnews.com/199506280039?m=1 (searched on March 5, 2021).

32 Interview with Kyung Chong-Min by Kim Dong-Won (September 23, 2020).

33 Digital Times, "Smartsensore nameun Yeonkuinsaeng geolgeot. IDECsojang Satoeuisa balkin Kyung Chong-Min-gyosu," *Digital Times* (November 7, 2011), www.dt.co.kr/contents.html?article_no=2011110802012069758002 (searched on December 18, 2020).

34 Kyung Chong-Min et al., *The Structure and Design of the High-Performance Microprocessor* (in Korean) (Seoul: Daeyoungsa, 2000); Youn-Long Lin, Chong-Min Kyung, Hiroto Yasuura, and Yongpan Liu (eds.), *Smart Sensors and Systems* (Dordrecht: Springer, 2015); Chong-Min Kyung, *Nano Devices and Circuit Techniques for Low-Energy Applications and Energy Harvesting: KAIST Research Series* (Dordrecht: Springer, 2016); and Chong-Min Kyung (ed.), *Theory and Applications of Smart Cameras: KAIST Research Series* (Dordrecht: Springer, 2016).

35 For more information about concurrent engineering, see Reference for Business, "Concurrent Engineering," www.referenceforbusiness.com/management/Comp-De/Concurrent-Engineering.html, and The European Space Agency, "What is Concurrent Engineering?" www.esa.int/Enabling_Support/Space_Engineering_Technology/CDF/What_is_concurrent_engineering (searched on March 7, 2021).

36 Kyung Chong-Min, *Keun Namuga jaraneun Ttang*, 217.

37 Kyung Chong-Min, *Igongkyega salaya Naraga sanda* (Seoul: Yas Media, 2004), 220–221.

38 For startup companies established by KAIST graduates, see Kim Dong-Won (ed.), *Miraereul hyanghan kkeunimeopneun Dojeon: KAIST 35nyeon* (Daejeon: KAIST, 2005), 145–169.

39 Kyung Chong-Min, *Changeopga*, 48–49.

40 For more information about Emcrafts, see www.emcrafts.com/en/products/se_main.php (searched on August 25, 2021).

41 Kyung Chong-Min, *Changeopga*, 51.

42 Hankuk Kyungje, "Dynalith System, kkomkkomhan Bandoche Seolgyegeomjeung," *Hankuk Kyungje* (June 18, 2014), www.hankyung.com/society/article/2014061887151 (searched on January 21, 2021).

43 Kyung Chong-Min, *Changeopga*, 56–58.

44 Ministry of Education, Science and Technology, "2011 'Global Frontier-Saeop' Gongo," (June 17, 2011), press release.

45 Yonhap News Agency, "Smart IT Yunghapsistemyeonkudan Gaeso," *Yonhap News Agency* (May 30, 2012), www.yna.co.kr/view/AKR20120529018700017 (searched on January 27, 2021).

46 CISS, "Agreement," http://ciss.re.kr/english/agreement (searched on January 27, 2021). By 2020, about 40 domestic and 20 foreign institutions/companies had made agreements to work with the CISS.

47 Park Bang-Ju, "Kyung Chong-Min Miraechangjokwahakbu ITyunghapsistemsaeopdanjang," *Science and Technology, 528* (2013), 66–69 on 68.

48 Kyung Chong-Min, *Changeopga*, 67–69. The full list of the 30 startups can be found on the CISS website, http://ciss.re.kr/english/introduction.

49 Kyung Chong-Min, *Changeopga*, 77–102. For more information about each company, see CISS website, http://ciss.re.kr/english/introduction.

50 WARP Solution, "What We Do," https://warpsolution.com/contact (searched on August 2, 2021).

51 OBELAB, "Application Areas," www.obelab.com/product/product_nirsit.php (searched on August 2, 2021).

52 Point2 Tech, "About Us," www.point2tech.com/product_ETube.php (searched on August 2, 2021).

53 Pico SERS, "Core Technology," www.picofd.com/technology/core-technology/ (searched on August 2, 2021).

54 Interview with Kyung Chong-Min by Kim Dong-Won (May 20, 2021).

55 Kyung Chong-Min, "Guknae Bandochesaneopui Miraewa Gisulinryeokui Yangseong Munje," *Jeonjajinheung, 6:5* (1986), 2–5.

56 Kyung Chong-Min, "Uri Cheomdansaneopi Kkot piryeomyeon," *Computer World* (January 1992), 81.

57 Kyung Chong-Min, "Jeonmunkaga shinnaneun Sahoereul mandeulja," *Hankyoreh* (January 29, 1994), www.hani.co.kr/arti/legacy/legacy_general/L291691.html (searched on February 2, 2021).

58 Kyung Chong-Min, "Populisme daehan Gyeongkye," *Gukmin-Ilbo* (November 25, 2004), http://news.kmib.co.kr/article/viewDetail.asp?newsClusterNo=01100201.20041125000002301; "Sangsikeul Jonjunghaneun Jidoja," *Gukmin-Ilbo* (April 21, 2005), http://news.kmib.co.kr/article/viewDetail.asp?newsClusterNo=01100201.20050421100000306 (searched on February 2, 2021).

59 Kyung Chong-Min, "Gukchaek Inryeokyangseong Bichaek," *Jeonja Shinmum* (August 23, 2004), www.etnews.com/200408220004 (searched on February 2, 2021).

60 Kyung Chong-Min, "Jeolmeun Gongdaegyosu 7gyemyeong," *Jeonja Shinmun* (July 1, 1993), 4; idem, *Keun Namuga jaraneun Ttang*, 199–203.

61 Kyung Chong-Min, "KAISTbaljeon mit Kyeonggyeongbanghyange daehan Sogyeon" (October 30, 2020). A later statement (January 20, 2021) was slightly toned down but delivered a similar message.

62 Kyung Chong-Min, "Daemanui Jeongbuchulyeon Yeonkuso," *Jeonja Shinmun* (June 7, 1994), www.etnews.com/199406070035 (searched on February 3, 2021).

63 Kyung Chong-Min, "Daeman Bandoche igiryeomyeon," *Jeonja Shinmun* (September 27, 2010), www.etnews.com/201009200041 (searched on February 3, 2021).

64 eeNews Analogue, "TSMC, Taiwan to Increase Foundry Market Share in 2022," *eeNews Analogue* (April 25, 2022), www.eenewsanalog.com/en/tsmc-taiwan-to-increase-foundry-market-share-in-2022/ (searched on July 10, 2022). In 2021, Taiwanese companies occupied total 64% share of the foundry market, while South Korean companies had 17%.

65 KBS News, "'Maenyeon 3000myeong Bujok' Bandoche Inryeoknan … Daechekeun?" *KBS News* (June 19, 2022), https://news.kbs.co.kr/news/view.do?ncd=5489649 (searched on July 14, 2022).

66 Hankuk Kyungje, "Kyeongjaengkuk Sahwalgeoneundae Jeongbuneun Dwitjim … Bandoche Saengtaegye, 40nyeonjeonboda Yeolak," *Hankuk Kyungje* (June 30, 2022), www.hankyung.com/economy/article/2022062292911 (searched on July 14, 2022).

67 Hankuk Kyungje, "Hynix-Micron Maegakilji," *Hankuk Kyungje* (April 22, 2002), www.hankyung. com/finance/article/2002042295058 (searched on February 27, 2021).

68 These opinions are reflected in the manifesto of June 7, 2002, his following newspaper articles, and an interview with Kyung Chong-Min by Kim Dong-Won (May 20, 2021).

69 MBC, "100-minutes Discussion" (May 9, 2002). See Jung-Ang Ilbo "Hynix Buhwal Dwitiyagi—500il," *Jung-Ang Ilbo* (March 14, 2007), www.joongang.co.kr/article/2661835#home (searched on June 29, 2022).

70 Chosun Ilbo, "[Gongdaegyosudeul] Hynix Jolsokmaegak Bandochesaneop Akhwa," *Chosun Ilbo* (June 9, 2002), https://biz.chosun.com/site/data/html_dir/2002/06/09/2002060970253.html (searched on July 10, 2022).

71 Kyung Chong-Min, "'Hynix Satae' Daeeung Yugam," *Jeonja Shinmun* (April 3, 2003), www.etn ews.com/200304020229; Chosun Ilbo, "Hynix Maekak Bandaehaetdeon Kyung Chong-Min-gyosu," *Chosun Ilbo* (February 28, 2004), https://biz.chosun.com/site/data/html_dir/2004/02/28/2004022870 039.html (searched on February 3, 2021).

72 Joong-Ang Ilbo, "Buhwalhan Hynix, nuga salryeotna," *Joong-Ang Ilbo* (November 26, 2014), https:// news.joins.com/article/16526233 (searched on February 3, 2021).

73 Kyung Chong-Min, *Changeopga*, 8.

74 Interview with Kyung Chong-Min by Kim Dong-Won (May 20, 2021).

4 Mr. NAND Flash
Lim Hyung-Kyu

Lim Hyung-Kyu is one of the most successful engineers-turned-top-executives at Samsung Electronics. He spent 34 years there, from 1976 to 2009 (except for five years' leave to obtain his master's and doctoral degrees), first as a talented research engineer and then as a brilliant executive. "A true Samsung man to the bone," as Kim Choong-Ki has called him, Lim and Samsung Electronics are inseparable, influencing and mutually benefitting each other.[1] The plaque Lim received in 2005, when he was named a University of Florida Distinguished Alumnus, nicely summarizes his achievements at Samsung:

> He spearheaded the development of EEPROM [electrically erasable programmable read-only memory], Flash Memory, SRAM [static random-access memory] and DRAM [dynamic random-access memory] products. As managing executive director and executive vice president of the Memory Division from 1996 to 1999, Samsung became the leading manufacturer of DRAM chips, the major memory components of PCs and laptop computers. He is particularly recognized for his work in the creation of the NAND ["not and"] Flash business at Samsung. In 2000, Lim was named president of the System LSI Business for Samsung and worked to make Samsung the worldwide leader on Home & Mobile System LSI Solution.[2]

Lim is, however, less well known to the public than Samsung's DRAM developers such as Chin Dae-Je and Hwang Chang-Gyu. Perhaps this is in part the result of his being "quiet, prudent, meticulous but excellent at managing organizations."[3] It may also spring from his deep loyalty to Samsung and its culture, which does not encourage individual heroism. He was, in fact, the most trusted "fireman" for Samsung's charismatic chairman, Lee Kun-Hee, who had sent Lim into the most troubled areas during the 1990s and 2000s and received the expected results.

EARLY YEARS

Lim Hyung-Kyu was born on February 4, 1953, on Geoje Island, the second largest island in Korea, located off the southern part of the peninsula. He was the fourth of six children and the second son in the family. His father was a local leader and paid special attention to the children's education. Since his father ran

DOI: 10.1201/9781003353911-7

the village's only book and stationery store, Lim was able to read as many books as he wished. The Korean War had ended just a few months after his birth, so living conditions were poor on the island, and Lim and his siblings had to work in the fields or collect firewood after school. His father gave them the choice, however, of either working in the fields or studying hard at home: Lim chose the latter and soon became one of the top students in his class.[4] After graduating from the middle school near his village, he moved to Busan, the second largest city in South Korea, to enter Kyeongnam High School, a prestigious school in the Busan area. There he studied hard and in 1972 entered the Department of Electronic Engineering at Seoul National University (SNU).

In the early 1970s, the Department of Electronic Engineering was one of the most popular and also most difficult departments to enter at SNU. Many of Lim's future colleagues at Samsung Electronics were also graduates of that department. The curriculum at the time, however, was not so impressive. There was no course on semiconductors, and most subjects were old ones, such as communication or circuit theory.[5] The students were supposed to study books and solve equations, but they had little opportunity to gain hands-on experience in electronics. Lim's most memorable class was the course on physical electronics offered by Min Hong-Sik, who taught about a p-n junction, the junction gate field-effect transistor (JFET), and other basics. This was his first contact with the semiconductor, and it fascinated him.

In his senior year, 1975, Lim saw an advertisement on the department bulletin board that Hankuk Semiconductor was looking for two highly qualified graduates: once selected, if they passed the entrance examination of the Korea Advanced Institute of Science (KAIS), the two would study semiconductors at KAIS as graduate students for two years, on the condition that they return to the company after receiving their master's degrees. Since the Samsung conglomerate had recently become the majority stockholder of the company with the plan of promoting semiconductors, its future seemed bright. Lim decided to apply for the post. He remembers the entrance examination for KAIS as the most competitive one of his entire life, but he passed it. In January 1976, he was hired by Hankuk Semiconductor, and he entered KAIS in March of that year.

Lim naturally chose semiconductors as his major, and Kim Choong-Ki became his advisor. It was Kim's second academic year teaching at KAIS, and he was busy building his semiconductor laboratory with his students. Lim's first job was to organize the laboratory as its secretary, but he also helped his colleagues mend pipes and purchase necessary parts in the street market. Kim taught his students how to become true engineers and serve the country as engineers, which deeply impressed Lim. He learned how to design and manufacture semiconductor chips under Kim and also gained valuable experience in building the small-scale fab (microchip fabrication plant) at KAIS (Figure 4.1). The subject of his master's thesis was the design and fabrication of a seven-segment decoder/drive with the p-type metal-oxide-semiconductor (PMOS) technology. He designed the circuit, made a layout and emulsion mask, manufactured chips through the PMOS process, and finally tested them with success.[6] Kim Choong-Ki appraised it as the first true semiconductor integrated circuit for which the full process had been carried out in South Korea, from the initial design to the final test.

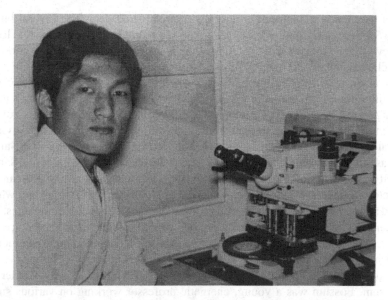

FIGURE 4.1 Lim Hyung-Kyu at Kim Choong-Ki's lab at KAIS (later KAIST) in 1977. Samsung sent Lim to KAIS to study semiconductors, and he became a specialist at Samsung Semiconductor in 1978, when he returned with his master's degree.

Source: Courtesy of Lim Hyung-Kyu.

In January 1978, after finishing his master's thesis, Lim returned to the company, whose name would soon be changed to Samsung Semiconductor.[7] Lim recalls that his years at KAIS made him the most experienced semiconductor specialist in the company when he returned there because he had worked with all the processes while at KAIS.[8] He was posted to the department of technology development, where he worked on developing the sound amplifier chip for TV sets. The process was one of reverse-engineering the similar Japanese chip, and he successfully accomplished the project within ten months. His boss then asked him to solve a problem in the production line of the complementary metal-oxide semiconductor (CMOS) chip, which was Samsung Semiconductor's main product at that time. Lim soon found the source of the trouble and suggested not only how to fix it but how to improve the whole process. He also led a development team that succeeded in "shrinking" the size of the chip, guaranteeing the double production of chips in the same size of the wafer. This was named the "low oxide step CMOS process," and his team later received a prize from the Samsung conglomerate for their work. Lim's bosses, especially Kim Kwang-Ho, were impressed by this continuous success and allowed him to lead a small independent research and development team. Lim organized the CMOS device team in order to develop new chips and their processes. He and two other engineers studied and tested various new chips and even reverse-engineered the 4Kb SRAM. But Lim didn't lead this ambitious project for long because he decided to go to the United States for advanced study.

This was not a sudden but a carefully contemplated decision. Several business trips to Silicon Valley and visits to American semiconductor companies had led Lim to believe that he needed more training at the world center of semiconductors. Many of his classmates at SNU and KAIS had already started their doctoral programs at American universities. The problem was whether Samsung Semiconductor would allow him a long-term leave for his advanced study. Kim Kwang-Ho, Lim's mentor, then suggested an option: the company would pay the expenses for his advanced study on the understanding that he would return to the company after receiving the doctoral degree. Lim accepted this generous offer and became the first Samsung overseas student. In retrospect, it was a very wise decision for Lim. After Samsung decided to enter the memory business in 1983, it hired many South Korean semiconductor engineers who had been educated at American universities and who had worked at prestigious research institutes or famous semiconductor companies. These engineers would become Lim's major rivals when he returned to the company. To compete with them, he needed his doctorate in semiconductors and experience in America.

In August 1981, Lim left for the University of Florida to study under Jerry G. Fossum. Fossum was a young, energetic professor working on various subjects related to semiconductors—CMOS integrated or memory circuits, metal-oxide-semiconductor field-effect transistors (MOSFETs), and so forth. He quickly recognized Lim's ability and made him his research assistant. Fossum also recommended charge-based modeling of the silicon-on-insulator (SOI) transistor as the subject of Lim's doctoral thesis.[9] Using his experience at Samsung Semiconductor, Lim completed the first step of the project within six months. Between 1983 and 1985, he published seven papers on the subject with Fossum and other students, all in the prestigious *IEEE Transactions on Electron Devices*.[10] One of them, Lim's 1983 "Threshold Voltage of Thin-Film Silicon-on-Insulator (SOI) MOSFET's," has been cited more than 500 times in other papers, books, and patents and became a classic paper on the subject.[11] Park Sung-Kye of SK Hynix, one of Kim Choong-Ki's doctoral candidates in the early 1990s (profiled in Chapter 7), remembers that he considered Lim's 1983 paper on threshold voltage to be the bible throughout his graduate years.[12] Fossum desperately wanted Lim to stay after graduation in 1984, but Lim had the obligation to return to Samsung. He also had no intention of working in academia. Both Bell Labs and Texas Instruments then offered him a job. He asked Samsung to allow him to work for a few years at one of the two in order to gain further experience, but the request was denied. Instead, the company instructed him to go to Samsung Semiconductor Inc. (SSI) in Silicon Valley.

A RISING STAR ENGINEER: FROM EEPROM TO NAND FLASH MEMORY

When Lim arrived in California's Silicon Valley in August 1984, Samsung Semiconductor Inc. was just a one-year-old research and development institute whose primary goals were to develop new very large-scale integration (VLSI) chips, to train Samsung engineers there, and to facilitate the importing of state-of-the-art semiconductor technology to South Korea.[13] Lee Il-Bok, who was one of the first engineers recruited by SSI in 1983, suggested that Lim work on an interesting subject: electrically erasable programmable read-only memory (EEPROM). A fairly new and promising technology developed by Motorola and Fujitsu, this nonvolatile memory (NVM) could be electrically erased and reprogrammed but was technically difficult to produce and also very expensive. Lim led a small research team to develop 64Kb EEPROM based on the 16Kb EEPROM. After struggling to solve many technical difficulties for almost a year, his team succeeded in developing the chip. He remembers that this project gave him very valuable experience in the development of memory.[14] The main factory line in South Korea tested his chip and confirmed the possibility of mass production. Samsung Semiconductor and Telecommunications (the former Samsung Semiconductor) then instructed Lim to return to Korea and lead the 64Kb EEPROM production. After spending more than four years in America, Lim finally returned to South Korea in November 1985. His ten-year apprenticeship was finally over.

Lim's first job after returning to South Korea was to lead the NVM team that aimed to prepare the mass production process of 16Kb and 64Kb EEPROM and to newly develop 256Kb EEPROM. At the age of 33, he became the department head responsible for the whole project. His department worked day and night to solve many problems, such as lowering the defect rate, and presented its results, "A 256Kb CMOS EEPROM with Enhanced Reliability and Testability," at the 1989 Symposium on VLSI Circuits.[15] Selling the product was, however, a totally different matter, and here Lim's team met a serious obstacle. Since the device was relatively new and its use was limited to printers, electronic scales, and some game consoles, most buyers preferred more reliable producers, not new ones like Samsung. Lim therefore traveled the globe to meet and persuade potential buyers to purchase his products. He recalls that he once visited a buyer in Edmonton, Canada, on a cold, snowy day and was delighted to sell his products.[16] The future of EEPROM was still unclear, and its sales couldn't cover even the development expenses. Even though Lim received a prestigious prize for his development of EEPROM in 1989 (Figure 4.2), that did not ensure his survival at Samsung Electronics.[17] The development of a high-quality charge-coupled device (CCD) for Samsung's camcorders, another project Samsung gave him, did not produce any satisfactory results, either.

In 1989, Lim chose mask (or masked) ROM (MROM) as the candidate to pursue. It is a NVM that can be "mass-produced with particular data bits already stored in it. Once the MROMs are made, the data can never be changed."[18] Since Lim's team had already successfully developed a much more difficult NVM—EEPROM—the development of 4Mb MROM was a relatively easy task. The popularity of the Nintendo game console and its replicas at the turn of the 1990s also guaranteed a wider market for MROMs across the world. Before Lim developed MROM at Samsung Electronics (which merged with Samsung Semiconductor and Telecommunications in 1988), its production was monopolized by a few Japanese companies, such as Toshiba and Sharp, and demand far exceeded the supply. As soon as the news broke about its successful development, Samsung Electronics received a rush of orders for MROM. By 1992, its sales had reached more than

FIGURE 4.2 **Left**: Lim Hyung-Kyu receiving a coveted prize for his contribution to the development of 256Kb EEPROM in 1989. **Right**: Lim's NAND development team often met even after they were scattered to different posts. In the picture, Shin Yoon-Seung (far right in the front row), another of Kim Choong-Ki's doctoral students, sits beside Lim (second from right). The woman in the back row is Yang Hyang-Ja, who rose to the director's level at Samsung Electronics and later became a member of Parliament.

Source: Courtesy of Lim Hyung-Kyu.

$400 million, with high profits. Samsung succeeded in developing 16Mb and 32Mb MROMs in 1991 and 1992, respectively.

The success of MROM was a turning point in Lim's career. First of all, though its sales and profits were much smaller than those for DRAM (Samsung Electronics' most profitable product at that time), its success was a meaningful victory for Lim. He was promoted to the director's level in December 1990 and could now compete with other rising star engineers at the company. Second, the success of MROM justified NVM research and development. Without producing any meaningful profit, the department might have been dissolved at any time, and the precious experience of several years spent on NVM might have evaporated. The following account of an episode in 1990 illustrates that not all senior executives were so sympathetic to the NVM project:

> In the summer of 1990, I had an opportunity to attend a meeting presided over by the vice president responsible for the entire semiconductor operation. During the meeting, he suddenly indicated a senior research engineer [Lim Hyung-Kyu] and asked him, somewhat nervously, "Should we continue the development of this product?" … The product that the vice president mentioned had no proper market yet, and everything from the process development to the market creation was new. From the point of view of a senior executive, the vice president wanted to know how long it would take to produce any profit from it. The senior research engineer [Lim] replied apologetically that Samsung must continue developing that product because other competitors were doing so, and that the future of it must be bright. So, Samsung continued developing the product, and that product—NAND flash memory—is now widely used in cell phones to store photos and information.[19]

The success of MROM not only rescued Lim but also enabled Samsung Electronics to continue accumulating valuable knowledge and experience on NVM that would later bear fruit in its NAND flash memory.

Lim's first task as director was to continue managing the development of NVM, but a new task was soon added: the development of static random-access memory (SRAM). SRAM is faster than DRAM and also uses less electricity in operation, but it can only store about one-quarter as many data bits as a DRAM. It was widely used as the instruction and data cache within microprocessor chips, and its market was growing rapidly. Nevertheless, it was much more difficult to design and manufacture SRAM than DRAM. Samsung Electronics had begun working on the development of SRAM in 1988 and in 1989 began to manufacture 1Mb SRAM with the speed of 70 nanoseconds.[20] Lim's team was not satisfied with this speed and aimed to develop 1Mb SRAM with the speed of 20 nanoseconds, which meant reconfiguring almost everything, from the design of the device to the method of testing it. In May 1991, his team succeeded in developing it, and by October 1992 it developed an even faster 1Mb SRAM with the speed of 10 nanoseconds. These fast SRAMs were very popular in the world market, and Lim's department developed a cache memory for Intel's CPU and another for Sun Microsystems. The rapid growth of the cell phone market from the early 1990s on also contributed to growing sales of his new fast SRAMs: Europe's GSM base system purchased a huge number of Samsung's fast 1Mb SRAMs because Samsung released them earlier than other companies. Impressed by this success, Samsung Electronics established a new marketing team to sell "non-DRAM" products such as MROM and SRAM. By the end of 1995, these non-DRAM memory products made up about 15% of all memory sales. The success of SRAM brought Lim another victory.

By the mid-1990s, Lim had emerged as one of four star semiconductor engineers within Samsung Electronics. These four—Chin Dae-Je, Kwon Oh-Hyun, Hwang Chang-Gyu, and Lim—shared many things common. They were almost the same age (born in 1952 or 1953) and had all graduated from SNU (from either the Department of Electrical Engineering or the Department of Electronic Engineering). Lim and Kwon had trained at KAIS under Kim Choong-Ki, and Chin had also attended Kim's graduate seminar before he left for America. All four had earned their doctoral degrees in semiconductors from American universities: Chin and Kwon from Stanford University, Hwang from the University of Massachusetts (Amherst), and Lim from the University of Florida. Two had worked at prestigious American semiconductor institutes before joining Samsung Electronics: Chin at IBM's Watson Research Center, and Hwang at Intel and Stanford University. With the exception of Lim, they had all been hired between 1985 and 1989 as chief research engineers in order to develop DRAM. In the early 1990s, these DRAM developers' names began to appear frequently in the mass media, where they were often praised as hero engineers: Chin as the developer of 4Mb and 16Mb DRAM, and Kwon and Hwang as the developers of 64Mb and 256Mb DRAMs. Lim, who had worked on NVM, a non-DRAM area, was a lesser known figure outside semiconductor circles. He had one advantage, however: unlike the other three, who had entered Samsung Electronics as middle-level research engineers, often with special invitations from the top, Lim had started his career there as a new recruit in 1976 and had climbed his way up, step by step. In Samsung's own words, Lim belonged to those who had "pure blood," that is, whose loyalty to Samsung no one could doubt. Samsung's high-ranking executives must have encouraged this competition among the promising semiconductor engineers, continuously testing their abilities and promoting them in different years. The real competition among them started after they were promoted to the director's level in the early 1990s. Lim desperately needed another big success to survive this severe competition.

In 1988, Lim read an interesting article in the technical digest of the International Electron Devices meeting in which Fujio Masuoka's research team presented "novel device technologies."[21] Masuoka was a renowned specialist on NVM, and first developed the floating gate technology that could be erased much faster—as fast as the "flash" of a camera.[22] Masuoka and his team had developed the first "not or" (NOR) EEPROM in 1984, and the first "not and" (NAND) EEPROM in 1987. As a fellow NVM specialist, Lim quickly recognized the merits of this new technology: despite some minor technical drawbacks, it could easily increase the level of integration of EEPROM in order to store large audio-video files. His team made some test chips and proved that they worked smoothly, but he was not sure of its commercialization, and it was put aside for a while. In early 1991, after he was promoted to director, Lim visited Samsung's research institute for home appliances to attend a meeting about CCD chips for camcorders. An engineer there demonstrated a Sony camcorder that used a floppy disk as storage.[23] An idea flashed into Lim's mind: could a NAND EEPROM memory, perhaps with 1Mb storage capacity, replace that bulky floppy disk for camcorders? His team immediately tried to develop 1Mb NAND EEPROM and also set a new unit of 256 bytes per page (or block) because of the structural restriction of NAND cells. However, potential buyers were reluctant to order this unproven device.

The opportunity for NAND EEPROM came from outside the firm in 1992. The news that Toshiba had decided to accelerate the production and sales of this new memory arrived at Samsung Electronics. Toshiba, the "inventor of flash memory," was the true pioneer in this device: it had first introduced NOR type EEPROM to the market in 1984, and NAND type "flash" memory in 1987.[24] Samsung Electronics had maintained a close relationship with Toshiba ever since its beginning, and the two companies even held regular semiconductor meetings. Lim asked his boss, Kim Kwang-Ho, to contact Toshiba about the possibility of cooperating on the production of the NAND EEPROM. Toshiba agreed to cooperate with Samsung Electronics, and the contract was signed in December 1992, within five months of floating the idea. Toshiba licensed the necessary patents to Samsung Electronics and even gave Samsung some blueprints of the memory block. Under their mutual agreement, either company could develop NAND EEPROM technology independently, but each firm was to check with the other every quarter on their compatibility.

There were several reasons Toshiba quickly agreed to cooperate with Samsung in 1992. Japanese companies such as Hitachi and Mitsubishi, as well as an American company, San Disk, had all adopted different flash memory technologies and were competing with Toshiba in the world market. Toshiba was therefore searching for a reliable ally with whom to share the same format of NAND EEPROM and chose Samsung Electronics as its partner. Toshiba certainly did not consider Samsung Electronics as a serious rival but did see it as a useful junior partner for the development of NAND EEPROM.

The cooperation with Toshiba on the NAND EEPROM (hereafter NAND flash memory) brought Lim's NVM team additional legitimacy within Samsung Electronics, as well as a great stimulus: it had finally discovered a next-generation memory device with great potential.[25] In 1992, Masuoka boldly and correctly predicted that NAND flash memory would replace both floppy and hard magnetic discs (and even DRAM) as the main storage device.[26] Samsung's development of 16Mb NAND flash memory was then accelerated and successfully finished in the second half of 1994, about six months after Toshiba did so. Since Samsung Electronics had accumulated as much know-how on EEPROM under Lim's leadership as Toshiba possessed, it then surpassed Toshiba on the development of 32Mb NAND flash memory. For this, Lim's team, which included the former students of Kim Choong-Ki such as Shin Yoon-Seung and Suh Kang-Deog, solved two vital technical difficulties that Toshiba's 16Mb NAND flash memory presented: (1) the high voltage (10V) within the memory cell, which imposed a limit for the larger size of memory, and (2) the problem of nonhomogeneity when seeking to program 256Kb cells simultaneously. Lim's team applied two new ideas to the design—self-boosting and incremental step programming pulse (ISPP)—and succeeded in producing a prototype of 32Mb NAND flash memory.[27] When the news of Samsung's success arrived at Toshiba, the latter immediately suggested the co-development of 32Mb NAND flash memory. Toshiba even loosened restrictions on the license of the NAND cell, while Samsung licensed its new design to Toshiba. In 1997, both companies presented the 32Mb NAND flash memory to the market simultaneously. From then on, the Toshiba-Samsung style NAND flash memory, which costs less and holds more memory, would dominate the ever increasing NAND flash market, rapidly surpassing other types of storage devices.

The sales of NAND flash memory were not smooth in the beginning, however. Two potential target markets were digital cameras and MP3 players, but these markets were still too small to consume large enough amounts of this type of memory. In 1994, the production line of the first commercial product, 16Mb NAND flash memory, was almost stopped because there were not enough orders. For some years Lim's major job was to encourage his NVM team to continue the research and development of NAND flash memory and to persuade his bosses to support the project. Samsung then developed a less sophisticated 4Mb NAND flash memory to replace a DRAM for the answering machine, which became an instant hit. Then the sudden boom in digital cameras, digital cell phones, and MP3 players from the mid-1990s on changed the situation completely. The sales of NAND flash memory doubled every year and became recognized as a major product of Samsung Electronics. Samsung continued developing bigger NAND flash memories—64Mb in 1997 and 128Mb in 1998—and led the market.

NAND flash memory continued to be Lim's beloved child even after he left to manage other aspects of Samsung's business. He later recalled:

[In 1999] I was in charge of the entire memory business. The overall situation was slowly improving, but the DRAM market was in stagnation. ... My great pleasure was to observe the rapid growth of the flash memory. Even though the market for flash memory was not big enough to attract much attention, it doubled its size every year and the sale of flash memory reached more than 100 million dollars in 1999. I was quite sure that the market for flash memory would be boundless. I felt like a father observing a rapidly growing child.[28]

Why, then, did Samsung Electronics support this seemingly unprofitable NVM development for so long without any noticeable profits? The answer may be closely related to the change in company leadership in the late 1980s. The new head of the Samsung conglomerate, Lee Kun-Hee, who succeeded his father Lee Byung-Chul in 1987, had a deep interest in semiconductors. It was he who had privately bought a majority of the stock in Hankuk Semiconductor in 1975 and then made it part of Samsung in 1977. Until the end of the 1980s, Samsung Electronics had been popularly known as a company that manufactured home appliances such as TVs, VCRs, refrigerators, and microwaves. The high management posts were usually filled by those who worked on these consumer products or by professional financiers. These executives were often envious and even suspicious of the up-and-coming young semiconductor engineers in the company. Under Lee Kun-Hee's new leadership, however, those engineers who had worked in semiconductor areas in the 1980s began to rise rapidly to higher management positions. These new leaders understood the importance of future technology like NVM and supported it using the huge profits from the sales of DRAMs.

Among the new high-ranking engineers-turned-executives, Kim Kwang-Ho and Lee Yoon-Woo were two key figures in the survival of Lim's NVM project. Kim Kwang-Ho is often praised as Samsung's first champion of semiconductors and as one who had worked on the area from the company's very early years. He first entered Tongyang Broadcasting Company (TBC), a part of the Samsung conglomerate, as an engineer in 1964. When Samsung Electronics was established in 1969, he was immediately transferred there and worked to manufacture TV sets. In 1979, he moved to Samsung Semiconductor as the principal executive of the company.[29] Kim was responsible for the management of the semiconductor business for the next eleven years: he led Samsung's first DRAM project (64Kb DRAM in 1983) and succeeding ones and became the president responsible for Samsung's entire semiconductor business in 1989. Between 1993 and 1996, he was the vice chairman and CEO of Samsung Electronics. While Kim was the first engineer-turned-executive for Samsung's semiconductor business, Lee Yoon-Woo was the first semiconductor-specialist-turned-executive. He had entered Samsung NEC (now Samsung SDI) in 1968 and moved to Hankuk Semiconductor (Samsung Semiconductor from 1977) in 1976. Since then, he has worked on the development of various semiconductors, including the first two DRAMs, 64Kb DRAM and 256Kb DRAM. He became a director of Samsung Semiconductor and Telecommunication in 1984, a senior executive director for the semiconductor business of Samsung Electronics in 1991, and a vice president and president of the entire semiconductor business in 1992. In 1996, he was promoted to the president and CEO of Samsung Electronics.

Lim and his NVM team were therefore very lucky to secure the support of these two key leaders of the company. Both leaders understood clearly the potential of NVM but also had plans to diversify to semiconductor products besides DRAMs, in order to prepare for the cycle of DRAM pricing: they never forgot the lesson of the collapse of DRAM's price in the mid-1980s.[30] Their trust was amply repaid when the sales of MROM, SRAM, and NAND flash memory rose rapidly from the mid-1990s on. As a reward for these successes, Lim was promoted to a junior managing executive director in December 1994. His role as an engineer developing specific technologies was almost over. From this time on, he was more of a professional executive manager supervising hundreds of engineers.

SYSTEM BUILDER: BECOMING A PROFESSIONAL EXECUTIVE

In 1995, Samsung Electronics reorganized the structure of its semiconductor operation. Under Lee Yoon-Woo's overall supervision, Chin Dae-Je, as vice president, became responsible for the entire memory operation. Lim was put in charge of the design of all the memory lines under Chin, while Kwon Oh-Hyun became responsible for the production of memory under Chin. Hwang Chang-Gyu, another star semiconductor engineer, was in charge of developing next-generation memory lines at the Samsung Semiconductor Research Institute. The competition among these four stars accelerated from then on as the slot for the next promotion became much narrower.[31]

As the new leader of memory design, Lim focused on Samsung Electronics' major product, DRAM. DRAM had been a cash cow for the company since 1988, when the prices of its 64Kb and 256Kb DRAMs had rebounded.[32] By 1992, Samsung's share of DRAM in the world market reached 13.6% and brought unprecedented profit to the company: for example, the sales of memory products in 1995, of which 87% consisted of the sale of DRAM, brought about USD 3.4 billion profit to the company.[33] However, the rapidly changing computer market in the mid-1990s demanded new standards for DRAMs. By 1995, Intel's central processing unit (CPU) dominated the whole computer market, and other types of CPUs developed by Hewlett Packard, DEC, and Sun slowly disappeared. To adapt Intel's more powerful new CPUs, new ideas and technologies were suggested to increase the speed of data processing: for example, new synchronous DRAMs rapidly replaced old asynchronous DRAMs. Lim's primary task was to prepare for the new standards for DRAMs. Under his leadership, Samsung adopted a double-data-rate (DDR) technology for DRAM and applied it to the design of 1Gb DRAM, which "result[ed] in 30% power reduction."[34] DDR was reported at the 1996 IEEE International Solid-State Circuits Conference: South Korean semiconductor engineers, for the first time, presented a paper on the design of DRAMs at a prestigious international conference, proving that they were no longer copycats of Japanese work. Samsung successfully applied both DDR and synchronous technologies to its DRAM designs and began to produce high-end DRAMs for PC servers and workstations. The development of the Rambus DRAM, which was widely considered in the late 1990s to be the next generation of memory, was also carried out under Lim's supervision. Samsung Electronics became one of the first semiconductor companies that supplied the Rambus DRAM and its succeeding models to the world market from 1999 on.[35]

Lim's other major job was to improve the design process. Until the mid-1990s, Samsung's chip designs were not properly standardized. Although the number of experienced designers was very limited, the number of semiconductor chips increased rapidly. Lim standardized the unit circuit and made it a block. He also demanded checklists to confirm every step of design. These process innovations proved very effective and helped increase the number of DRAM designs from an average of five per year to an average of twenty. Lim proudly recalled that these innovations contributed greatly to the company's increasing ability to manufacture several different kinds of chips, which became vital during the economic crisis of the late 1990s.[36] In December 1995, Lim was promoted to a senior managing executive director, "just a year after he was promoted to a junior executive director ... due to his excellent achievement."[37]

In early 1997, Lee Kun-Hee, chairman of Samsung conglomerate, reshuffled Samsung Electronics' management to respond both to the sharp drop in DRAM price and to the general economic recession. Chin became responsible for the System Large-Scale Integration (System LSI) operation, and Kwon Oh-Hyun was also transferred there. This was Lee's ambitious move to strengthen this "nonmemory" operation with these two star engineers. Samsung Electronics had manufactured various System LSI chips in small scale since the mid-1980s, but their sales and profits had been much lower than those of memory chips. Since the System LSI market is much bigger than the memory market and the price of this line of chips is less volatile than that of memory chips, the company could not ignore this promising area any longer. Samsung's continuous success in the memory business over the preceding fourteen years also led Lee and Samsung engineers to believe that they could win this market without much difficulty.[38]

With the departure of Chin and Kwon, Lim became responsible for the entire memory business from January 1997 to December 1999. His primary job as the head of the memory operation was to raise Samsung's competitiveness in the world's memory market. He proposed two goals to achieve this: product differentiation and "design for reliability, productivity and testability." Product differentiation aimed to cultivate new markets for DRAMs. The major target of Samsung's DRAM had been personal computers, but Lim wanted to sell diverse DRAMs to diverse markets, such as those for servers and graphics. To win these new markets, his second goal of "design for reliability, productivity and testability" had to be reached. Lim worked hard to recruit talented semiconductor designers and trained them accordingly. He also expanded the product planning part of the company so as to contact and persuade potential buyers more effectively. His hard work paid off well: IBM, Hewlett Packard, and Sun Microsystems soon began to purchase Samsung's DRAM for their servers.[39]

The period when Lim was in charge of the entire memory business largely overlapped with the years when South Korea experienced economic hardships. The price of DRAMs began to drop suddenly at the beginning of 1996 (from $44 to $8 within a few months) and recovered slightly only from 1999 on. The Asian Financial Crisis, which started in Southeast Asia, hit South Korea in the fall of 1997. The South Korean government received emergency relief from the International Monetary Fund (IMF) on the condition that it would restructure its whole economic system according to the IMF's suggestions. The government that was newly elected in December 1997 carried out the economic reforms from 1998 on, which included several mergers and acquisitions of big companies. Samsung Electronics survived, but LG Semiconductor was forced to merge with Hyundai Electronics. Actually,

FIGURE 4.3   **Left**: Lim Hyung-Kyu (right) attended the press conference in 1999, along with Lee Yoon-Woo (center) and Hwang Chang-Kyu (left), where Samsung Electronics announced its first mass production of 256Mb DRAM. **Right**: A launching ceremony of the S3C2410 chipset for small devices such as cell phones or handheld game machines. From 2000 to 2004, Lim was responsible for the System LSI operation (first as vice president and then as president) and contributed greatly to the development of System LSI at Samsung Electronics.

Source: Courtesy of Lim Hyung-Kyu.

Samsung expanded its share of memory in the world market during the economic crisis, though its profits decreased sharply. In 1998, it even started manufacturing 256Mb DRAM and 128Mb NAND flash memory (Figure 4.3). Lim's contribution to weathering the storm was duly noted by the top management: he was promoted to a vice president in January 1999.

In January 2000, Lim was transferred to the System LSI operation. Chin Dae-Je was promoted to the president to supervise "Information and Home Appliances." When Chin had been moved to System LSI in 1997, he had confidently declared that he would make Samsung Electronics one of the top ten companies in the world in nonmemory areas by 2002: three teams under his supervision were the alpha chip (CPU) and merged DRAM with logic (MDL) group, the application-specific integrated circuit (ASIC) group, and the microcomputer and power semiconductor device group.[40] He made a similar assertion two years later but narrowed his targets to two: alpha chips and ASIC.[41] Although Chin's System LSI operation succeeded in developing the fast alpha chip, it failed to secure a meaningful market share.[42] Lee Kun-Hee did not seem satisfied with the progress of the System LSI operation under Chin and decided to assign the task to Lim.

Once again, Lim became the "problem solver" for a promising but troubled area. He later recalled that he welcomed this new challenge:

> When I moved from the memory to the non-memory operation, many people around me worried about my future career: Samsung's memory was the first in the world, while its non-memory was around the twentieth. However, I liked this new task. I loved to do this undeveloped but promising field rather than the already fully developed one. Moreover, as CEO of the System LSI operation, I had full power to manage not only research and development but also production, sales, planning, personnel administration, and finance. It was the best opportunity for me to learn the role of CEO.[43]

Lim immediately rearranged the main targets of the System LSI operation, deleting less promising or unprofitable areas, while the LCD Driver IC (LDI), micro control unit (MCU), smart card chip, and radio frequency (RF) chips for communication were selected to nurture.[44] The results of this arrangement soon became apparent. By the end of 2000, Samsung Electronics had sold USD 1.8 billion worth of nonmemory products—about 80% growth over sales of the previous year.[45] Lim was duly promoted to a president in 2001 but still supervised the System LSI operation. Bolstered by this success, he enthusiastically expanded the System LSI operation from 2001 on, constructing new production lines and adding eleven nonmemory chips, including seven system-on-a-chip (SoC) integrated circuits, to the production list.[46] At the same time, Lim dropped as many as twenty targets, including analog chips and MCU, that could not produce any promising results. The restructuring worked well: the 2002 annual report of Samsung Electronics proudly reported that "[The System LSI operation] continued its high growth owing to the growing sales of both the display drive IC and new SoC products."[47]

Lim ambitiously aimed to make Samsung Electronics number five in the nonmemory market and also to raise the System LSI operation to the equal of the memory operation in both sales and profits by 2007.[48] To achieve this goal, he set about recruiting talented engineers not only in South Korea but around the world. He spent many days a year traveling to Silicon Valley for this task. Fortunately for him, at the turn of the new century, both KAIST and SNU had begun to produce the necessary nonmemory specialists. Many of those whom Lim recruited during his tenure as the head of the System LSI rose to the director's level.

Lim soon realized that Samsung Electronics could not singlehandedly meet all the diverse demands of System LSI buyers. He confessed in a press conference, "We have a lot of things to do but don't have enough time and those who can do them."[49] The solution was to search for partner companies, especially small and medium-sized startups. When Lim told the reporters his difficulties, Samsung's website posted the news under the headline "Samsung's System LSI Business Introduced Strategic Business Opportunities with Local Design Houses."[50] It was truly a win–win cooperation between a giant company like Samsung Electronics and small or medium-sized startup companies. Many among the selected fifteen grew to be competent medium- or large-sized companies, with their stock listed in the Korea Securities Dealers Automated Quotation (KOSDAQ) index. Samsung Electronics was certainly the major beneficiary of this cooperation: for example, in 2004, the LCD Driver IC (LDI), the product that Lim had selected to nurture, became the first System LSI chip to achieve USD 1 billion in annual sales.[51]

Lim's other major contribution to the System LSI operation was to expand the foundry.[52] The foundry was a weak point of Samsung Electronics, which had focused on the memory business since 1983. Although Lim had wanted to expand the foundry business from the beginning, he had inherited outdated facilities and weak technology when he became head of the operation. The immediate mission was to replace the old 8-inch production line with a new 12-inch production line. This involved a serious dilemma: construction of a 12-inch line required a great deal of investment at a time when there was little demand for Samsung's nonmemory products, yet without a larger production line, Samsung could not produce the top-notch nonmemory products that the market demanded. In 2002, Samsung Electronics finally decided to build the new production line under Lim's supervision. In terms of this expansion of the foundry business, Samsung was lucky to join the IBM-led consortium in 2003 to develop the new logic process technology so as to compete with Intel and Taiwan Semiconductor Manufacturing Company (TSMC). Lim remembers that Lee Jae-Yong, the son and heir apparent of Samsung's chairman Lee Kun-Hee, played an important role in the construction of the 12-inch production line as well as in the connection with IBM.[53]

By the end of 2003, the System LSI operation had successfully constructed a state-of-the-art 12-inch production line and had begun to produce eight top-quality nonmemory chips. Its overall sales reached USD 3 billion. An executive director of Samsung Electronics later appraised Lim's contribution to System LSI: "If Lim had not cultivated the System LSI operation of Samsung Electronics [in the early 2000s], neither Samsung smartphones nor smart TVs would have ever existed."[54] This was only a slight exaggeration, for the 2004 annual report of Samsung Electronics states that "The System LSI operation developed the display driver IC, CMOS image sensor, and the most efficient mobile processor" and that the System LSI was chosen as one of eight "growth engines" of the company.[55]

In December 2003 Lee Kun-Hee met with Lim alone, and they had a long talk on various subjects, including the future of flash memory.[56] A few days later, Lim was appointed chief technology officer (CTO), the position that supervises all of Samsung Electronics' technology and also prepares for future technology. For the first time since being hired at Samsung in 1976 (except for his leaves to attend graduate schools), Lim was no longer in research or production. He recalls that he was bewildered and disappointed to leave the System LSI division to become CTO. He was replaced by Kwon Oh-Hyun, who successfully cultivated the System LSI operation and later became CEO of the company.

What kind boss was Lim during the 1990s and early 2000s, when he personally managed semiconductor engineers to design, develop, and produce chips? A Samsung engineer who had worked under him in the memory operation remembered that Lim had emphasized details but had also been open-minded:

> My first impression from his appearance was that he must be "tough." Since he was very meticulous and checked every detail of the work, his nickname among engineers was "assistant manager." However, he was open-minded on private occasions and understood the various difficulties of engineers. ... His office was literally open to engineers with any new ideas and good suggestions. I asked his secretary whether I really could enter his office without any appointment, and she told me that the opened door of his office was the answer. ... This attitude greatly stimulated engineers under him to work harder by themselves and to search for new issues. He was famous for preparing the ground for the new business wherever he moved among different operations.[57]

Another example that illustrates his style as an executive director is provided by a female clerk-turned-engineer. Yang Hyang-Ja was hired by Samsung Electronics as an assistant to semiconductor engineers after graduating from a commercial high school. Without a college degree, her major tasks were making photocopies for engineers, serving cups of coffee, and even cleaning tables. Yet she wanted to become a semiconductor engineer, so she studied the subject by herself. Slowly her colleagues and superiors recognized her effort and began to help her. She remembers Lim's help as follows:

> Lim was the supervisor of my department. He knew of my efforts to become an engineer and accepted me as a member of a development team. Once a director was searching for a new secretary who had graduated from a commercial high school and could read and speak Japanese. He spotted me, but Lim declined the request, saying, "She does not fit that post." He believed that I must grow as a research engineer, not a secretary.[58]

Yang worked as chief engineer in SRAM, DRAM, and flash memory departments and in 2013 became a director at Samsung Electronics. In 2020, she was elected a member of Parliament representing her own hometown and became the leading figure in Parliament supporting the development of semiconductors.[59]

Lim's new position of CTO had existed at Samsung Electronics since the early 2000s, although a president or a vice chairman of the company had often occupied it as an additional position. Lim was therefore the first CTO who was independent of any other executive posts. His main tasks were "to carry out basic research, to prepare the roadmap for the future, to manage patents and standards, and to coordinate R&D within the company."[60] This was a new "experiment" by Lee Kun-Hee, who intended to reduce the overlapping investments within Samsung Electronics. Ten virtually "independent" operations within the company were competing with one another to develop their own products first, and the overlapping investments in R&D had become prevalent and serious. The choice of Lim for the job thus reflected Lee Kun-Hee's trust in him: Lim had enough knowledge and experience to deal with this difficult task, and his loyalty to the company was beyond question (Figure 4.4).

However, it soon became apparent that the task was Herculean. As CTO, Lim could not dictate to or even question other presidents of operations who had different interests and goals: for example, he did emphasize coordination between the semiconductor operation and other finished goods operations, yet he had little direct authority to impose it.[61] His tenure as CTO was relatively short, just one year. Nonetheless, when South Korean media mentioned his name thereafter, the title "CTO" was frequently affixed. In 2007, the Korea Industrial Technology Association selected Lim as "CTO of the Year."[62] After retirement in 2010, he served as a secretary of the CTO club and a principal liaison between the government and the club. His experience as CTO also contributed to his future career at the SK conglomerate between 2014 and 2016.

In January 2005, Lim became head of the Samsung Advanced Institute of Technology (SAIT), while the vice chairman Lee Yoon-Woo replaced Lim as CTO while continuing his existing task of handling external relations. SAIT had been established in 1987 by Lee Byung-Chul, the founder of the Samsung conglomerate, as "Samsung Group's R&D Hub," with the aim of becoming "the incubator for cutting-edge technologies under the founding philosophy of 'Boundless search for breakthroughs.'"[63] As the head of the Samsung's most important research institute, Lim's first task was to reexamine its basic roles within the Samsung conglomerate and set up more realistic goals. He soon found that the institute had not played its

FIGURE 4.4   Samsung's chairman Lee Kun-Hee (second from left) and his wife (far left) greet Lim Hyung-Kyu (second from right) and his wife (far right) at a party.

Source: Courtesy of Lim Hyung-Kyu.

proper roles as originally intended and immediately began to reorganize the struc-ture of the institute. Almost half of the existing projects were dropped or merged with similar ones. Many researchers at the institute were transferred to the produc-tion lines in Samsung's related companies.[64] Research related to Samsung's major business areas at the time (semiconductors, liquid crystal displays, etc.) was reduced, while research dedicated to the future (green energy, the environment, and bioengi-neering) was greatly enlarged, so that the ratio between them changed from 8:2 to 5:5. Lim boldly predicted that green energy, the environment, and bioengineering would become Samsung's major business areas soon.[65] He later recalled that he truly enjoyed his job at SAIT.[66]

In October 2007 Lee Kun-Hee appointed Lim the head of a task force to search for "new growth power" for the future, from an office located at the Samsung conglomerate's headquarters. The team was rather small but had the power to use any specialists in the Samsung conglomerate. When the team's office moved to Samsung Electronics in May 2008, Lim resigned as the president of SAIT and concentrated on leading the task force. He continued emphasizing the need to nur-ture the areas of new energy, the environment, bioengineering, and health care. Five specific fields were selected: biosimilars, medical appliances, secondary batteries for the electric car, LED lighting, and solar panels. The task force was put in charge of the biosimilar operation, which was a completely new business for Samsung, and the other four operations were distributed to either Samsung Electronics or Samsung SDI. Under Lim's supervision, the biosimilar operation laid a solid foundation for the future Samsung Biologics (established in 2011) and Bioepis (established in 2012). Biologics became partners with GlaxoSmithKline and other global pharmaceutical companies to produce biological and medical products. Bioepis is a joint company with Biogen, an American multinational biotechnology company that pioneered the research and development of treatments for neurological diseases, including Alzheimer's. Some members of Lim's task force later became leaders in both companies.

In January 2009, Samsung Electronics announced large-scale organizational and personnel changes to respond to the global economic crisis that had started in the United States in the fall of 2008. Many of the old guard at the company, including Lim and Hwang Chang-Gyu, would step down before they reached 60 years old (the age limit for executives that Lee Kun-Hee had set).[67] As head of the task force, however, Lim had important unfinished work and was asked to remain until the end of 2009, with a special one-year contract. The 2009 shakeup was widely considered to be a step toward turning over the crown of the Samsung conglomerate to Lee Kun-Hee's son, Lee Jae-Yong. Lim had been Lee Kun-Hee's man through thick and thin. Of the four star semiconductor engineers at Samsung during the 1990s and 2000s, only Kwon Oh-Hyun remained, continuing to play important roles as president, vice chairman, and chairman of the company until early 2019. Lim finally ended his 34-year career at Samsung in 2009 but remained a "special counsellor," as was customary, until the end of 2011. He later recalled that he had done all he could do at Samsung and that he enjoyed freedom and leisure for the first time, without any heavy burden or pressure, for the next four years.[68]

AFTER SAMSUNG ELECTRONICS

Lim, in his early sixties, was too young to spend all his free time traveling and playing golf with his family and friends. He became the president of the KAIST Alumni Association in 2011 and served in this post for the next three years. He did his best to revitalize this inactive organization by promoting contacts among the alumni and achieved some success. Along with 30 other alumni, he established the KAIST Alumni Scholarship Foundation to support talented undergraduate students. Since he was one of the most accomplished semiconductor specialists in South Korea, there were several calls from government and from political parties after his retirement, but he declined them. Lim was also a popular lecturer and gave many impressive talks at universities, professional societies, and public meetings. In June 2020, for example, he was invited to give a special talk on "New Trends in Global Industry and Opportunity for South Korea" at a study group in Parliament.[69]

In January 2014, four years after Lim had retired from Samsung Electronics, the SK conglomerate announced that it would appoint Lim the vice chairman responsible for the overall management of its information and communications technology (ICT), including semiconductors, cell phone services, and the search for new growth powers for the future (Figure 4.5).[70] South Korean media reported that Chey

SK, 'ICT 총괄직' 신설…삼성 DNA 심는다

입력 2014-01-22 18:41 | 수정 2014-01-23 10:40

삼성종합기술원장 출신 임형규 전 사장 부회장으로 영입

SK그룹이 그룹의 미래 성장동력으로 정보통신기술(ICT)을 선정하고, 신사업 발굴을 위한 전담조직을 신설한다. 이를 위해 최태원 SK 회장이 추천한 임형규<사진> 전 삼성종합기술원장을 영입하며, 삼성의 1등 DNA 옮겨 심기에 나섰다.

FIGURE 4.5 Left: In January 2014, South Korean media reported that the SK conglomerate—which had acquired Samsung Electronics' chief rival in the memory business, Hynix, two years earlier—would appoint Lim Hyung-Kyu vice chairman of SK. **Right**: Lim helped SK conglomerate's chairman Chey Tae-Won (second from left) reconstruct SK Hynix, and the company soon hit its stride.

Sources: *etoday*, January 22, 2014; and courtesy of Lim Hyung-Kyu.

Tae-Won, chairman of the SK conglomerate, had "personally recommended" Lim and that the Samsung conglomerate "understood" Lim's move to SK, even though SK Hynix had been Samsung's major Korean rival in the semiconductor business ever since SK had taken over Hynix in 2012.[71] A newspaper reported Lim's move to SK under the following headline: "Samsung's First [Overseas] Scholarship Student, 'Mr. NAND,' Becomes SK Hynix's Board Director."[72] Lim concentrated on building proper systems to strengthen the memory operation, as well as to enlarge the System LSI operation at SK Hynix, and on coordinating information technologies among SK's related companies. He quickly realized the basic difference between Samsung Electronics and SK Hynix: unlike Samsung, few engineers had been promoted to CEO positions at SK Hynix, so that they were somewhat passive when receiving orders from above. Lim tried hard to change this culture into one in which semiconductor engineers would play more important and broader roles in management and therefore grow as true entrepreneurs. A newspaper reported in November 2014 that the recent "success after success" of SK Hynix was due to Lim's "invisible assistance."[73] He served as SK's vice chairman until the end of 2016 and then stepped down to become an advisor.

Although Lim has reached his seventies, he remains active in preaching the importance of semiconductors to South Korean industry in his clear and direct manner. In the spring of 2021, when tension between the United States and China over semiconductors intensified, South Korean media expressed serious worries about the future of the South Korean semiconductor industry: South Korea still depends heavily on American technology in manufacturing semiconductors, yet China has been the number one importer of South Korean semiconductors since the early 2000s. So how can South Korea maintain a balance between the two giant countries? Or, in the worst case, which country should South Korea side with? In a newspaper article published on April 20, 2021, Lim presents a clear answer to the dilemma:

I don't agree with the opinion that South Korea's semiconductor industry is sandwiched between the US and China. The choice is clear: South Korea must closely cooperate with the US and need not worry about China's pressure. … We must remember that it is the US that has created and developed semiconductors. It is also the US that still dominates core technologies, necessary software, and the market for semiconductors. … China is just South Korea's major market, no more, no less. … The development of semiconductors in South Korea in the past and the present is the result of the US–South Korea alliance.[74]

The contributions of "Mr. NAND Flash" to the development of South Korean semiconductors are not yet over.

NOTES

1   Interview with Kim Choong-Ki by Kim Dong-Won (September 14, 2020).
2   This description is inscribed in the plaque Lim received in the spring of 2005, when he was named a University of Florida Distinguished Alumnus. See https://fora.aa.ufl.edu/docs//21//Distinguis hed%20Alum%20through%20May%202022.pdf (searched on December 1, 2020).
3   Joong-Ang Ilbo, "Samsungjeonjareul Umjikineun Saramdeul," *Joong-Ang Ilbo* (February 23, 2002), https://news.joins.com/article/673226 (searched on November 13, 2020).
4   Interviews with Lim Hyung-Kyu by Kim Dong-Won (February 14 and September 17, 2020).
5   See Seoul National University, *Seouldaehakgyo 50nyeonsa* [*Fifty-Year History of Seoul National University*], Vol. 2 (Seoul: Seoul National University Press, 1996), 147–155. By the end of the 1970s, most professors in the department had trained at Japanese universities, and few had received their doctoral degrees from US universities. It was quite the opposite at KAIS (later KAIST), where most professors had received their doctorates from US or other Western universities.
6   Hyung-Kyu Lim, "Design and Fabrication of a Seven-Segment Decoder/Drive with PMOS Technology" (master's thesis, Seoul: KAIS, 1978). His name in the thesis was recorded as Rim Hyung-Gyu. It was later published in *Jeonjagonghakhoeji, 15:3* (July 1978), 11–17.
7   For the history of Hankuk Semiconductor and Samsung Semiconductor, see Samsung Semiconductor and Telecommunications, *Samsung Bandochetongsin 10nyeonsa* [*Ten-Year History of Samsung Semiconductor and Telecommunications*] (Seoul: Samsung Bandochetongsin, 1987), 170–186. Samsung Semiconductor became Samsung Semiconductor and Telecommunications in 1982, which merged with Samsung Electronics in 1988.
8   Park Sang-In et al., *Uri Kim Choong-Ki Seonsaengnim* [*Our Teacher, Kim Choong-Ki*] (Daejeon: Privately printed, 2002), 29.
9   Hyung-Kyu Lim, "Charge-Based Modeling of Thin-Film Silicon-on-Insulator MOS Field-Effect Transistors" (Ph.D. thesis, University of Florida, 1984).
10 Some selections are Hyung-Kyu Lim and Jerry G. Fossum, "An Analytic Characterization of Weak-Inversion Drift Current in a Long-Channel MOSFET," *IEEE Transactions on Electron Devices, 30:6* (1983), 713–715; Hyung-Kyu Lim and Jerry G. Fossum, "Current Voltage Characteristics of Thin-Film SOI MOSFET's Strong Inversion," *IEEE Transactions on Electron Devices, 31:4* (1984), 401–408; and Hyung-Kyu Lim and Jerry G. Fossum, "A Charge-Based Large-Signal Model for Thin-Film SOI MOSFET's," *IEEE Transactions on Electron Devices, 32:2* (1985), 446–457.
11 Hyung-Kyu Lim and Jerry G. Fossum, "Threshold Voltage of Thin-Film Silicon-on-Insulator (SOI) MOSFET's," *IEEE Transactions on Electron Devices, 30:10* (1983), 1244–1251. For the citation of the paper, see https://ieeexplore.ieee.org/document/1483183 (searched on October 30, 2020).
12 Interview with Park Sung-Kye by Kim Dong-Won (June 9, 2022).
13 Samsung Electronics, *Samsungjeonja 30nyeonsa* [*Thirty-Year History of Samsung Electronics*] (Seoul: Samsung Electronics, 1999), 202–203. See also Lee Im-Sung's recollections of SSI in Joong-Ang Ilbo, "30nyeonjeon Eoryeopge ppurin Ssiat, Yeon 300eok Dollar Yeolmaero," *San Francisco Joong-Ang Ilbo* (August 1, 2013), http://m.koreadaily.com/news/read.asp?art_id=1877832&referer (searched on October 23, 2020).

14 Interview with Lim Hyung-Kyu by Kim Dong-Won (September 17, 2020).

15 Jae-Young Do et al., "A 256K EEPROM with Enhanced Reliability and Testability," *1988 Symposium on VLSI Circuits,* held in Tokyo (1988), 83–84. This was the first work, based on the efforts made at Samsung Electronics's research facility, that was presented at and then published by an international conference.

16 Interview with Lim Hyung-Kyu by Kim Dong-Won (September 17, 2020).

17 Maeil Kyungje, "Jeongjinki Eonronmunhwasang Sisang," *Maeil Kyungje* (July 13, 1989), http://m.mk.co.kr/onews/1989/999213#mkmain (searched on October 26, 2020).

18 Jim Turley, "Masked ROM," in Jim Turley (ed.), *The Essential Guide to Semiconductors* (Upper Saddle River, N.J.: Prentice Hall, 2003), 151.

19 Lee Jae-Hyun, *Samsung iraeseo ganghada: Samsungjeonja Bandoche Memorisaeopbu Iyagi 1* (Seoul: Barunbooks, 2017), 276.

20 Samsung Electronics, *Samsungjeonja 30nyeonsa*, 298–299. The book, however, does not mention Lim's name as the developer of fast SRAMs. In contrast, those who developed DRAMs, Chin Dae-Je (16Mb), Kwon Oh-Hyun (64Mb), and Hwang Chang-Gyu (256Mb), are specifically named.

21 Masaki Momodomi et al., "New Device Technologies for 5 V-only 4Mb EEPROM with NAND Structure Cell," *International Electron Devices Meeting (IEDM) Technical Digest* (1988), 412–415.

22 For more information about Masuoka Fujio and his work, see Koji Sakui, "Professor Fujio Masuoka's Passion and Patience toward Flash Memory," *IEEE Solid-State Circuits Magazine, 5:4* (2013), 30–33; Forbes editorial, "Unsung Hero," *Forbes* (June 23, 2002), www.forbes.com/global/2002/0624/030.html#299c6b903da3; and Masuoka Fujio, "Oral History of Fujio Masuoka" interviewed by Jeff Katz on September 21, 2012, stored in Computer History Museum (reference number: X6623.2013), http://archive.computerhistory.org/resources/access/text/2013/01/102746492-05-01-acc.pdf (searched on October 28, 2010).

23 Interview with Lim Hyung-Kyu by Kim Dong-Won (September 17, 2020).

24 Toshiba, *Flash Memory: Semiconductor Catalog March 2016* (2016), 2 and 4.

25 Samsung Electronics, *Samsungjeonja 30nyeonsa*, 387 and 509. The development of NAND flash memory is only briefly mentioned in it.

26 Fujio Masuoka, "Technology Trend of Flash-EEPROM. Can Flash-EEPROM Overcome DRAM?" *1992 Symposium on VLSI Technology Digest of Technical Papers* (1992), 6–9.

27 Kang-Deog Suh et al., "A 3.3V 32Mb NAND Flash Memory with Incremental Step Pulse Programming Scheme," *Proceedings of 1995 International Solid-State Circuits Conference* (1995), 128–129.

28 Interview with Lim Hyung-Kyu by Kim Dong-Won (May 29, 2021).

29 Kim Kwang-Ho later remembered that he had considered the transfer to Samsung Semiconductor as a demotion. When he first arrived at Samsung Semiconductor, the company was almost bankrupt. For his recollections of the early days of Samsung's semiconductor business, see Kim Kwang-Ho, "Samsung Bandocheui Oneuli itgikkaji," in Korea Electronics Association, *Gijeokui Sigan, 50 (1959–2009): The Miraculous Time* (Seoul: Korean Electronics Association, 2009), 179–182.

30 Samsung Electronics, *Samsungjeonja 30nyeonsa*, 211–212.

31 The number of high-ranking officers at Samsung Electronics in 1999, for example, was as follows: 3 presidents, 16 vice presidents, 26 senior executive directors, 46 junior executive directors, and 192 directors. About one-third of them worked on the semiconductor business. See Samsung Electronics, *Samsungjeonja 30nyeonsa*, 552–571. The total number of high-ranking officers had more than doubled by the time Lim retired in 2009.

32 Samsung Electronics, *Samsungjeonja 30nyeonsa*, 211–212.

33 Ibid., 400.

34 Jei-Hwan Yoo et al., "A 32-bank 1 Gb DRAM with 1 GB/s Bandwidth," *1996 IEEE International Solid-State Circuits Conference, Digest of Technical Papers* (1996), 378–379.

35 Samsung Electronics, *Samsungjeonja 30nyeonsa*, 510. However, the Rambus DRAM, which Intel adopted as standard for its CPUs in the late 1990s, faded out and disappeared in the market from 2002.

36 Interview with Lim Hyung-Kyu by Kim Dong-Won (May 29, 2021).

37 Maeil Kyungje, "Samsung-group 468myeong Choedaegyumo Imwoninsa Danhaeng," *Maeil Kyungje* (December 8, 1995), www.mk.co.kr/news/business/view/1995/12/56489/ (searched on November 11, 2020).

38 *Samsungjeonja 30nyeonsa* reflects this optimistic view in the late 1990s. For more information about Samsung's System LSI during the 1980s and 1990s, see Samsung Electronics, *Samsungjeonja 30nyeonsa*, 388–392, 511–515.

39 Ibid., 507.

40 Maeil Kyungjae, "Bimemorido Samsung apjang ... Chin Dae-Jedaepyo," *Maeil Kyungje* (February 27, 1997), www.mk.co.kr/news/it/view/1997/02/11540/ (searched on November 17, 2020).

41 Joong-Ang Ilbo, "Samsungjeonja, 'Bimemori Seungbu' ... 2001nyeonkkaji 12eokbul Tuip," *Joong-Ang Ilbo* (June 22, 1999), https://news.joins.com/article/3792150 (searched on November 17, 2020). *Samsungjeonja 30nyeonsa*, published in 1999, reflects Chin Dae-Je's rosy view on System LSI similarly.

42 Samsung Electronics, *Samsungjeonja 40nyeon: Dojeonkwa Changjoui Yeoksa* [*Forty-Year History of Samsung Electronics: The Legacy of the Challenge and Creativity*] (Suwon, Kyeongkido: Samsung Electronics, 2010), 83–88.

43 Interview with Lim Hyung-Kyu by Kim Dong-Won (May 29, 2021).

44 Hankuk Kyungje, "Samsung, 4gae Bimemori Jipjung Yukseong ... LDI, MCU Segye 1wi Mokpyo," *Hankuk Kyungje* (October 3, 2000), www.hankyung.com/news/article/2000100204661 (searched on November 17, 2020). This rearrangement was made early in 2000 but announced much later, in October 2000.

45 Hankuk Kyungje, "Samsung, Bimemori Daeyakjin, ... Olmaechul 80% Geupjeung 18eokbul," *Hankuk Kyungje* (December 11, 2000), www.hankyung.com/news/articles/2000121074891 (searched on November 17, 2020).

46 Hankuk Kyungje, "Samsungjeonja, Bimemori Daedaejeok Tuja ... Gaebalbi 2000eok Chuga," *Hankuk Kyungje* (April, 15, 2002), www.hankyung.com/news/article/2002041562761; Joong-Ang Ilbo, "System-on-chipdeung 11gae Chumdan Bimemory Bandoche, Samsungjeonja Guksanhwa Wanryo," *Joong-Ang Ilbo* (July 26, 2002), https://news.joins.com/article/4317696 (searched on November 17, 2020).

47 Samsung Electronics, *2002 Annual Report* (in Korean) (Suwon, Kyeongkido: Samsung Electronics, 2003), 47.

48 Hankuk Kyungje, "Bimemori Ganghwa ... Samsungjeonja, System LSIsaeop Yukseong Uimi," *Hankuk Kyungje* (August 27, 2002), www.hankyung.com/news/article/2002082731911; Joong-Ang Ilbo, "Bimemori Maechul 5nyeonhu 5wiro," *Joong-Ang Ibo* (August 28, 2002), https://news.joins.com/article/4334076 (searched on November 17, 2020).

49 Joong-Ang Ilbo, "Samsungjeonja Bencheowa Sonjapgo Bimemori Bakcha," *Joong-Ang Ilbo* (July 30, 2001), https://news.joins.com/article/4110902 (searched on November 17, 2020).

50 Samsung, "Samsung's System LSI Business Introduced Strategic Business Opportunities with Local Design Houses," www.samsung.com/semiconductor/newsroom/news-events/samsungs-system-lsi-business-introduces-strategic-business-opportunities-with-local-design-houses/ (searched on November 17, 2020).

51 Joong-Ang Ilbo, "Samsung LDI Bandoche Maechul 10eokbul Dolpa," *Joon-Ang Ilbo* (May 10, 2005), https://news.joins.com/article/44630 (searched on November 17, 2020).

52 For the foundry business in the early 2000s, see Samsung Electronics, *Samsungjeonja 40nyeon*, 89–91.

53 Interview with Lim Hyung-Kyu by Kim Dong-Won (May 29, 2021).

54 Joong-Ang Ilbo, "'Bandoche Shinhwa Rival' Lim Hyung-Kyu, Hwang Chang-Gyu 2 Round," *Joong-Ang Ilbo* (January 23, 2014), https://news.joins.com/article/13712601 (searched on November 13, 2020).

55 Samsung Electronics, *2004 Annual Report* (in Korean) (Suwon, Kyeongkido: Samsung Electronics, 2005), 49; and idem, *2005 Annual Report* (in Korean) (Suwon, Kyeongkido: Samsung Electronics, 2006), 10.

56 Interview with Lim Hyung-Kyu by Kim Dong-Won (May 29, 2021).

57 Lee Jae-Hyun, *Samsung iraeseo ganghada*, 199–200. Lee notes four different types of leaders under whom he had worked at Samsung Electronics. Though specific names are not mentioned, they must be Chin Dae-Je, Kwon Oh-Hyun, Lim Hyung-Kyu, and Hwang Chang-Gyu.

58 Joong-Ang Ilbo, "Jukgetda sipdorok Gongbuhaetda ... Miss Yangseo Yang Hyang-Jassi doetda," *Joong-Ang Ilbo* (September 17, 2018), https://news.joins.com/article/22975746 (searched on November 23, 2020).

59 Yang Hyang-Ja became chair of the special committees for the development of the semiconductor industry in two different political parties in 2020 and 2022.

60 Maeil Kyungje, "Samsungjeonja 'Dream Team'e Haeksimproject matgyeo," *Maeil Kyungje* (September 23, 2004), www.mk.co.kr/news/economy/view/2004/09/333443/ (searched on November 18, 2020).

61 Interview with Lim Hyung-Kyu by Kim Dong-Won (May 29, 2021).

62 Joong-Ang Ilbo, "Lim Hyung-Kyu Samsung Gisulwonjangdeung 5myeong 'Olhae Gisul Gyeongyeonginsang," *Joong-Ang Ilbo* (February 22, 2007), https://news.joins.com/article/2641809 (searched on November 18, 2020).

63 Samsung Institute of Advanced Technology, "About SAIT," www.sait.samsung.co.kr/saithome/about/who.do (searched on November 18, 2020).

64 Interview with Lim Hyung-Kyu by Kim Dong-Won (September 17, 2020).

65 Maeil Kyungje, "Samsung Saesaup Yeonkubijung 50%ro," *Maeil Kyungje* (July 12, 2007), www.mk.co.kr/news/business/view/2007/07/368243/ (searched on November 18, 2020).

66 Interview with Lim Hyung-Kyu by Kim Dong-Won (May 29, 2021).

67 Hankuk Kyungje, "Lee Yoon-Woo 'Bupum', Choi Jisung 'Jepum' … '2gaeui Samsungjeongja' Chulbum," *Hankuk Kyungje* (January 17, 2009), www.hankyung.com/news/article/2009011640701 (searched on November 19, 2020).

68 Interview with Lim Hyung-Kyu by Kim Dong-Won (May 29, 2021).

69 Yonhap News, "Minju, jeon Samsungjeonjasajang buleo 'Global Sanup' Yeolgong," *Yonhap News* (June 30, 2020), www.yna.co.kr/view/AKR20200630191400001 (searched on April 15, 2021).

70 Hankuk Kyungje, "Samsungchulsin Lim Hyung-Kyu, SK Miraesanup Chonggwalhanda," *Hankuk Kyungje* (January 23, 2014), www.hankyung.com/news/article/2014012228711 (searched on November 19, 2020).

71 Chosun Ilbo, "SK, Samsung CTOchulsin Lim Hyung-Kyu ICT Gisul-Seongjang Chonggwalbuhoejang Yeongip," *Chosun Ilbo* (January 22, 2014), https://biz.chosun.com/site/data/html_dir/2014/01/22/2014012203125.html; Insight Korea, "Samsungchulsin Lim Hyung-Kyu Buhoejang Yeongjp," *Insight Korea* (June 26, 2015), www.insightkorea.co.kr/news/articleView.html?idxno=13394 (searched on November 19, 2020).

72 Joong-Ang Ilbo, "Samsnug 1ho Janghaksaeng 'Mr. NAND,' SK Hynix Deungkiisae," *Joong-Ang Ilbo* (March 5, 2014), https://news.joins.com/article/14075010 (searched on November 15, 2020).

73 etoday, "'Seungsengjangku' SK Hynix sumeun Joryeokja Lim Hyung-Kyu Buhoejang," *etoday* (November 6, 2014), www.etoday.co.kr/news/view/1013860 (searched on November 17, 2020).

74 Maeil Kyungje, "K Bandoche, Mieopsineun Jonripbulga … Jung Nunchibol Pilyoeopseo," *Maeil Kyungje* (April 20, 2021), www.mk.co.kr/news/economy/view/2021/04/380924/ (searched on July 10, 2021).

5 Exploring New Semiconductor Areas

Cho Byung-Jin

Cho Byung-Jin has been a star engineering professor ever since he began to teach semiconductors at the National University of Singapore (NUS) in 1997. He started his career as an industrial researcher at Hyundai Electronics (later SK Hynix) and then continued it as a star professor at NUS and KAIST. He is one of the most productive researchers among Kim Choong-Ki's former students, publishing about 300 papers in prestigious international journals, presenting more than 320 papers at professional conferences, and filing more than 50 patents. He is a specialist in complementary metal-oxide semiconductor (CMOS) technology and one of the pioneers in graphene technology and flexible thermoelectric devices. Among his many published works over the last 35 years, two papers, "Quasi-Breakdown of Ultrathin Gate Oxide under High Field Stress" (1994) and "A Wearable Thermoelectric Generator Fabricated on a Glass Fabric" (2014), are considered breakthrough basic works in each corresponding subject.[1] Cho himself is proud of these two, saying:

> Professor Kim Choong-Ki repeatedly emphasized to us that we must prepare the dinner table for others [i.e., do something new for future research] rather than just join the table with our spoons [i.e., join the research that others have already opened]. I believe these two works belong to the former.[2]

It is therefore no wonder that Cho was hired as a professor at NUS three years after the publication of his 1994 paper and that he became the first South Korean to receive the coveted Netexplo Grand Prix in 2015 for his 2014 work.

EARLY YEARS

Cho Byung-Jin was born on March 16, 1964, in Busan, the second largest city in South Korea, and was brought up there for the next seventeen years. His father was a local businessman. Cho longed to become a teacher from his very early years: he recalls that he wanted to become an elementary school teacher when he was in elementary school and a mathematics teacher when he was a middle school and high school student.[3]

In 1981, Cho came to Seoul to enter the Department of Electrical Engineering at Korea University. After Samsung Electronics began its memory business in 1983, he and his classmates were intrigued by semiconductors and wanted to study them

DOI: 10.1201/9781003353911-8

more. But he had little opportunity to learn about or conduct experiments in the subject during his four undergraduate years. In his senior year, he decided to apply to the Korea Advanced Institute of Science and Technology (KAIST) because it had first-class professors and facilities for studying the semiconductor, and also offered a unique exemption from compulsory military service. He successfully passed the entrance examination and entered the Department of Electrical Engineering at the beginning of 1985. Korea University, though one of the top universities in South Korea, had not sent many graduates to KAIST: Cho remembers that few Korea University graduates were in the department when he entered it and that they worked in other fields.[4]

When Cho applied for Kim Choong-Ki's laboratory at the end of his first semester at KAIST, Kim himself was not on campus but at the Microelectronics Center of North Carolina as a visiting researcher. Cho was one of seven students whom Kim accepted that year without personal interviews. He recalls:

> I asked seniors and other students about the professors, including Professor Kim Choong-Ki. Many praised Kim as an excellent researcher and teacher. Even the janitor at the gate praised him as "a good-natured person." So, I decided to become Kim's student. I consider my decision to become Kim's student as one of the most fortunate blessings in my life. During the next six years, I really learned a lot from him—not only semiconductors but also how to become a good engineer and a teacher. He was never an easy teacher to satisfy. He often told us that "My primary job as a professor is to criticize you!" Indeed, he criticized us a lot, and I was afraid of him during my student years, not even answering clearly what I knew. However, we students knew that his criticism was correct in most cases and did not include any personal attack. I also realized, later, that it took a lot of observation and time to make any correct criticism of students. So, I emulated Kim and criticized my students a lot at the National University of Singapore and KAIST.
>
> I also learned a lot about the management of the laboratory from Kim during my student days. The students of neighboring laboratories, or even those from other departments, asked and used the instruments that we had made. They often asked us if they could use the precious reagents that we had prepared or bought. We complained a lot to Kim about this, but he dismissed our protests, saying that "They are like our brothers or guests. Do you deny your brothers or guests at home the use of your pens or tables? Try to be benevolent to and considerate of others." He actually treated the instruments and facilities in his laboratory as "common-use" ones, not his own property. Another important lesson I learned from Kim was his dictum for education: "It is good for students' education to construct the instruments by themselves, though perhaps not so good for the research. So, don't try to buy them if you can make them." Since I agreed with him, I later applied both these management methods to my laboratories at NUS and KAIST.[5]

For the next six years, Cho underwent Kim's typical hard training (Figure 5.1).

Under Kim's supervision, Cho studied the rapid thermal process and published its early results in the *Proceedings of the Annual Meeting of the Korea Electrical Engineering Society* in the spring of 1986, which became the subject of his 1988 master's thesis.[6] The rapid thermal process was a popular research subject in Kim's laboratory during the late 1980s, and several graduate students, such as Kim Kyeong-Tae, Kim Jeong-Gyoo, and Park Sung-Kye, worked on the topic. It was Cho, however,

FIGURE 5.1 Cho Byung-Jin has respected his mentor Kim Choong-Ki since his student years and made Kim his model as a teacher. The photo was taken when (from left to right) Cho, Kim, and Lee Dong-Yup (who came from another university to work temporarily at Kim's laboratory) climbed Mount Seorak in the winter of 1989.

Source: Courtesy of Cho Byung-Jin.

who rose to become the master of the rapid thermal process, publishing nine more papers on the subject, including three in international journals, before completing his doctorate in 1991.[7] In fact, Cho was the most productive student-author in Kim's laboratory. His doctoral thesis, "Modeling of Rapid Thermal Diffusion of Phosphorous into Silicon and Its Application to VLSI Fabrication," was a practical application of the rapid thermal process to the production of semiconductor devices: he predicted, in the conclusion, that this promising technology would become the industry standard "in the near future."[8] As he predicted, it soon became the industry standard.[9]

With his distinguished achievements at KAIST, Cho was hired after graduation as a postdoctoral researcher at the Interuniversity Microelectronics Centre (IMEC) in Leuven, Belgium. Founded in 1984, IMEC has been the "world leading R&D and innovation hub in nanoelectronics and digital technologies."[10] The center has various advanced research facilities, including two upscaled cleanrooms. During his eighteen-month stay at IMEC, Cho met many renowned semiconductor specialists and studied various new subjects and instruments. He and his colleagues at IMEC developed a new rapid thermal process system that "uses a vertical cylindrical quartz tube, while the wafer is placed horizontally" (Figure 5.2).[11] When his tenure at IMEC ended in early 1993, Cho returned to fulfill his three-year obligation to work in South Korea in return for having been exempted from military service. There were two choices for him: the academy or industry. Since he wanted to gain practical experience and the industry had better facilities for semiconductor research, he chose Hyundai Electronics (later named Hynix and, from 2012 on, SK Hynix).

FIGURE 5.2 The period between 1991 and 1997 was the critical time when Cho Byung-Jin grew as a mature, competent semiconductor researcher. He spent about two years as a postdoctoral fellow at IMEC in Belgium and four years at Hyundai Electronics as team manager of the memory R&D division. **Left**: Cho standing beside the rapid thermal processing (RTP) equipment that he had designed and made at IMEC. **Right**: At Hyundai, Cho not only published his first breakthrough paper, "Quasi-Breakdown of Ultrathin Gate Oxide under High Field Stress" (Seok-Hee Lee et al., 1994), but also mentored junior engineers such as Pi Seung-Ho (on right).

Source: Courtesy of Cho Byung-Jin.

In April 1993, Cho became the team manager of the memory division at Hyundai's Semiconductor Research and Development Laboratory. His research interests broadened to include not only rapid thermal process and equipment design but also gate materials, thin dielectrics for high-density DRAM, and flash EEPROM. Most notably, his team succeeded in developing the 256Mb synchronous DRAM in 1996, and he received a special award from the company for this achievement. He remained prolific, producing five research papers for international journals, two for domestic ones, and several other conference papers during his tenure at Hyundai between 1993 and 1997. All these papers, written with his Hyundai colleagues, focused on the improvement of the production or process of various semiconductor devices.

Among these publications, Cho's 1994 paper with his teammate Lee Seok-Hee and others, "Quasi-Breakdown of Ultrathin Gate Oxide under High Field Stress," deserves special attention.[12] Cho and Lee Seok-Hee discovered a new phenomenon while they worked on the ultrathin oxide reliability, and Cho named it "quasi-breakdown." This paper was not only a monumental achievement in the field but also the first paper by Hyundai Electronics employees to be published in the prestigious *IEEE International Electron Devices Meeting (IEDM) Technical Digest*. Owing to this success and Cho's help, Lee Seok-Hee was able to attend Stanford University for his advanced studies on material engineering. He then worked at Intel for eleven years, returning to South Korea in 2010 to become a professor at KAIST. Cho, who had attracted Lee to KAIST, nevertheless advised him that he could contribute more by working in industry than at the academy and recommended that he go to SK Hynix. Lee took Cho's advice, moved to SK Hynix in 2013, and climbed the ladder quickly, becoming CEO in 2018.[13]

Lee Seok-Hee was not Cho's only mentee at Hyundai Electronics; there were many more. Pi Seung-Ho, who joined the company in 1995, after receiving his doctorate in material engineering at KAIST, remembers his debt to Cho:

> I usually asked anyone what I would like to know or learn. … This tendency doubled after I entered Hyundai Electronics. My mentor then was Cho Byung-Jin, who really taught me a lot. Once he brought me to his office and taught me how to measure and analyze the device's signals with a parameter analyzer. He also taught me how to find the problems and solve them when I used it. Usually no one taught such things at the company. I had learned some things about the instrument from articles or heard about them from other engineers but hadn't experienced them myself until then. Thanks to Cho, I gained some insight into the semiconductor process that few other engineers could get.[14]

As Cho's three-year obligation period came to an end in 1996, he began to seriously consider his future career again. He could have remained at Hyundai Electronics and climbed the ladder of promotion quickly, but he wanted to teach what he had learned in the industry—he had dreamed of becoming a teacher since childhood. From Cho's point of view, there were only two universities in South Korea at the time that had proper facilities and good students for semiconductor education: KAIST and Seoul National University (SNU). Neither of them, however, seemed likely to hire him: KAIST had already hired two of Kim's former students, Kyung Chong-Min (profiled in Chapter 3) and Han Chul-Hi, both of whom were senior to Cho, and SNU preferred either its own doctoral degree holders or those who had earned their advanced degrees from prestigious American universities. Cho therefore turned his eyes to foreign universities, sending applications to the Hong Kong University of Science and Technology and the National University of Singapore (NUS).[15] Fortunately, NUS was then seeking a specialist in CMOS technology and was eager to hire him. NUS had an excellent reputation for supporting its professors' research and had a cleanroom and other high-quality facilities to support semiconductor research. Its Department of Electrical Engineering already had more than twenty semiconductor specialists by 1997. Cho remembers that when he informed Kim of his decision to go to Singapore, his mentor worried about Cho's command of English.[16]

SINGAPORE

Singapore, an independent state since 1965, is one of the "Four Asian Tigers," along with Hong Kong, Taiwan, and South Korea. Owing to its strategically important location, the logistics, shipping, finance, and oil-refinery industries flourished there even before this city-state gained its independence from Malaysia. Nevertheless, Lee Kuan Yew (1923–2015), the country's ambitious and energetic leader, wanted more and tried hard to attract foreign manufacturing companies such as General Electric, Hewlett Packard, NEC, and Fujitsu to Singapore.[17] Computers and semiconductors were added to the list in the early 1980s, and Chartered Semiconductor Manufacturing (CSM) was founded in 1987 with financial investment from the government. CSM rose to be the world's third largest semiconductor foundry in the early 2000s and became a major rival of Taiwan Semiconductor Manufacturing Company (TSMC) and United Microelectronics Corporation (UMC), but it was jointly bought out in 2009 by the United States' Advanced Micro Devices (AMD) and the United Arab Emirates' Advanced Technology Investment Company (ATIC). Although the focus of the Singapore government had begun to change from "Intelligent Island" to "Biopolis" at the turn of the new century, research on semiconductors continued in order to achieve this new goal.[18] The National University of Singapore, the foremost and oldest university in the country, has been the center of semiconductor research there since the early 1980s.

Cho arrived in Singapore in April 1997 with his wife and two sons. It was not easy for his family to settle down in this tropical region, where the average daily temperature is over 30°C (86°F) year-round and seldom drops below 25°C (77°F) at night, even during the winter: South Korea, in contrast, has four distinct seasons. Singapore's multicultural and multiracial environment was also a serious obstacle for them to overcome: about 75% of Singaporeans are of Chinese descent; 13%, Malaysian; 9%, Indian; and 3%, of other heritage. Although English is the most frequently used language, especially in academia, Malay, Mandarin, and Tamil are also official languages of Singapore. Cho soon realized that his superiors were observing him and his family carefully to see whether they would be temporary visitors or become permanent Singaporeans. Cho and his family decided to pursue the latter course in order to thrive in this tightly interwoven city-state. He therefore sent his children to a local school where the majority of students were of Chinese origin, whereas most foreign professors, including other Koreans, sent their children to either private or international schools. Cho himself tried hard to learn Singapore's unique culture and to become friends with his new colleagues. When he first arrived in Singapore, his English was not good enough to communicate or teach fluently. He remembers that several students complained about his poor English in the beginning.[19]

As the first South Korean professor in the Faculty of Engineering, as well as the first specialist in CMOS in his department, Cho was a figure of great interest from the outset. He did not disappoint those who observed him closely and received the full confidence of the department within three years. In 2000, he received a large research fund (equivalent to USD 10 million) from the Singaporean government to build a new, state-of-the-art cleanroom for CMOS research. Using his experience at KAIST and Hyundai Electronics, he completed the cleanroom within a year. It soon evolved into the Silicon Nano Device Laboratory (SNDL), with the full capacity of a semiconductor fabrication facility, and Cho became its first director.[20] The major goal of SNDL was "to develop scientific and technological bases to meet the most critical needs for future generations of very fast CMOS silicon nanoelectric devices."[21] It collaborated closely with the Institute of Microelectronics and Institute of Material Research and Engineering in the Faculty of Engineering and aimed to "become one of the world's leading research laboratories in the field."

Laboratories at NUS were set up according to subjects, and the directors of laboratories had full power to manage them, including hiring new professors or researchers.

Cho later mentioned that this unique rule might have been born of Singapore's authoritarian culture as well as its high regard for effectiveness, and that it greatly helped a newcomer like him.[22] He worked hard to attract talented researchers and professors to his laboratory. He also adopted some new rules for SNDL to increase its productivity: all facilities and funds in the laboratory, regardless of who made, purchased, or attracted them, were not specific members' property but for the use of all members of the laboratory, and each member was required to assist either other members' research or the laboratory's projects. These rules were especially helpful for newly hired professors or researchers who had many new ideas but few funds or connections. These new rules soon proved to be very effective not only in increasing the number of research papers but also in attracting more funds from out-side. At the 2004 International Electron Device Meeting (IEDM) in San Francisco, for example, Cho's SNDL presented six papers, the "largest number among all the university laboratories in the world."[23] Its researchers also published sixteen papers in *IEEE Electron Device Letters* and filed four US patents in 2004 alone. SNDL also succeeded in securing large research grants in 2004, including SGD 3 million from Singapore's prestigious A*STAR (Agency for Science, Technology and Research) Thematic Strategic Research Programme and the equivalent of USD 2.2 million from Jusung Engineering, South Korea's largest manufacturer of semiconductor pro-cess equipment.[24] Cho became one of the top professors at NUS in attracting large research grants from outside.

It is therefore no wonder that in 2005, when NUS celebrated its hundredth anniver-sary, the *Straits Times* selected Cho as one of the four most creative minds at NUS in its special essay "Creating the Future" (Figure 5.3):[25]

Associate Professor Cho Byung-Jin, 43, from the electrical and computer engineering department, had only praise for the university's support.

The supervisor of NUS' Silicon Nano Device Laboratory (SNDL), which focuses on nano-sized electronic device technology development, said his faculty helped him source for funding to start the SNDL in 2000.

He said: "NUS' support for my research was beyond my imagination. For nano-electronic research, a good research facility is critical."

The SNDL has attracted foreign interest. In a collaboration with South Korea's largest maker of semiconductor manufacturing equipment, the SNDL is working on developing faster and smaller transistors that measure as small as 30 nanometres (nm), compared to the current industry standard of 90 nm. A millimetre equals one million nanometres.

Prof. Cho said such a tiny high-performance transistor is the basic building block for the chips necessary for the next generation of electronic devices, such as multi-functional mobile phones.

At NUS, Cho continued to be a prolific writer, publishing more than 90 papers between 1997 and 2007.[26] Most were team papers based on his research at SNDL with his students and colleagues and were about new ideas or techniques that could be applied to manufacturing various semiconductor devices. Examples (often with a first author other than Cho)[27] include "Evolution of Quasi-Breakdown Mechanism in Thin Gate Oxides" (2002), "HfO_2 and Lanthanide-doped HfO_2 MIM Capacitors for

FIGURE 5.3 At the National University of Singapore (NUS), Cho Byung-Jin emerged as a star engineering professor who successfully operated a state-of-the-art semiconductor laboratory with a large research fund. He was chosen as one of the four most creative minds at NUS when Singapore's *Straits Times* published a special essay, "Creating the Future," to celebrate the university's hundredth anniversary. Cho holds a silicon wafer in the upper photo.

Source: *The Strait Times*, July 1, 2005, 3.

RF/Mixed IC Applications" (2003), "A Novel Approach for Integration of Dual Metal Gate Process Using Ultra Thin Aluminum Nitride Buffer Layer" (2003), "Analysis of Charge Trapping and Breakdown Mechanism in High-K Dielectrics with Metal Gate Electrode Using Carrier Separation" (2003), "Engineering of Voltage Nonlinearity in High-K MIM Capacitor for Analogue/Mixed Signal ICs" (2004), "High-K HfAlO Charge Trapping Layer in SONOS-type Nonvolatile Memory Device for High Speed Operation" (2004), "Substituted Aluminum Metal Gate on High-K Dielectric for Low Work-Function and Fermi-Level Pinning Free" (2004), "High Capacitance Density (>17fF/μm^2) Nb_2O_5-based MIM Capacitors for Future RF IC Application" (2005), "Dual Metal Gate Process by Metal Substitution of Dopant-Free Polysilicon on High-K Dielectric" (2005), "Electron Mobility Enhancement Using Ultrathin Pure Ge on Si Substrate" (2005), "SiGe on Insulator MOSFET Integrated with Schottky Source/Drain and HfO/TaN Gate Stack" (2006), and "Tensile-Strained Germanium CMOS Integration on Silicon" (2007).

Between 1997 and 2007, Cho trained 25 graduate students—fifteen doctoral and ten master's students.[28] Many came from China, Singapore, and South Korea, but others also came from India, Indonesia, and other countries. He later recalled that some of them, especially those from China, were truly exceptional.[29] After graduation, his students scattered around the world, with Singapore, South Korea, Taiwan, and the United States being four major countries where they chose to work. Most of his graduates went into the semiconductor industry, including CSM, TSMC, and Samsung Electronics: for example, Kim Sun-Jung, who was Cho's doctoral candidate at NUS, entered Samsung Electronics after graduation and climbed to junior director at Samsung Electronics in 2021, responsible for the development of the 3-nanometer gate-all-around filed-effect-transistor (GAAFET) device.[30] Three went to SEMATECH, a US-based consortium that concentrates on research and development of advanced chip manufacturing: this connection would be particularly helpful to Cho when he returned to South Korea and searched for a new research subject.

Cho paid a lot of attention to his students to train them properly. Following his mentor's dictum, he criticized his students a lot but in a gentler way. His students recognized Cho's efforts on their behalf and often expressed their sincere gratitude in their acknowledgments:

- "The seven years I have worked with him [Cho] as a research engineer, a master's student, and a doctoral candidate have changed the direction of my life and laid the cornerstone of my future endeavor." (Kim Sun Jung)
- "I really thank him [Cho] for offering me the opportunity to be a doctoral candidate. He is not only an experienced advisor for me but also an elder who gives me confidence and blessing. I will remember his inculcation in my life." (Yang Weifeng)
- "I also greatly appreciate Prof. Cho from the bottom of my heart for his knowledge, expertise, and foresight in the field of semiconductor technology. Without his guidance, it would be impossible for me to have completed this." (Zhang Lu)

- "He gave me a lot of encouragement and wisdom through several projects during the course of my research and always provided me with valuable insight, ensuring that I did not lose sight of my primary research objectives." (Park Chang Seo)
- "In particular, it is Assoc. Prof. Cho who has inspired me to reach for the highest standard in my research and who has tirelessly reviewed and guided me in all my publications. It is with his help that I am able to produce credible results in the area of oxide and high-K reliability." (Loh Wei Yip)

Several doctoral candidates at SNDL, who were technically not Cho's students, also expressed their thanks to him in their acknowledgments: "I am also deeply grateful to Professor Byung-Jin Cho, who provided me the opportunity to join the research team in the NUS. The knowledge and experience that I have gained from the team will benefit me greatly throughout my career" (Sun Zhiqiang).

In the fall of 2006, at the height of his career at the NUS, Cho began to reconsider his future. After relocating to Singapore in 1997, it had seemed that NUS would be his permanent destination. He was highly respected there as a star professor and received a much higher salary than other professors at the university. There was no specific reason for him to leave what he had constructed at SNDL over the preceding decade. At the age of 42, however, he wondered whether he would live out his career as a professor and researcher in this tropical city-state. He sent a long letter to Kim Choong-Ki describing the management and success of his laboratory. Kim showed the letter to Lee Yong-Hoon, the dean of the College of Engineering at KAIST at that time, and Lee contacted Cho, through another professor, to ask whether he was interested in returning to South Korea and specifically to KAIST. Cho did not reply but did begin to seriously consider the possibility of returning to Korea.

However, too many factors were involved. The education of Cho's children was the most serious question. His two sons had grown up in Singapore and were now high school students there: moving back to South Korea would make their lives not just different but very difficult. Also, the average salary of KAIST professors was much lower than Cho's salary at NUS. The result was that Cho decided to visit KAIST before making any final decision. In December 2006, he traveled to KAIST to give a seminar at the Department of Electrical Engineering. The seminar was well attended by many of his former teachers and colleagues, as well as graduate students. It was very successful, and Lee Yong-Hoon and other senior professors strongly encouraged Cho to apply to KAIST. Cho noticed an atmosphere in his old department that he had not experienced at NUS: warmth and cordiality. He returned to Singapore and struggled as to what he should choose for himself and his family. He finally decided to return to South Korea and sent his application to KAIST. When the president of KAIST, Suh Nam-Pyo, visited Singapore to attend a meeting, Cho gave him a tour of SNDL and had a short job interview. Suh was very satisfied with the interview and promised to speed up the hiring process. The decision to hire Cho soon met an unexpected barrier, however, when some professors in the Department of Electrical Engineering objected to Cho's appointment for the following reason: the department had already hired too many professors (three!) who had been trained by Kim Choong-Ki.[31] This objection was overruled by the president of KAIST, the dean of the College of Engineering, and other senior professors in the department, on the grounds that Cho was a star engineering professor whom any university in the world would happily hire. In August 2007, Cho and his family returned to South Korea to start another new life.

BACK TO KAIST

Once again, it was not easy for Cho and his family to settle down in their new environment. The expectations Cho faced were very high, not only within KAIST but also within the South Korean semiconductor community. Moreover, financial support from KAIST, although generous by South Korean standards, was far less than that provided by NUS. In the beginning, Cho therefore depended on his South Korean friends and his old network to start his own laboratory. Hynix (the former Hyundai Electronics), where he had worked in 1994–1997, provided him with some research funds. Jusung Engineering (which had provided Cho's SNDL with a grant equivalent to USD 2.2 million in 2004), donated USD 1 million worth of equipment for his research. It was, however, the National NanoFab Center (NNFC) on KAIST's campus that enabled him to continue the research he had started in Singapore as well as to start new research projects. Opened in 2005, the center has three major missions: "1) to offer nanotechnology equipment/facility/process services to all kinds of users including academia, research institutes and industries, 2) to provide nanotechnology education to students, researchers, and experts with hands-on experience, and 3) to promote nanotechnology commercialization of the R&D outcomes."[32] Lee Hee-Chul, the president of the NNFC at that time, greatly helped Cho settle down successfully at the center.[33]

Cho's first serious task at KAIST was to search out proper research topics for himself and his new graduate students. He soon realized that he could not continue his CMOS-related research in South Korea to the extent that he had at NUS because the South Korean government had severely reduced its funding for CMOS and its related subjects. As a result, 2008 and the first half of 2009 were a transitional period during which he finished up his old research projects with his NUS graduate students and/or KAIST students, publishing with them, for example, "Carbon-Doped Polysilicon Floating Gate for Improved Date Retention and P/E Window of Flash Memory" (2008) and "Metal Carbides for Band-Edge Work Function Metal Gate CMOS Devices" (2008). His team paper in the *Journal of Applied Physics*, entitled "Monolayer Graphene Growth on Sputtered Thin Film Platinum" (Byung-Jin Kang et al., 2009), signaled that Cho had finally discovered a promising new research subject to pursue: graphene.[34]

Graphene is "a single layer of carbon atoms, tightly bound in a hexagonal honeycomb lattice."[35] It was first isolated and characterized in 2004 by Andre Geim and Konstantin Novoselov at the University of Manchester, which led them to share the Nobel Prize in Physics in 2010. Graphene is "many times stronger than steel, yet incredibly lightweight and flexible, ... electrically and thermally conductive but transparent," and "one million times thinner than the diameter of a single human hair."[36] Owing to its amazing characteristics, scientists and engineers tried to apply graphene to "transport, medicine, electronics, energy, defense, and desalination." As the University of Manchester's website suggests, "the potential of graphene is limited only by our imagination."[37]

Cho was one of the first South Korean semiconductor engineers to recognize graphene's importance and applicability. He was the prime mover in organizing the first study group of graphene in South Korea, which evolved into the Korea Graphene Society. He served as the second president of the society. The research fund from SEMATECH in 2008 finally enabled him to concentrate on this uncultivated field.[38] Cho's 2009 team paper on graphene spelled out both the optimistic prospect of graphene's applications to semiconductors and some of the technical difficulties that needed to be overcome:

> Graphene, a two-dimensional (2D) honeycomb lattice of sp^2-bonded carbon atoms, is a promising candidate for post-CMOS electronics because of its excellent electrical properties such as long-range ballistic transport and very high carrier mobility. For electronic device applications, it is necessary to form a large-scale graphene layer on a substrate with good uniformity. Thermal decomposition of SiC at a high temperature, chemical-vapor deposition (CVD)-like graphene growth on Ni Substrate, and chemical reduction in graphite oxide film on substrate are now being widely studied for large-scale graphene synthesis.[39]

Between 2009 and 2018, graphene became Cho's major research topic, and about half of his published papers over these years were devoted to this subject. Some examples are "Monolayer Graphene Growth on Sputtered Thin Film Platinum" (2009), "Flexible Resistive Switching Memory Device Based on Graphene Oxide" (2010), "Graphene Gate Electrode for MOS Structure-Based Electronic Devices" (2011), "Direct Measurement of Adhesion Energy of Monolayer Graphene As-Grown on Copper and Its Application to Renewable Transfer Process" (2012), "Electro-Magnetic Interference (EMI) Shielding Effectiveness of Monolayer Graphene" (2012), "Synthesis of Monolayer Graphene Having Negligible Amount of Wrinkles by Stress Relaxation" (2013), "High Performance Graphene Field Effect Transistors on an Aluminum Nitride Substrate with High Surface Phonon Energy" (2014), "Large-Area, Periodic, Hexagonal Wrinkles on Nanocrystalline Graphitic Film" (2015), "Hybrid Integration of Graphene Analogue and Silicon Complementary Metal-Oxide-Semiconductor Digital Circuits" (2016), "Direct Graphene Transfer and Its Application to Transfer Printing Using Mechanically Controlled, Large Area Graphene/Copper Freestanding Layer" (2018), and "A High-Performance Top-Gated Graphene Field-Effect Transistor with Excellent Flexibility Enabled by an iCVD Copolymer Gate Dielectric" (2018).

The South Korean media quickly realized the importance of Cho's research on graphene because his new ideas could be applied directly to manufacturing semiconductors. His team paper "Graphene Gate Electrode for MOS Structure-Based Electronic Devices" (Jong-Kyung Park et al., 2011), for example, suggests replacing the conventional TaN metal with a monolayer graphene as a gate electrode in charge-trap flash (CTF) memory devices (Figure 5.4).[40] Cho and his coauthors claimed that this idea could be "applied to nonvolatile Flash memory devices, whose performance is critically affected by the quality of the gate dielectric. CTF memory with a graphene gate electrode shows superior data retention and program/erase performance that current

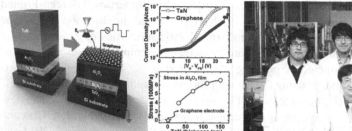

FIGURE 5.4 Cho Byung-Jin was one of the first South Korean researchers who paid attention to a new material—graphene—and its applications to semiconductor chips. **Left**: Cho's team paper "Graphene Gate Electrode for MOS Structure-Based Electronic Devices" suggests that a graphene gate electrode can greatly improve the performance of flash memory and mass-producible electronic devices based on MOS structure. **Right**: Cho (center front) and his team who participated in the research.

Sources: Jong-Kyung Park et al., "Graphene Gate Electrode...," 5383; and *Daejeon Ilbo*, November 22, 2011.

CTF devices cannot achieve."[41] South Korean newspapers reported, rather brashly, that Samsung Electronics, the number one manufacturer of NAND flash memory in the world, was seriously considering employing this new technology in order to widen the gap between it and its competitors.[42] Cho's team paper "Direct Measurement of Adhesion Energy of Monolayer Graphene As-Grown on Copper and Its Application to Renewable Transfer Process" (Taesik Yoon et al., 2012), which suggests not only economical but also environment-friendly technology to separate the graphene monolayer from the metal substrates, was also highly praised by the media: "South Koran research team advances the practical use of graphene for electronics."[43] Foreign journals also paid attention to Cho's graphene research: *Interference Technology*, for example, reported on his other team paper published that year, "Electro-Magnetic Interference (EMI) Shielding Effectiveness of Monolayer Graphene" (Seul-Ki Hong et al., 2012), noting that Cho and his coauthors were "successfully demonstrating that single-layer graphene may be the ideal choice of material for high-performance EMI shielding."[44] Hence it is not surprising that Cho emerged during the 2010s as the representative of graphene technology in the South Korean media.

The thermoelectric device was the other new research subject that Cho pursued after he returned to KAIST. The thermoelectric effect had been independently discovered by three scientists—Thomas Johann Seebeck, Jean C. A. Peltier, and William Thomson—in the early nineteenth century, but the development of the thermoelectric device, which converts temperature difference into electricity, had been painfully slow, and the devices were usually bulky and expensive. Thanks to the South Korean government's concern about reusable energy in the early 2010s, Cho was able to attract a large amount of research funding to find a new way of connecting semiconductors with reusable energy. He targeted the development of self-powered, wearable thermoelectric devices using "flexible" electronics.

Cho's first team paper on the subject, "Thermoelectric Properties of Screen-Printed ZnSb Film" (Heon-Bok Lee et al., 2011), clearly indicated the direction of this line of research in the coming years: "The thermoelectric properties of ZnSb thin film prepared by screen-printing technique are investigated, aiming to achieve a low-cost and eco-friendly thermoelectric power generator module. ... The feasibility of a *flexible* thermoelectric module using the screen-printing technique is also demonstrated" (emphasis added).[45] The development of economical and eco-friendly as well as flexible thermoelectric devices therefore became major goals of his research. Cho and his students together published more than 25 papers on the subject between 2011 and 2020, many of which are highly practicable. Some examples are "Improvement of Thermoelectric Properties of Screen-Printed Bi_2Te_3 Thick Film by Optimization of the Annealing Process" (2013), "A Wearable Thermoelectric Generator Fabricated on a Glass Fabric" (2014), "Synthesis on Ultrathin Polymer Insulating Layers by Initiated Chemical Vapour Deposition for Low-Powered Soft Electronics" (2015), "Post Ionized Defect Engineering of Screen-Printed $Bi_2Te_{2.7}Se_{0.3}$ Thick Film for High Performance Flexible Thermoelectric Generator" (2016), "Material Optimization for a High Power Thermoelectric Generator in Wearable Application" (2017), "High-Performance Self-Powered Wireless Sensor Node Driven by a Flexible Thermoelectric Generator" (2018), "Self-Powered Wearable Electrocardiography Using a Wearable Thermoelectric Power Generator" (2018), "Mechanical and Electrical Reliability Analysis of Flexible Si CMOS Integrated Circuit" (2019), "Two-Dimensional Array Thermal Haptic Module Based on a Flexible Thermoelectric Device" (2020), and "Thermal Display Glove for Interacting with Virtual Reality" (2020). The National Research Foundation in South Korea recognized his leading role in the field and in 2015 appointed his laboratory, Nano Electronics and Energy Device Laboratory, the "Engineering Research Center (ERC) for the Development of Flexible Thermoelectric Semiconductor Device," with a seven-year research grant equivalent to about USD 12 million.[46] Cho, however, chose to shorten the recipient period to five years in order to input more funds per year.

The South Korean media competed to report on Cho's development of flexible thermoelectric devices when his team published the paper "A Wearable Thermoelectric Generator Fabricated on a Glass Fabric" (Su-Jin Kim et al., 2014) in *Energy and Environmental Science* (Figure 5.5).[47] Yonhap News Agency, for example, reported that Cho's team had developed a light and wearable thermoelectric device using glass fiber and that it could generate enough electricity to charge a cell phone battery using body temperature, although the latter statement is certainly much exaggerated.[48] And on April 11, 2014, Korea Net, the official website of South Korea, posted a detailed story of Cho's invention under the headline "Charge Your Battery by Wearing It."[49] On February 4, 2015, the Ministry of Science, ICT, and Future Planning (formerly the Ministry of Science and Technology) proudly announced that Cho had won the Netexplo Grand Prix, which is cosponsored by UNESCO, for his wearable thermoelectric device.[50] He was the first Korean engineer who was chosen in the top ten, not to mention to receive the grand prize, since Netexplo had begun giving the award in 2008. The South Korean media, including two public TV stations, publicized his award along with interviews of Cho and a rosy picture of his technology.[51] A joint 2015 paper on the synthesis of ultrathin polymer insulating layers, written

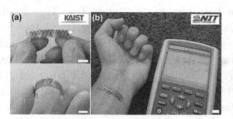

Fig. 6 Demonstration of band-type flexible TE generator for harvesting thermal energy from human skin: (a) photos of band-type flexible TE generator and (b) electricity generation measured on human skin at an air temperature of 15 °C. Scale bar, 1 cm.

FIGURE 5.5 The research and development of the flexible thermoelectric device is Cho Byung-Jin's most famous achievement among the South Korean public. **Left**: Figures from Cho's team paper "A Wearable Thermoelectric Generator Fabricated on a Glass Fabric." **Right**: Cho received the 2015 Netexplo Grand Prix on February 4, 2015.

Sources: Su-Jim Kim et al., "A Wearable Themoelectric Generator...," 1963; and courtesy of Cho Byung-Jin.

with Im Sung-Gap's team at KAIST's Department of Chemical and Biomolecular Engineering, also attracted much media attention due to the hope that this research would accelerate the commercial use of wearable electric devices.[52]

TEGway, a startup company that Cho and another KAIST graduate, Lee Kyung-Soo, co-founded in the fall of 2014 to commercialize thermoelectric devices, soon attracted much attention in media and business circles. A business news agency reported on the company under the provocative title "Five-Month-Old-Start-up TEGway wins Grand Prix at UNESCO's Netexplo Award."[53] As CEO of TEGway, Lee aimed for the company's technology on thermoelectricity to become the industry standard.[54] Cho served as chief technology officer (CTO) until 2019. The company grew rapidly and successfully developed several applications for flexible thermoelectric devices. Its most successful product, ThermoReal, for example, can be applied to virtual reality (VR) and augmented reality (AR) so that the user can actually feel warmth or coldness.[55] It was chosen as one of 464 honorees that received the Innovation Award at the 2020 Consumer Electronics Show (CES).[56] By 2020, TEGway held "over 90 intellectual property rights including patent registration, licenses, and trademarks" and worked with "major corporations in the gaming market, automobile, beauty, health, and wearables."[57]

Besides graphene and thermoelectric devices, Cho worked on various other subjects, including his beloved specialty, CMOS. He has always believed that CMOS knowledge and experience are fundamental for semiconductor engineers and has therefore paid a lot of attention to CMOS research. He worked on these subjects not only with his academic colleagues from KAIST, NUS, and other South Korean universities but also with research engineers from Samsung Electronics and SK Hynix, publishing more than twenty team papers between 2008 and 2020. Some examples are "Endurance Reliability of Multilevel-Cell Flash Memory Using ZrO_2/Si_3N_4 Dual

Charge Storage Layer" (2008), "Improvement of Memory Performance by High Temperature Annealing of the Al_2O_3 Blocking Layer in a Charge-Trap Type Flash Memory Device" (2010), "Lanthanum-Oxide-Doped Nitride Charge Trap Layer for a TANOS Memory Device" (2011), "Origin of Transient Vth Shift after Erase and Its Impact on 2D/3D Structure Charge Trap Flash Memory Cell Operations" (2012), "Surface-Controlled Ultrathin (2nm) Poly-Si Channel Junctionless FET toward 3D NAND Flash Memory Applications" (2014), "Improved Electromagnetic-Resistance of Cu Interconnects by Graphene-Based Capping Layer" (2015), "Valley-Engineered Ultra-Thin Silicon for High-Performance Junctionless Transistors" (2016), "Very Low Work Function ALD-Erbium Carbide (ErC_2) Metal Electrode on High-k Dielectrics" (2016), "Surface-Localized Sealing of Porous Ultralow-k Dielectric Films with Ultrathin Polymer Coating" (2017), "Large Grain Ruthenium for Alternative Interconnects" (2018), "H_2 High Pressure Annealed Y-Doped ZrO_2 Gate Dielectric with an EOT of 0.57nm for Ge MOSFETs" (2019), "Capacitance Boosting by Anti-Ferroelectric Blocking Layer in Charge Trap Flash Memory Device" (2020), and "Method to Achieve the Morphotropic Phase Boundary in $Hf_xZr_{1-x}O_2$ by Electric Field Cycling for DRAM Cell Capacitor Applications" (2021). In 2018, alarmed by the Japanese government's sudden restriction on the export of basic materials and devices to South Korea, the South Korean government resumed funding for CMOS. Cho's team was one of the major beneficiaries of this changing policy. More research results on CMOS by Cho and his students are expected to come out in the 2020s.

From 2007 to 2021, Cho produced 37 master's and 18 doctoral degree holders at KAIST. The most popular subject has been graphene, followed by thermoelectric devices, flash memory, CMOS, among others. Most of Cho's former students work at either Samsung Electronics or SK Hynix as semiconductor researchers and developers, and some of them, though relatively young, have climbed to the director's level at their companies. As many South Korean universities expanded their existing or newly established semiconductor programs from the beginning of the 2020s, three of Cho's former doctoral candidates became professors in 2022 alone: two of them had worked at Samsung Electronics and the third at SK Hynix before they moved to the universities.[58] Since Cho's former students are relatively young compared to those of Kim Choong-Ki and Kyung Chong-Min (Chapters 2 and 3), their future careers and contributions to the development of the South Korean semiconductor community are still in progress.

At KAIST Cho has been as influential and eccentric as his mentor Kim Choong-Ki. His teaching style is somewhat different from Kim's, however. He allows his graduate students a great deal of freedom. Cho, for example, encourages students to choose their own research topics but is also always ready to suggest one for them. He insists that his students consider other people in the laboratory, on campus, and in society: if students ignore this, they receive a lot of sharp scolding from him—a legacy inherited from his mentor but one that Cho has somewhat augmented. He also requires his students to read classic literature or social science books and to submit a monthly book report on them, which is certainly a nightmare for many of his students: for example, the website of Cho's laboratory has a unique section, the "Culture Column," which is filled with students' essays on various books on romance, music, science, theology, and history.[59]

Some quotes from the acknowledgments of his former students at KAIST hint at Cho's deep and unique influence on them:

- "Professor Cho was humorous and always gave me a lot of encouragement. He also taught me how to become an engineer with a cool head and scientific insight." (Seo Jun Ho, 2009)
- "When I visited him to discuss my future plans as a graduate student in the first year, I was deeply impressed by his saying that he would like to work on the subjects that can help make the world better. He accepted me as his student and taught me various requirements to become an ideal engineer in addition to professional knowledge." (Oh Joong Kun, 2012)
- "Thanks to Professor Cho's careful training, I can grow as a fine adult." (Song Seung Min, 2015)
- "Professor Cho emphasized that engineers should have broad perspectives. He demanded that we become engineers with liberal arts knowledge and carefully monitor the needs of society. I am deeply grateful that he taught me how to express my thoughts clearly in writing and how to cultivate my own philosophy as a human." (Kim Sun Jin, 2017)
- "I learned from Professor Cho how to overcome various difficulties as well as how to do research. Whenever I thought that I couldn't do this or that, he always persuaded me that there is no problem without solutions, which was a great encouragement for me." (Lee Sang Jae, 2019)

The process by which Cho selects new graduate students for his laboratory is equally unique. He seldom pays attention to their undergraduate grades because he thinks there is little correlation between students' grades in the past and their ability to do research in the future. He instead asks them the following three questions: Do they like sports, and are they good at them? Are they avid readers, and how good are their book reports? And are they good at dating?[60] He asks aspiring semiconductor engineers these questions because those who like and are good at sports may have stronger bodies and will overcome difficulties they will encounter while doing research; those who read a lot and write book reports well may have the ability to organize big projects better by themselves; and those who are good at dating may pay more attention to others and try harder to understand others' minds. He has summarized his principles:

> Semiconductor research requires a lot of equipment and facilities, so that it can't be pursued alone. The production of a single semiconductor device, for example, needs more than twenty different instruments. So, it is essential for semiconductor engineers to have the ability to communicate with others and be considerate of them in order to ask for their assistance.[61]

Although it is debatable how accomplished Cho himself was in these three categories when he was a graduate student, his point seems clear and even reasonable.

In early 2017, Cho suddenly had trouble with his eyes, and local eye doctors could not identify the cause. He went to a larger hospital in Seoul and found that he had a brain tumor. He was immediately hospitalized and had a long and serious operation

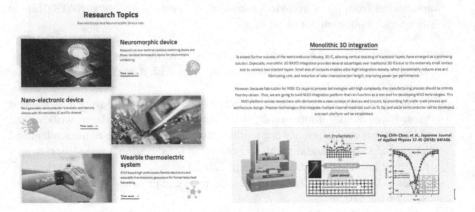

FIGURE 5.6 Cho Byung-Jin continues exploring new research areas. The monolithic three-dimensional (M3D) integration, which is a part of the nanoelectronic device division, is one topic to which he is paying special attention.

Source: KAIST's NAND Lab website.

in March. The tumor was successfully removed, but it took almost six months for him to return to the KAIST campus. This unexpected turn of events was a great shock to Cho, his family, and his colleagues and students, but it gave him an opportunity to plan the future as well as review the past. He decided to quit his CTO position at TEGway since he could not simultaneously hold down both that job and his KAIST professorship with his poor health. It was not an easy decision for him or for the company because he was a founder, chief engineer, and major stockholder of TEGway. He finally left the company in 2019. Although he must take medication daily and has substantially reduced his research and teaching, Cho has continued educating his students and producing important research results. In that respect, his winning the 2019 KAIST Grand Prize in Research gave him special "consolation and encouragement," which may accelerate his recovery.[62]

In 2022, Cho's Nanoelectronic and Neuromorphic Device Lab (NAND Lab) worked on three research subjects: (1) developing the ferroelectric-based device for neuromorphic computing; (2) developing the next-generation semiconductor processing technology using monolithic 3D (M3D) integration; and (3) developing the next-generation memory device technology.[63] Among them, M3D integration, which "enables revolutionary digital system architectures of computation immersed in memory" by stacking "layers of logic circuits and memories" vertically "with nanoscale inter-layer vias," deserves a special attention (Figure 5.6).[64] This new technology certainly has many advantages over the old ones because of "the extremely small contact size to connect two stacked layers" that enables "ultra-high integration density." However, as the NAND Lab notes,

> because fabrication for M3D ICs requires process technologies with high complexity, the manufacturing process should be entirely foundry-driven. Thus, we are going to build a[n] M3D integration platform that can function as a test-bed for developing

M3D technologies. This M3D platform serves researchers who demonstrate a new concept of devices and circuits, by providing full-wafer scale process and architecture design. Process technologies that integrate multiple channel materials such as Si, Ge, and oxide semiconductor will be developed, and each platform will be established.[65]

Who knows whether Cho can produce another, third, groundbreaking paper for this new area before his retirement in 2029? Cho's enthusiasm for searching and cultivating new research areas certainly has not yet ended.

Kim Choong-Ki recently recounted an interesting episode about Cho.[66] In May 2022, Kyung Chong-Min and Cho visited Kim to celebrate "Teacher's Day" (May 15 in South Korea) and had a long conversation on several topics over lunch. Kim remembers that "when Cho started talking about the recent trend in semiconductor chips, I barely understood what he was talking about." This is quite surprising because Kim has been closely monitoring his former students' work as well as the development of semiconductors, yet it is understandable if the subject itself is very new. In the twenty-first century semiconductor community, where changes occur at lightning speed, Cho is without doubt one of the engineers who makes those changes, not just a fast follower.

DUAL MISSIONS

The development of semiconductor engineering has been a lifelong mission that Cho has carried out very successfully ever since he entered KAIST in 1985. His award of the 2015 Netexplo Grand Prix for the flexible thermoelectric device and his recent award of the 2023 Kahng Dawon Prize (device/processing area) only partially illustrate his profound contributions to the development of semiconductor engineering in South Korea.[67] Cho has, however, also had another "mission" to complete: Christian missionary work. He was not born a Christian but first encountered Christianity after he entered Korea University as an undergraduate. His belief strengthened as time went on, became much more serious after he entered KAIST, and posed a difficult dilemma. He once seriously considered abandoning his doctoral work and starting to study theology, wondering whether it was more appropriate for him to serve as a missionary in undeveloped countries.[68] The struggle continued when he was a postdoctoral researcher at IMEC and an industrial researcher at Hyundai Electronics. He even organized a Bible study group while he worked at Hyundai, which grew from three members in the beginning to 30 by the time he left. Cho finally resolved the dilemma when he decided to go to NUS, where he believed he would be able to carry out both missions together. At NUS, he organized another Bible study group, mostly for Chinese students; he even edited an introductory book in English for the group. His missionary work continued when he returned to KAIST in 2007. He discovered many Christian students on campus from China, Indonesia, Mongolia, Thailand, and other countries, but there was no church service for them. He therefore organized a KAIST International Church (KIC) to offer a regular service in English.[69] KIC grew rapidly and was later registered as an official "club" at KAIST.[70] When asked in 2021 what he would like to do in the future, Cho replied that he would like to continue pursuing his two missions until his retirement.[71] Cho is certainly the rare figure who could manage these two very different missions quite successfully for a long time.

As Cho becomes a senior in the South Korean semiconductor community, however, a third role seems to be slowly emerging—that of speaking for the South Korean semiconductor community. Unlike Kyung Chong-Min, a fellow KAIST professor, Cho has not spoken publicly on broader subjects, such as policies for the semiconductor industry or general education, nor has he published newspaper articles or books on nonengineering topics. In an interview with a newspaper reporter in August 2021, Cho openly criticized government policy for the first time, pointing out the grave obstacles it posed that would block the development of the South Korean semiconductor industry in the future:

> The South Korean government has not supported the R&D of semiconductors in the university for a long time. It thought that there is no need for such support since South Korean industry has been doing well. The Ministry of Science and Technology did not include the silicon semiconductor in its list of research subjects for many years [between the mid-2000s and the late 2010s]. Therefore, those who work on silicon semiconductors had to work on other subjects instead. Universities stopped selecting those who majored in that subject as professors, and the number of graduate students in that area also decreased. The overall result was a rapid drop in the number of silicon semiconductor engineers. Samsung Electronics, for example, could not secure a sufficient number of semiconductor engineers because there were so few. Therefore, South Koran companies had no choice but to hire anyone [of] those who had studied any engineering and then re-train them. There are some cases of graduates from the department of nuclear engineering applying to Samsung Electronics.[72]

When the reporter noted that South Korea is still number one in the world memory market, Cho's criticism became sharper, emphasizing that quality is as important as quantity:

> Those who are leading the South Korean semiconductor community now, who are mostly in their fifties or sixties, were the most talented ones when they entered the university [in the 1970s and 1980s]. Unfortunately, there is a big gap between those who majored in semiconductors in the 1980s and those [who are doing so] now. Now, almost all the top high school students apply to medical schools throughout the country, and then those remaining apply to the engineering schools of either Seoul National University or KAIST. This will make a big difference 20 or 30 years later. The future of the South Korean semiconductor industry cannot be guaranteed if this situation continues.[73]

It appears that Cho's third mission has just been launched.

NOTES

1   Seok-Hee Lee, Byung-Jin Cho, Jong-Chul Kim, and Soo-Han Choi, "Quasi-Breakdown of Ultrathin Gate Oxide under High Field Stress," *IEEE International Electron Devices Meeting (IEDM) Technical Digest* (December 1994), 605–608; and Su-Jin Kim, Ju-Hyung We, and Byung-Jin Cho, "A Wearable Thermoelectric Generator Fabricated on a Glass Fabric," *Energy and Environmental Science, 7* (2014), 1959–1965.
2   Interview with Cho Byung-Jin by Kim Dong-Won (May 21, 2021).
3   Interview with Cho Byung-Jin by Kim Dong-Won (September 28, 2020).

4 Ibid.

5 Ibid.

6 Cho Byung-Jin, Kim Kyeong-Tae, and Kim Choong-Ki, "Experimental Results of a Prototype Rapid Thermal Annealing System," *Daehanjeonjagonghakhoe Haksulbalpyohoe Nonmunjip, 5:1* (1986), 71–73; Byung-Jin Cho, "Design Study of a Rapid Thermal Annealing System and Temperature Control" (master's thesis, Seoul: KAIST, 1988).

7 The three papers published in international journals are Byung-Jin Cho and Choong-Ki Kim, "Elimination of Slips on Silicon Wafer Edge in Rapid Thermal Process by Using a Ring Oxide," *Journal of Applied Physics, 67* (1990), 7583–7586; Jeong-Gyoo Kim, Byung-Jin Cho, and Choong-Ki Kim, "AES Study of Rapid Thermal Baron Diffusion into Silicon from a Solid Diffusion Source in Oxygen Ambient," *Journal of the Electrochemical Society, 137* (1990), 2857–2860; and Byung-Jin Cho, Sung-Kye Park, and Choong-Ki Kim, "Estimation of Effective Diffusion Time in a Rapid Thermal Diffusion Using a Solid Diffusion Source," *IEEE Transactions on Electron Devices, 39:1* (1992), 111–117. The last paper was published after Cho's graduation but was the result of his research at KAIST.

8 Byung-Jin Cho, "Modeling of Rapid Thermal Diffusion of Phosphorous into Silicon and Its Application to VLSI Fabrication" (doctoral thesis, Seoul: KAIST, 1991), 174.

9 In 1993, when the Samsung conglomerate awarded Kim Choong-Ki the Hoam Prize, the rapid thermal process was mentioned as one of his major contributions to the development of the South Korean semiconductor industry. See www.hoamfoundation.org/eng/award/part_view.asp?idx=10 (searched on February 19, 2021).

10 For more information about the IMEC, see www.imec-int.com/en/about-us#about (searched on February 19, 2021).

11 Byung-Jin Cho, Peter Vandenabeele, and Karen Maex, "Development of Hexagonal-Shaped Rapid Thermal Processor Using a Vertical Tube," *IEEE Transactions on Semiconductor Manufacturing, 7:3* (1994), 345–353.

12 Seok-Hee Lee, Byung-Jin Cho, Jong-Chul Kim, and Soo-Han Choi, "Quasi-Breakdown of Ultrathin Gate Oxide under High Field Stress," 605–608. This paper has been cited 182 times by other papers since its publication (see https://ieeexplore.ieee.org/document/383337). Quasi-breakdown is also known as "soft-breakdown."

13 Interview with Cho Byung-Jin by Kim Dong-Won (May 21, 2021).

14 Technology and Innovation, "Choegogisulgyeongyeongin Interview—SK Hynix Miraegisulwon R&D Gongjeongdamdang Pi Seung-Hobusajang," *Technology and Innovation, 437* (January 2020), http://azine.kr/m/_webzine/wz.php?c=71&b=103404&g= (searched on March 10, 2021).

15 Christian Forum in Science and Engineering, "Cho Byung-Jin (1)," (July 19, 2015), www.sciengineer.or.kr/board_kVuz18/19005?ckattempt=1 (searched on March 24, 2021).

16 Interview with Cho Byung-Jin by Kim Dong-Won (September 28, 2020).

17 Ezra F. Vogel, *The Four Little Dragons: The Spread of Industrialization in East Asia* (Cambridge, Mass.: Harvard University Press, 1991), 78.

18 Gregory Clancey, "Intelligent Island to Biopolis: Smart Minds, Sick Bodies and Millennial Turns in Singapore," *Science, Technology and Society, 17:1* (2012), 13–35.

19 Interview with Cho Byung-Jin by Kim Dong-Won (September 28, 2020).

20 For more information about SNDL, see S. Ramakrishana and T. C. Lim, "Overview of the NUS Nanoscience and Nanotechnology Initiative and Its Available Facilities," presented at South East Asia Materials Network Meeting, IMRE, Singapore (November 14–16, 2005), www.icmr.ucsb.edu/progr ams/archive/documents/Kim.pdf, 29–33 (searched on March 1, 2021).

21 NUS Faculty of Engineering, *A Vision for Tomorrow: Annual Report 2004–2005* (Singapore: NUS, 2005), 16–17 on 16. www.eng.nus.edu.sg/wp-content/uploads/2019/02/NUS_AR_2004.pdf (searched on February 2, 2021).

22 Interview with Cho Byung-Jin by Kim Dong-Won (September 28, 2020).

23 NUS Faculty of Engineering, *A Vision for Tomorrow*, 17.

24 Ibid.

25 Dawn Wong, "Creating the Future," *The Straits Times* (July 1, 2005), 3.

26 Cho published ten more papers with his SNDL students after he left NUS in 2007.

27 Except in cases where they are quoted or otherwise highlighted, these and subsequent team works coauthored by Cho Byung-Jin and his students and colleagues are not listed in full in the References or in a note.

28 Seven of Cho's graduate students received their degrees after he left Singapore.

29 Interview with Cho Byung-Jin by Kim Dong-Won (May 21, 2021).

30 Samsung Newsroom, "Samsungjeonja 2022nyeon Jeongki Imwon Insa" (December 9, 2021), https://news.samsung.com/kr/삼성전자-2022년-정기-임원-인사 (searched on September 1, 2022).

31 Interviews with anonymous KAIST professors.

32 National NanoFab Center, "Mission and Vision," www.nnfc.re.kr/eng/pageView/322 (searched on March 11, 2021).

33 Interview with Cho Byung-Jin by Kim Dong-Won (September 28, 2020).

34 Byung-Jin Kang, Jeong-Hun Mun, Chang-Yong Hwang, and Byung-Jin Cho, "Monolayer Graphene Growth on Sputtered Thin Film Platinum," *Journal of Applied Physics*, *106* (2009), 104309, https://doi.org/10.1063/1.3254193 (searched on March 11, 2021).

35 Jesus de la Fuente, "Understanding Graphene," www.graphenea.com/pages/graphene#.YEqvTS2cYgp (searched on March 11, 2021).

36 The University of Manchester, "Graphene: Applications," www.graphene.manchester.ac.uk/learn/applications/ (searched on March 11, 2021).

37 Ibid.

38 Interview with Cho Byung-Jin by Kim Dong-Won (May 21, 2021).

39 Byung-Jin Kang et.al, "Monolayer Graphene Growth on Sputtered Thin Film Platinum," 104309.

40 Jong-Kyung Park, Seung-Min Song, Jeong-Hun Mun, and Byung-Jin Cho, "Graphene Gate Electrode for MOS Structure-Based Electronic Devices," *NANO Letters* (November 2011), 5383–5386.

41 Ibid., 5386.

42 Hankuk Kyungje, "Samsungjeonja Graphene Doip …Memory Gyeongjaengryeok Ganghwa," *Hankuk Kyungje* (January 4, 2012), www.hankyung.com/finance/article/2012010469456; eDaily, "1/5ro julin Memory Chip … Samsung, Ilboneul apdohada," *eDaily* (January 5, 2012), www.edaily.co.kr/news/realtime/realtime_NewsRead.asp?newsid=01088966599394768 (searched on March 12, 2021).

43 Taesik Yoon, Woo-Cheol Shin, Taek-Young Kim, Jeong-Hun Mun, Taek-Soo Kim, and Byung-Jin Cho, "Direct Measurement of Adhesion Energy of Monolayer Graphene As-Grown on Copper and Its Application to Renewable Transfer Process," *NANO Letters* (February 2012), 1448–1452. For the South Korean media's report on this paper, see Joong-Ang Ilbo, "Guknaeyeonkujin, Jeonjagiyong 'Graphene' Silyonghwa apdangkyeo," *Joong-Ang Ilbo* (February 29, 2012), https://news.joins.com/article/7497201 (searched on March 12, 2021).

44 Seul-Ki Hong, Ki-Yeoung Kim, Taek-Yong Kim, Jong-Hoon Kim, Seoung-Wook Park, Joung-Ho Kim, and Byung-Jin Cho, "Electromagnetic Interference Shielding Effectiveness of Monolayer Graphene," *Nanotechnology*, *23* (2012), 455704; Interference Technology, "New Research Suggests Monolayer Graphene Is the Most Effective Material for EMI Shielding," *Interference Technology* (October 25, 2012), https://interferencetechnology.com/new-research-suggests-monolayer-graphene-is-most-effective-material-for-emi-shielding/ (searched on March 14, 2021).

45 Heon-Bok Lee, Ju-Hyung We, Hyun-Jeong Yang, Kukjoo Kim, Kyung-Cheol Choi, and Byung-Jin Cho, "Thermoelectric Properties of Screen-Printed ZnSb Film," *Thin Solid Films*, *519* (2011), 5441–5443 on 5441.

46 National Research Foundation, "Engineering Research Center," www.nrf.re.kr/cms/page/main?menu_no=131 (searched on March 15, 2021).

47 Su-Jin Kim, Ju-Hyung We, and Byung-Jin Cho, "A Wearable Thermoelectric Generator Fabricated on a Glass Fabric," *Energy and Environmental Science*, *7* (2014), 1959–1965.

48 Yonhap News, "Cheoneuro Jeonkisaengsan … 'Wearable Battery' Gisulgaebal," *Yonhap News* (April 7, 2014), www.yna.co.kr/view/AKR20140407081500017 (searched on March 15, 2021).

49 Korea Net, "Charge Your Battery by Wearing It," *Korea Net* (April 11, 2014), www.korea.net/NewsFocus/Sci-Tech/view?articleId=118808 (searched on July 22, 2022).

50 Ministry of Science, ICT, and Future Planning, "Wearable Che-on Jeonryeoksaengsan Gisul, 2015 UNSECO 10dae Gisul 1wi Grand Prix Susang," unpublished report for the press. Cho's device had been already selected as one of top ten IT innovations in January by 200 specialists from around the world.

51 Chosun Ilbo, "KAIST Wearable Baljeonki, UNESCOga bbopeun Sesangeul bakkul Choegogisule Seonjeong," *Chosun Ilbo* (February 4, 2015), https://biz.chosun.com/site/data/html_dir/2015/02/04/2015020403318.html; Jeonja Shinmun, "Cho Byung-Jin KAISTgyosu 'Wearable Baljeonsoja', UNESCO Netexplo Award Daesang," *Jeonja Shinmun* (February 4, 2015), www.etnews.com/201 50204000275?m=1; KBS, "Hankuk Wearable Baljeonjangchi, UNESCO Choegoui Gisul Seonjeong," *KBS News* (February 4, 2015), https://news.kbs.co.kr/news/view.do?ncd=3014902; and Maeil Kyungje, "Segye Choecho, Che-oneuro Jeonkireul Saengsanhaneun 'Wearable Baljeonsoja'," *Maeil Kyungje* (February 21, 2015), www.mk.co.kr/news/home/view/2015/02/167666/ (searched on March 15, 2021).

52 Hanul Moon et al., "Synthesis of Ultrathin Polymer Insulating Layers by Initiated Chemical Vapour Deposition for Low-Powered Soft Electronics," *Nature Materials*, *14* (2015), 628–635, online version published on March 9, 2015; and Maeil Kyungje, " 'Chakyonghyeong Jeonjagigi' Jejak Pilsumuljilin Gobunja Jeolyeonmak Gaebal," *Maeil Kyungje* (March 12, 2015), www.mk.co.kr/news/it/view/2015/03/227417/ (searched on March 17, 2012).

53 Pulse News, "5-Month-Old-Start-Up TEGway wins Grand Prix at UNESCO's Netexplo Award," *Pulse News* (February 5, 2015), https://pulsenews.co.kr/view.php?year=2015&no=118570 (searched on March 15, 2021).

54 Hello DD, "Bunjaeng Gajeonghaji anko sseun Teuheoneun Ssregi," *Hello DD* (March 18, 2015), www.hellodd.com/news/articleView.html?idxno=52495 (searched on March 15, 2021).

55 Hello DD, " 'Jakge mandeuni mot mandeulge eopseo' … Yeoljeonsoja sae Yeoksa sseunda," *Hello DD* (September 22, 2020), www.hellodd.com/news/articleView.html?idxno=72943 (searched on March 15, 2021).

56 Consumer Technology Association, "CES 2020 Innovation Award Product, ThermoReal," www.ces.tech/Innovation-Awards/Honorees/2020/Honorees/T/ThermoReal-Made-with-Flexible-Thermoelectric-De.aspx (searched on August 10, 2021).

57 TEGway, "Company Overview," in http://tegway.co/tegway/ (searched on March 18, 2021).

58 This information was supplied by Cho Byung-Jin (October 15, 2022).

59 Nanoelectronic and Neuromorphic Device Lab, "Culture Column" in Culture Activity, https://nand.kaist.ac.kr:54856/bbs/board.php?bo_table=sub6_1&page=1 (searched on July 25, 2022).

60 Hello DD, "Hulryunghan Gwahakjaga doeryeomyeon Yeonaesoseol mani ilgeoyajo," *Hello DD* (December 15, 2015), www.hellodd.com/news/articleView.html?idxno=56176 (searched on March 22, 2021).

61 Ibid.

62 Interview with Cho Byung-Jin by Kim Dong-Won (May 21, 2021).

63 Nanoelectronic and Neuromorphic Device Lab, "Research Topics," https://nand.kaist.ac.kr:54856 (searched on July 19, 2022).

64 For more information about monolithic three-dimensional (M3D) integration, see Max M. Shulaker, Tony F. Wu, Mohamed M. Sabry, Hai Wei, H. S. Philip Wong, and Subhasish Mitra, "Monolithic 3D Integration: A Path from Concept to Reality," *2015 Design, Automation & Test in Europe Conference & Exhibition* (2015), 1197–1202.

65 Nanoelectronic and Neuromorphic Device Lab, "Monolithic 3D Integration," https://nand.kaist.ac.kr:54856/sub2_1_b.php (searched on July 19, 2022).

66 Interview with Kim Choong-Ki by Kim Dong-Won (May 31, 2022).

67 Kahng Dawon is a Korean American semiconductor engineer who is best known as the inventor of MOSFET with Mohamed M. Atalla in 1960, when they worked at the Bell Labs. In 2009 Kahng and Atalla were inducted into the National Inventors Hall of Fame (see www.invent.org/inductees/dawon-kahng). To commemorate the most famous Korean semiconductor engineer into the world, the Korean Conference on Semiconductors established Kahng Dawon Prizes for the device/processing area and the circuit/system area and began to award them in 2017. For more information about Kahng Dawon Prizes, see the 30th Korean Conference on Semiconductors, "Shisang Gaeyo," http://kcs.cosar.or.kr/2023/about_award1.jsp (searched on February 1, 2023). See also Jeonja Shinmun, "Yoo Hoi-Jun and Cho Byung-Jinkyosu, 6hoe Kahng Dawonsang Susang … 'AImit Goyujeonche Gisulbaljeon Giyeo'," *Jeonja Shinmun* (February 15, 2023), www.etnews.com/20230214000122 (searched on February 15, 2023).

68 Christian Forum in Science and Engineering, "Cho Byung-Jin (1)."
69 Christian Forum in Science and Engineering, "Cho Byung-Jin (2)," (July 20, 2015), www.sciengineer. or.kr/board_kVuz18/18998 (searched on March 24, 2021).
70 For this activity, Cho received a special ("community service") award from the KAIST president on February 15, 2023.
71 Interview with Cho Byung-Jin by Kim Dong-Won (May 21, 2021).
72 Joong-Ang Ilbo, "Bandoche Misegongjeong 2-3nanoga Hangye, ijen 3D-soja Yeonku," *Joong-Ang Ilbo* (August 27, 2021), www.joongang.co.kr/article/24132626#home (searched on July 20, 2022).
73 Ibid.

6 Master of TFT-LCD and OLED
Ha Yong-Min

Ha Yong-Min is the master of the thin-film-transistor liquid-crystal display (TFT-LCD) and organic light-emitting diode (OLED) at LG Display. Tenacious, optimistic, and energetic, he is an excellent example of the Korean proverb, "If you want a well, only dig in one place": ever since beginning to work at GoldStar (renamed LG Electronics in 1995) in the summer of 1994, Ha has focused solely on raising low-temperature poly-Si (LTPS) TFT-LCD and, later, OLED technologies to the next level. On October 14, 2014, the South Korean government awarded him a presidential citation for his contributions to

> the development of the ultra high-resolution panels for smartphones, tablet PCs, notebooks, and monitors. He also succeeded in developing 20-inch AMOLED [active matrix organic light-emitting diodes] for the first time in the world (2004), developing and commercializing AMOLED for mobile phones (2008), and developing flexible AMOLED. All these contributed to LG Display's leading role in the display industry in the world.[1]

At LG Display, Ha has always been the vanguard when the company started development of new technologies or their commercialization according to the market's various needs. He has also been a competent and reliable "fireman" whenever the company encountered important problems to resolve in terms of technology.

Ever since the 1990s, LG has been strong in the large (TFT-LCD and later OLED) panel market for TV sets and in the market for medium-sized panels for laptops and desktops, continuously competing with Samsung to be the number one manufacturer in the large-panel world market but remaining relatively weak in the small-sized panel market. From the very beginning, Ha was a pioneer and a specialist within the company to develop both LTPS TFT-LCD and OLED panels for this weak but profitable market. Thanks to the endless efforts of its engineers such as Ha, LG Display has provided Apple, Nokia, Huawei, Google, and LG Electronics with its TFT-LCD or OLED panels for "mobile" (i.e., small-sized) devices since the mid-2000s, often challenging Samsung Display in the market.[2] Although in the late 2010s the rapid growth of Chinese companies, such as Beijing Oriental Electronics (BOE), began to threaten the hegemony of the two South Korean companies in display markets, LG Display has remained a major player in OLED panel markets in the 2020s, even cultivating the new market of OLED display for automobiles.[3]

DOI: 10.1201/9781003353911-9

EARLY YEARS

Ha Yong-Min was born on December 10, 1966, near Jinju City, which has been a major center in the southern part of the Korean peninsula throughout its history. His father, a farmer, expected his youngest and brightest son to become a lawyer and was disappointed to find that he wanted to become an engineer instead. Ha liked mathematics a lot during his school years in Jinju and heard that electrical engineers would easily get jobs after graduating from college.[4] He applied to the Department of Electronic Engineering at Seoul National University (SNU), and successfully entered it in the spring of 1984.[5] The Department of Electronic Engineering was the most popular department in the College of Engineering at SNU in the early 1980s, thanks to the rapid development of the electronics industry in the 1970s and 1980s. Ha remembers that he found many truly brilliant minds in the department and studied hard so as not to get left behind. He also remembers that the courses on semiconductor devices and physical electronics taught by Min Hong-Sik were the most interesting and memorable and that he learned a great deal about the theory of semiconductor devices.[6] In his senior year, he became interested in acoustics because his advisor was a specialist in the subject. Since he did not expect any financial support for his stay in Seoul from his parents, he offered private tutoring to high school students from time to time. As his graduation approached, Ha decided to apply to KAIST to study semiconductor engineering in earnest. The exemption from compulsory military service that KAIST offered was a major draw for most applicants, including Ha, but the full financial support at KAIST was another important factor for him. He passed the entrance examination and entered the Department of Electrical Engineering at KAIST in early 1988, along with fifteen of his SNU classmates.

The life and work at KAIST were quite different from those at SNU, even in Ha's first semester. Two things especially surprised him: the first was that he hadn't realized how diligently he must study there, and the second was that most professors at KAIST were much younger than those at SNU and more enthusiastic about teaching.[7] He decided to major in semiconductor engineering and chose Kim Choong-Ki as his advisor. The discipline at the laboratory was much stricter than he had originally expected, and some seniors drilled junior members like Ha strictly. He survived in this harsh environment without much difficulty, partly because he was accustomed to physical hardship, which he had experienced working on his father's farm. Although some of his classmates complained about being assigned trivial duties or about physical hardship in the laboratory, he quietly carried out such tasks and realized that they were real "learning." He remembers that when Kim heard the complaints of his students about physical hardship in the laboratory, he shot back, "Have you hammered a nail or sawed a log? Engineering is not just book learning and is often learned more by actual experience. Some SNU graduates often forget it!"[8] Ha also recalls that he learned some other important lessons from Kim that he later tried hard to keep in mind:

> I still remember that when Professor Kim got angry, he was really scary. He lost his temper with students whenever they were irresponsible or selfish, or, sometimes, when he wanted to teach them a lesson. However, he forgave many mistakes that certainly

deserved sharp blame. Once, for example, I broke a quartz tube for LPCVD [low-pressure chemical vapor deposition] during an experiment and confessed my mistake to him. I expected sharp scolding from him but, to my surprise, he just said, "A mistake in the school is good medicine for the future. Don't worry about it." He was forty-eight years old at that time. I haven't reached that level yet: when a TFT-LCD panel that my team [at LG Philips] had constructed for two months didn't work properly because of some minor mistakes, I lost my temper and scolded my juniors severely. I may reach Kim's level only ten years from now. ... Around 1990, when I attended a party to celebrate a senior's [Kim Jung-Kyu's] successful doctoral defense, Professor Kim told us that we should contribute to the engineering output after graduation because we would each become responsible for the earnings of at least a thousand people.[9]

Ha's master thesis was on the design and implementation of the halogen-lamp-heated LPCVD system for the deposition of polysilicon and silicon dioxide.[10] LPCVD was employed "to decrease any unwanted gas phase reactions and to increase uniformity across the substrate," so that it guarantees uniform film on the wafer.[11] By skillfully controlling temperature and input gas, Ha could successfully deposit multiple films without removal of the wafer from the vacuum chamber. The result was presented at the fall conference of the Korea Institute of Electrical Engineering in 1991 and became Ha's first published paper.[12] He remembers that he employed several instruments and facilities that his lab colleagues had built for their own research.[13] As his master's work was coming to a close, he realized he wanted to study semiconductors more and decided to advance to the doctoral program (Figure 6.1).

Ha started his doctoral studies in the spring of 1990 under the same advisor, Kim Choong-Ki, but he had little time for his own research in the beginning. KAIST was moving from its Seoul campus to the new Daejeon campus, and Ha and his colleagues had to use a large portion of their time on this relocation in 1992. As Ha's classmate Park Sung-Kye (profiled in Chapter 7) recalls, they "had done all the moving process and setup, from piping for the water supply and drainage to setting up all the facilities.

FIGURE 6.1 Left: Ha Yong-Min poses in front of an instrument in Kim Choong-Ki's laboratory during his student years at KAIST. **Right**: Ha and his colleagues in Kim's laboratory. Park Sung-Kye, one of Ha's close friends at KAIST, stands beside him.

Source: Courtesy of Ha Yong-Min.

I injured my little finger while moving the furnace. After finishing the move, we had a drinking party with the professors."[14]

Ha's doctoral thesis was on the structure and fabrication of "High Performance Polysilicon Thin Film Transistors (TFT)."[15] It is partly an extension of his master's thesis (on poly-Si and LPCVD), but it is aimed to design and fabricate state-of-the-art TFT at relatively low temperature using several new techniques, including rapid thermal chemical vapor deposition. For a long time thereafter, "poly-Si TFT" was Ha's trademark. TFT was a relatively new but rapidly developing field whose application to the liquid crystal display (LCD) was one of the hottest agendas in the semiconductor community during the 1980s. TFT-based LCD was widely considered to replace the old cathode ray tube display, and some Japanese companies, most notably Sharp, have developed TFT-LCD for laptop computer screens since the late 1980s.[16] Kim Choong-Ki's laboratory also worked on the subject beginning in the mid-1980s; Kim Nam-Deog, for example, published his master's thesis on the fabrication of hydrogenated amorphous silicon TFT for flat panel display in 1986 and became the pioneer of the field at Samsung Electronics.[17] Ha remembers that when he started his poly-Si TFT project in 1990, the subject was still in its infancy in South Korea, even at KAIST.[18]

In the introduction to his 1994 doctoral thesis, Ha considered the poly-Si TFT to be much better than the currently available amorphous silicon TFT because the poly-Si TFT has

> a great potential to be commercialized due to the possibility of integration of driver circuits which include shift registers and logic gate with pixels on the same glass substrates. In addition, ... poly-Si TFT will have a larger aperture ratio, for a given pixel size, than amorphous Si (a-Si) TFT. Thus poly-Si TFT will be applied to LCD with smaller pixels, higher resolutions, and higher aperture ratio, and especially, will be viable for LCD with a small pixel pitch because of the difficulties in mounting external driver ICs. ... In the near future low temperature poly-Si TFT LCD will be applied to TV projectors and view finders. An ultimate application of poly-Si TFT might be the high definition TV (HDTV) which requires large area and high resolution LCDs.[19]

This summary turned out to foreshadow his work over the next twenty years at LG. Ha is one of the rare figures who have actually applied what they learned at school to the real world after graduation, not just for the first few years but for almost their entire careers. From 1990 to 1994, he solved many technical difficulties and achieved satisfactory results, some of which were published a few months before he received his doctorate in the summer of 1994.[20]

During his six years at Kim Choong-Ki's laboratory, Ha also learned how to manage a project in order to produce the targeted results, which often included non-research-related jobs. Ha remembers that "when I entered LG, I realized that there were few differences between Kim's laboratory and the company. Both demanded close cooperation with other researchers to accomplish the intended project. At both places I also had to manage the facilities, human relationships, and sometimes even the finances."[21] One negative legacy of his experience at Kim's laboratory was to emphasize perfection. Ha's subordinates at LG Display often tell him, "You are a perfectionist and expect too much from us. Please lower your expectations and standards."[22]

It is interesting to compare Ha with his classmate Park Sung-Kye. Both entered KAIST in the spring of 1988 as master's candidates and joined Kim Choong-Ki's laboratory, advanced to the doctoral program together in 1990, and graduated in the summer of 1994. Their research fields were different, however: Ha worked on poly-Si TFT, while Park worked on a more efficient method for fabricating SOI MOSFET (silicon-on-insulator metal-oxide-semiconductor field-effect transistor). In 1992, following Kim's advice, both became industry-sponsored scholarship students funded by the GoldStar group: Park was supported by GoldStar Electron (renamed LG Semiconductor in 1995) and Ha by GoldStar (renamed LG Electronics). After receiving their doctorates in 1994, Ha went to GoldStar's LCD laboratory to work on poly-Si TFT, while Park went to GoldStar Electron to develop SRAMs and DRAMs. Their destiny began to diverge sharply in 1999, when LG Semiconductor was merged with Hyundai Electronics (renamed Hynix in 2001). Both Park and Ha became *the* leading figures in their companies and regarded each other very highly: Park expressed his special thanks to Ha in the acknowledgments to his doctoral thesis, writing, "I would like to express my deepest gratitude to Yong-Min, who has given me true friendship, a great deal of assistance, and consolation from the very beginning of our laboratory life. He was the model student in the laboratory"; Ha, in turn, later praised Park as "the treasure of Hynix."[23]

LG ELECTRONICS AND LG PHILIPS LCD: LTPS TFT-LCD AND AMOLED

GoldStar was the South Korean electronics company that attained many "firsts": it began to produce radio sets on November 15, 1959, for the first time in South Korea; it became the first South Korean electronics company to export its radio sets to the United States in 1962; and in the 1960s, it was the first South Korean company to manufacture "monochrome TVs, air conditioners, refrigerators, etc."[24] Established in 1958 by Koo In-Hwoi, who had started the Lucky Chemical Industrial Company (renamed LG Chem in 1995) to manufacture toothpastes, toothbrushes, soaps, and detergents for the first time in South Korea, GoldStar was the most popular brand name in the country during the 1960s and 1970s and dominated the South Korean electronics market. Lee Byung-Chul's Samsung Electronics (established in 1969) was a direct challenge to the hegemony of GoldStar in the domestic market, but GoldStar maintained its predominance until the mid-1980s, when Samsung suddenly started the development of DRAMs and scored a striking success. From the mid-1990s on, LG Electronics and Samsung Electronics became two worldwide brand names in home electronic appliances, as well as bitter rivals in both the domestic and the world markets.

A big difference between GoldStar (and LG Electronics) and Samsung Electronics was their company cultures, which reflected the two distinctive philosophies of their founders. Whereas Samsung Electronics encouraged competition among its employees and emphasized effectiveness and final results, GoldStar emphasized harmony among members of the company. At Samsung Electronics, an ambitious and talented engineer could climb the promotion ladder very quickly, often ignoring seniority, but this kind of sudden jump was not possible at GoldStar. The conservative culture at GoldStar did not change much even after GoldStar changed its name to LG Electronics and had a new boss, Koo Bon-Moo, who also became the head of the LG conglomerate in 1995.

GoldStar was not a stranger to the semiconductor business.[25] In the 1970s, the company had started developing several key electronic components, including transistors and some integrated circuits. Then, in 1979, GoldStar acquired a small semiconductor company from Taihan Electric Wires and changed its name to Goldstar Semiconductor, though its development of semiconductors remained slow and conservative compared to that at Samsung Electronics. It was the young Koo Bon-Moo who paid a lot of attention to the growth of the semiconductor business at GoldStar. In 1989, he established GoldStar Electron (renamed LG Semiconductor in 1995) and moved most of the semiconductor parts from GoldStar Semiconductor to the new company in order to accelerate the development of semiconductors. Koo achieved only limited success, however, and the gap widened between Samsung and GoldStar in both memory and other semiconductors.[26] Nevertheless, Koo was firmly determined to advance the semiconductor business, and supported the company even more after he became the head of the LG conglomerate in 1995.

Unfortunately, Koo Bon-Moo met an unpredictable obstacle in 1998, when a new government decided to reduce the number of semiconductor companies to solve the economic crisis that had hit the South Korean economy in the fall of 1997. Under strong government pressure, Koo lamentably agreed to sell his beloved LG Semiconductor to Hyundai Electronics in 1999. This forced merger badly demoralized Koo, and he boycotted the Federation of Korean Industries meetings for many years because it had pressed him to give in to the government's demands.[27] This 1999 merger greatly influenced Ha Yong-Min and his friend Park Sung-Kye's lives—but very differently: after losing its semiconductor business to Hyundai, LG began to pay more attention to the development of next-generation displays, especially TFT-LCD, which was already Ha's specialty; whereas Park Sung-Kye, who had entered GoldStar Electron (LG Semiconductor) in the summer of 1994, had to move to Hyundai Electronics in 1999, where he struggled to survive in the beginning but soon emerged as the key memory developer there.

From the very beginning, in the summer of 1994, Ha was given the task of developing TFT-LCD at GoldStar's main research center in Anyang, near Seoul. He was appointed the chief researcher to lead a team to develop LTPS TFT-LCD. It was quite natural for LG Electronics to assign Ha to this future technology because low-temperature technology is required to manufacture large-sized flat panel LCDs for both TV screens and computer monitors, and TVs had been one of LG Electronics' major products since the 1960s. The choice of polycrystalline silicon over amorphous silicon was also an important decision: first, the former provides better resolution and quicker response time than the latter, although it requires more complicated processes, such as laser annealing, for the formation; and second, Ha had worked on the subject at KAIST. LG had already manufactured TFT-LCD screens for notebook computers since 1995, using the amorphous silicon technology, but its market share was not significant.[28] Nevertheless, LG Electronics started developing LTPS TFT-LCD for the future and supported it continuously.

Ha didn't disappoint his superiors. His team made a meaningful achievement in 1999 and wanted to apply it to actual production. With permission from Ha's superiors, his team moved from the Anyang Research Center to Gumi's production line, which is about 150 miles away. Some noticeable results came out soon: his team succeeded in developing and commercializing the 10.4-inch LTPS TFT-LCD for aviation in 2001, which guaranteed its high resolution and high reliability.[29] In addition, during this early period, Ha worked on in-plane switching (IPS) technology and filed a patent, "IPS Type LCD," in 1997 with his LG colleague Moon Buhm-Jin.[30]

In 1999, while Ha was busy commercializing his LTPS TFT-LCD in the Gumi production line, his affiliation was changed from LG Electronics, via LG LCD, to the newly founded LG Philips LCD. LG Philips LCD (renamed LG Display in March 2008) was established in September 1999, with the aim of becoming "the World's No. 1 LCD Company."[31] LG had developed TFT-LCD since 1987, but its research and development had subsequently been divided between GoldStar (LG Electronics) and GoldStar Electron (LG Semiconductor). Its rival Samsung, in contrast, had developed and manufactured all kinds of displays, including TFT-LCD at Samsung Display and Device Inc. (SDI), since the late 1980s, which had enabled Samsung to develop new technologies first and to present its products more quickly than LG.[32] In November 1998, the LG conglomerate decided to establish a separate company for LCD development and production, naming it LG LCD and having it absorb two different R&D teams from LG Electronics and LG Semiconductor. After losing its semiconductor business (mostly memory business) to Hyundai in 1999, LG had only one semiconductor-related area—TFT-LCD —and hence began to focus on this area. By establishing a joint company with Philips, LG could expect to use Philips' worldwide marketing network, as well as to improve its finances for the development of future technology, while Philips could, in turn, use LG's production technology and productivity.[33] This marriage was quite successful, and LG Philips LCD soon grew into a major global manufacturer of TFT-LCD, even defeating its rival Samsung Electronics to become the top maker of TFT-LCD in the second quarter of 2001.[34]

Between 2000 and 2007, Ha made several important technological breakthroughs. The first achievement was the development of the 10.4-inch LTPS TFT-LCD in January 2001. More successes followed. Ha's team developed a 20.1-inch LTPS quad ultra-extended graphic array (QUXGA) LCD in October 2002, and a small, narrow bezel (2.4q VGA) LTPS TFT-LCD in 2007.[35] Ha also started working on the active matrix organic light-emitting diode (AMOLED) in the early 2000s and succeeded in developing the "world's largest" 20.1-inch LTPS AMOLED panel in October 2004.[36] Owing to these technological leaps in the early 2000s, LG Philips LCD's world market share rose steadily during this period—especially for notebook panels and TV panels, which in 2006 reached 26.2% and 23.6%, respectively—threatening its archrival Samsung in the TFT-LCD market.[37]

Among Ha's many achievements in the early 2000s, the development of OLED requires special attention. OLED is "a flat light emitting technology, made by placing a series of organic thin films between two conductors. When electrical current is applied, a bright light is emitted. OLED are emissive displays that do not require a backlight and so are thinner and more efficient than LCD displays (which do require a white backlight)."[38] First developed in the 1960s, OLEDs began to be commercialized by Pioneer, TDK, Samsung-NEC, and Eastman Kodak–Sanyo at the turn of the twentieth century.[39] LG Philips LCD was not a front-runner in this hotly contested future area but soon achieved remarkable success under Ha's leadership. Ha's team had developed OLED technology since the early 2000s and possessed some of the source technologies for it.[40] The South Korean Ministry of Industry and Commerce even selected Ha's 20.1-inch LTPS AMOLED panel as one of the "Ten New Technologies in 2004": the ministry praised this new technology for adopting a "creative but simple process" to make "the largest and highest resolution LTPS AMOLED in the world," which could be applied to both TV and mobile phone screens.[41]

Ha remembers the period between 2002 and 2003 as one of the most difficult years in his career at LG, largely because the top management of LG Philips LCD paid little attention to small displays for mobile devices and focused on developing larger LCD panels for TVs. Although LG Philips LCD recognized the importance of LTPS-TFT technology for both LCD and OLED and added its development to the goals of the company from 2004 on, it was too conservative to advance this new technology aggressively and apply it to other emerging areas.[42] Worse, Philips objected to LG Philips LCD's entering the mobile phone screen market because Philips itself had already invested a lot there. As a result, Ha and his team had no choice but to just focus on the technology development. Without enough large investment, however, this new technology could not produce any meaningful business results. As the company could not give Ha's team any clear visions for their careers, some team members returned to the Anyang Research Center to work on other subjects. Others made a more drastic decision—to move to Samsung SDI to continue the development of AMOLED there.

Ha himself was so disappointed in LG Philips LCD's disinterest in OLED that he seriously considered leaving the company to become a professor.[43] There was an attractive offer from a university, but after some consideration, he decided to remain at the company. He remembers that there were two reasons for this decision. First, he could not abandon the remaining members of the team: "They were looking at me. If I moved to a university, what would have happened to them? I made up my mind that I would stay and work at LG until it closes."[44] The second reason was, as he recalls, the lessons that he had learned from Kim Choong-Ki, who had always pushed his students to go into industry after graduation, encouraging them to "each become responsible for the earnings of at least a thousand persons."[45]

Fortunately for both LG Philips LCD and Ha, in late 2004 the CEO of the company, Koo Bon-Joon, finally decided to build a fab (display fabrication plant) for LTPS TFT in Gumi, which would manufacture not only LCD but also AMOLED in small quantities. Koo Bon-Joon—a younger brother of Koo Bon-Moo, head of the LG conglomerate—had special affection for the LG Philips LCD. Under his strong leadership and aggressive management between 1999 and early 2007, LG Philips LCD grew rapidly, expanded its facilities, and built a new Paju compound near Seoul.

Ha then became busy recruiting new researchers who would work to fill the vacancies, successfully accomplished that mission, and then resumed his research and development of LTPS TFT-LCD and AMOLED. Some results began to appear. Ha's team and a research team from SNU together presented a summary of their recent work on AMOLED and also published two papers in 2005 and 2006.[46] Since the demand for OLED still remained low and AMOLED is driven by LTPS TFT, Ha concentrated more on the development of LTPS TFT-LCD and its production. He remembers that he really learned a lot from his new responsibilities for the production of TFT-LCDs and enjoyed the new role, even though it was stressful and hard.[47]

In 2006, Ha received a Master of Business Administration degree from the Helsinki School of Economics (HSE). Established in 1911, HSE was the "largest and top-ranking business school in Finland, consistently listed among the best business schools worldwide by the *Financial Times*."[48] The school offered a special program to earn an "MBA for Executives" in several countries in conjunction with its partner universities. In South Korea, the Seoul School of Integrated Sciences and Technologies (aSSIST) has offered this joint MBA program since 1995.[49] The LG conglomerate has encouraged some of its junior executives, who had little experience in business management, to enter HSE's degree programs, and several LG Display directors have received their MBA degrees from it.

There was another big event in this period for the development of LG Philips LCD. In February 2003, the firm announced that it had decided to add several new production lines in Paju, near Seoul. The company had originally planned to extend its current production lines in Gumi but failed to negotiate with the Gumi City. It instead chose Paju, which offered various incentives to win the project. Paju is less than an hour's drive from Seoul, which seems to easily attract more talented workers from the Seoul area to the company.[50] From the mid-2000s on, as the new production lines were nearing completion in Paju, manpower began to move from Gumi to Paju. From 2010 on, Paju replaced Gumi as LG Display's major production site.

RISING STAR ENGINEER AT LG DISPLAY: APPLE, POLED, AND BEYOND

On New Year's Day 2008, Ha was promoted to junior director at LG Philips LCD (which was then renamed LG Display in March 2008, after LG purchased most of its stock from Philips). Ha was one of ten new directors promoted that day and the youngest among them. In fact, he was the second youngest director among 68 high-ranking officers whose names are registered in the 2008 annual report (the youngest being a lawyer newly employed by the company in November of that year).[51] Analysis of the list of directors in the 2008 annual report indicates several interesting and important aspects of LG Philips LCD (and LG Display). First of all, about half of the directors were over 50. Second, about one-third of the directors had business management, law, or social science backgrounds. Third, few directors had earned doctoral degrees in semiconductors or related areas from reputable universities. These three aspects demonstrate quite a different company milieu at LG Philips LCD compared to that at Samsung Electronics or its sister companies, where most directors were much younger, more engineering-oriented, and more highly educated in the sciences. Ha's entry into the conservative LG Philips LCD's directors' club not only confirmed the company's trust in him but also signaled top management's more aggressive approach going forward.

FIGURE 6.2   Ha Yong-Min was the central figure to develop AMOLED at LG Philips Display and LG Display. **Left**: Ha (second from left in the back row) attended the AMOLED kickoff ceremony on November 30, 2007. **Right**: At a launching ceremony in 2008 (date unknown), Ha (left) held aloft a poster inscribed with the slogan "We will certainly succeed in the development of OLED," along with the signatures of researchers.

Source: Courtesy of Ha Yong-Min.

Ha's first task as junior director was to develop OLED, which he had worked on since the early 2000s (Figure 6.2). LG Display established a new business division for OLED in June 2008, "in order to strengthen this new growth engine," and put Ha "in charge of development and production of OLED" in the division.[52] That year, Ha's team succeeded in developing a 3-inch AMOLED with a nonlaser advanced solid phase crystallization (A-SPC) method: for this, they developed a new technology using a rapid thermal anneal method that improves the image quality of the screen and lowers the cost.[53] Unfortunately, Ha could not concentrate on this subject for long because, at the beginning of 2009, the company ordered him to take up a much broader task, "the product development of mobile LTPS TFT-LCD," which proved critical for both LG Display and Ha himself. From the beginning of 2009 to mid-2010, he therefore focused on the product development of LTPS TFT-LCD for Apple's Retina display. He was also responsible for the overall product development of LTPS TFT-LCD and AMOLED for small-sized panels.

In April 2008, a South Korean newspaper reported that LG Display would supply 1 million TFT-LCD panels per month to Apple for its iPod beginning later that year, and that "[LG Display] is negotiating with Apple to supply its panels for iPhone, too."[54] On January 11, 2009, Reuters provided more detailed information about the relation between LG Display and Apple (Figure 6.3):

> South Korea's LG Display said it had signed a deal to supply LCD panels to Apple Inc. for five years. The world's second-largest maker of LCD screens did not disclose the total size of the deal but said in a filing to the Korea Exchange that it would receive a $500 million advance from Apple this month. "Although LG Display already had a relationship with Apple, the deal's duration and the size of the advance show that the two companies are involved in a long-term, strategic alliance," said Son Young-Jun, a spokesman for LG Display. ... "LG Display's panels already represent more than

GLOBAL MARKETS   JANUARY 11, 2009 / 8:48 PM / UPDATED 13 YEARS AGO

LG signs LCD supply deal with Apple

By Reuters Staff   2 MIN READ   f   y

A show attendee passes by the LG Electronics booth during the 2009 International Consumer Electronics Show
(CES) in Las Vegas, Nevada January 9, 2009. REUTERS/Steve Marcus

FIGURE 6.3   Beginning in 2008, LG Display became a supplier for Apple's various products,
including the iPod, iPad, and iMac series. Reuters reported it in detail on January 11, 2009.

Source: *Reuters*, **"LG Signs LCD Supply Deal with Apple," January 11, 2009.**

70 percent of Apple's notebooks and monitors," said Park Sang-Hyun, an analyst at HI
Investment & Securities.[55]

LG Display's business with Apple quickly extended to providing TFT-LCD panels
for many other Apple products, including its new iPhone 4.

The iPhone 4, which was introduced in June 2010, was "the biggest leap since the
original iPhone" and was equipped with many new technologies, such as "FaceTime
Video Calling, Retina Display, 5 Megapixel Camera & HD Video Recording."[56]
Among these new features, Retina display deserved special attention from the
beginning:

Apple's stunning 3.5 inch Retina display has 960×640 pixels—four times as many
pixels as the iPhone3GS and 78 percent of the pixels on an iPad. The resulting 326
pixels per inch is so dense that the human eye is unable to distinguish individual pixels
when the phone is held at a normal distance, making text, images and video look sharper,
smoother and more realistic than ever before on an electronic display.[57]

The news that LG Display had supplied Apple with its TFT-LCD panels for iPhone 4
spread quickly. Many newspapers competitively reported it: "LG Becomes Partner of

Apple for the Latter's New Weapon iPhone 4," "Disassembled iPhone 4 Shows CPU from Samsung, LCD from LG," "LG Display's Panel Is the Most Expensive Part in iPhone 4," and "LG Display's CEO Kwon said, 'We Can't Meet All Demands after Steve Job's High Praise [of the Retina Display].'"[58]

As LG Display became a major supplier of display panels for Apple from 2008 on, many of the company's developments and products became closely correlated with Apple's trajectory. As the head of the Development of Mobile LTPS LCD and Glass Mobile OLED (January 2009~August 2010), of the Development of Oxide TFT Technology (September 2010~March 2012), and of the Advanced Display Group (April 2012~August 2016), Ha (Figure 6.4) was a major figure managing the technical challenges that Apple presented: he was the driving force in developing the 3.5-inch high-resolution panel using both LTPS and IPS that eventually became Retina. From 2010 on, LG Display focused on further developing the IPS technology. The development of the advanced high-performance IPS (AH-IPS) panels for the iPhone series, iPad series, and various sizes of MacBook series were all concentrated between 2010 and 2016.

Among these many technical challenges, the development of oxide TFT deserves special attention. Oxide TFT has many desirable properties that exceed those of incumbent hydrogenated amorphous silicon (a-Si:H) TFT by guaranteeing higher resolutions and speedier switching, which are perfect for the display. In the early 2010s, several companies, most notably Sharp, began to adopt this technology for display drivers, and LG Display also started developing it under Ha's leadership. Two

FIGURE 6.4 As the most experienced specialist in LTPS TFT-LCD, Ha Yong-Min frequently escorted VIPs who visited LG Display's production lines in Paju. Here Ha (far right) gives a tour to two "strategic clients" (third and fourth from right).

Source: Courtesy of Ha Yong-Min.

patents filed in 2012 and two papers published in IEEE journals in 2013 and 2014 illustrate his team's efforts to develop high-quality oxide TFTs in the early 2010s.[59] For example, Ha's 2013 team paper on "Temperature Sensor Made of Amorphous Indium-Gallium-Zinc Oxide TFTs" claims that his sensor will "provide a simple, low-cost, and easy manufacturing solution for temperature measurement and thus can be integrated with display on glass together with a IGZO (indium gallium zinc oxide) backplane."[60] The first real result came out in 2013, when Ha's team successfully commercialized a 5.0-inch HD (high-definition) oxide TFT for smartphone screens that greatly reduces the consumption of electricity.[61] In 2014, he also developed the 27-inch HD display that applied both oxide TFT and liquid crystal alignment technologies for the first time in the world (Figure 6.5).[62] Thanks to this effort, LG Display became a core supplier of displays for Apple's iMac, MacBook, and iPad series, and others.

With his brilliant and continuous contributions to the company, Ha was promoted to senior director in January 2015 and to executive vice president in January 2018. This was quite a rapid promotion at LG Display (and within the LG conglomerate), where "harmony" and "seniority" still prevailed: when Ha was promoted in 2015 and 2018, he was the youngest of the seventeen senior directors and also of the six executive vice presidents.[63] His name, however, barely appeared in the South Korean mass media, being only briefly mentioned when reporting the promotions at LG Display. For example, his promotion to senior director in 2015 was noted as follows: "Ha Yong-Min was promoted from junior to senior director for his contributions to strengthening several businesses through development of new innovative technologies."[64] The announcement of his promotion to executive vice president was also short: "Executive Vice President Ha has played a core role in developing LTPS

FIGURE 6.5 On "The Fifth Display Day," October 6, 2014, Ha Yong-Min received a presidential citation for his contributions to the development of ultra-high-resolution panels for small and medium-sized electronic devices such as smartphones and tablets.

Source: Courtesy of Ha Yong-Min.

technology and its mass-production system, as well as successfully leading the display group for the strategic client."[65]

A temporary setback in the relationship between LG Display and Apple was on the horizon, however. While LG Display concentrated on the development of IPS technology, its rival Samsung SDI (and Samsung Display from 2012 on) focused on developing the new technology—AMOLED. During the 2000s, Samsung had originally adopted vertical alignment (VA) technology rather than the IPS for its liquid crystal alignment. After the unprecedented success of the iPhone 4 in 2010, Samsung quickly changed its policy: the company began to supply Apple with Retina-like panels using IPS technology and aggressively developed and commercialized its AMOLED display panels for small devices (such as mobile phones) and for medium-sized devices (such as tablets). Samsung's mobile phones began to employ AMOLED displays as early as 2007, and Samsung rapidly expanded their uses beginning in 2010, when it launched its Galaxy S series. Since the Galaxy S series became very popular in both the domestic and world markets beginning in the 2010s, even challenging Apple's iPhones, the adoption of AMOLED panels for the Galaxy series meant the mass production of small-sized AMOLED panels, which eventually made Samsung the undisputed number one manufacturer in the small-sized OLED panel markets since the early 2010s.[66]

Although LG Display had possessed OLED technology since the early 2000s, it was painfully slow to recognize the winds of change in the marketplace. The company provided LG Electronics with its advanced IPS TFT-LCD panels for LG Electronics' various series of smartphones in the 2010s but only supplied a small number of OLED panels for its G Flex series between 2013 and 2015. The G Flex series, however, soon proved a catastrophic failure.[67] LG Display had to wait two more years before resuming supplying OLED panels to LG Electronics for its more successful V30 smartphones in 2017. This neglect of OLED panels for smartphones in the early 2010s is well reflected in several interviews given by LG Display's top managers. In his interview with South Korean media in 2010, for example, CEO Kwon Young-Soo said: "LG Display is interested in OLED. Though we are a little bit behind the competitor [Samsung] on it, we started developing OLED because its market is expanding rapidly. … Since OLED guarantees crisper coloring, we will focus on manufacturing high-end TV sets rather than mobile devices using this technology."[68] In another interview the next year, Kwon emphasized IPS's superiority over OLED, especially in mobile phone screens, and even called LG Display's IPS panel for mobile phones "Smart Display."[69] This attitude was similarly reflected in LG Display's annual reports between 2008 and 2012, which repeatedly state that "OLED business makes up a very small part of the company's assets and sales; only LCD business is considered for public announcement in the report."[70]

The display market was changing rapidly and dramatically from the mid-2010s on. Although the small-sized OLED panel markets were expanding rapidly during the 2010s, especially from 2016 on, the growth of the LCD panel market for this category was not as impressive.[71] Between 2010 and 2017, it was Samsung that absolutely dominated these OLED panel markets, and LG Display's market share was almost negligible until 2016.[72] The real threat, however, came from China, whose companies aggressively sold their LCD panels at a much lower price from the mid-2010s on and even began to occupy considerable market share in the OLED market beginning in the late 2010s.[73] LG belatedly realized the danger and constructed new production lines for OLED panels for small devices, such as the smart watch, but its production was delayed because of several technical difficulties.

Even Apple began to seriously consider adopting OLED for its iPhones in the mid-2010s: *Forbes* reported in November 2015 that Apple was considering adopting OLED for its future iPhones and that both Samsung and LG Display would become "likely suppliers."[74] There was a widely circulated conjecture that Apple's new iPhone 8, which was planned to debut in 2017, would adopt the OLED panel: CNET, for example, reported in February 2017 that the iPhone 8 might be equipped with the curved OLED screens, based on the large purchase of OLED panels from Samsung.[75] A South Korean financial newspaper had already reported similar news a few months before, under the sensational headline, "LG Display Lost Apple, Its Customer for Last Ten Years."[76] The rumor, however, soon proved only a half truth: both the iPhone 8 and the iPhone 8 Plus, which were released in September 2017, were equipped with the same Retina display based on the IPS TFT-LCD panels, but the iPhone X, which became available a few months after the release of the iPhone 8, adopted the OLED screen (dubbed "Super Retina").[77] The future of the iPhone's display was clearly settling down.

This 2016–2017 shock, however, became strong medicine for LG Display, which began to aggressively develop OLED panels for small and medium-sized devices from then on. Beginning in 2017, the tone of voice about OLED changed drastically in the company's annual reports, definitely becoming more positive: that year, for instance, its report noted that the company "provided customer mobile phone companies with pOLED panels in large quantities thanks to the mass production of pOLED in the 6th generation line," that it had "endeavored to lower the cost during the production process, especially in the new kinds of panels like OLED," and that "by opening the 6th generation line and [making] additional investment, the company strengthened the foundation to expand OLED business for small and medium devices."[78] Meanwhile, Apple wanted neither to abandon its old partner LG Display— which still supplied most of the TFT-LCD panels for its iPhones, iPads, MacBooks, and iMacs—nor to depend too much on Samsung, its bitter rival in the smartphone market. As a result, Apple decided to keep LG Display as its "second" supplier of OLED panels and to invest in its production; as business headlines put it, "Apple Is Trying to Rely Less on Samsung by Having LG Make Some of Its iPhone OLED Screens" and "Apple Reportedly Adds LG as Second OLED Display Supplier for iPhone XS and XS Max."[79]

Yet despite Apple's strong desire to partner with LG Display for the supply of OLED panels, there were some technical difficulties for LG Display to solve. For example, in April 2018, *The Wall Street Journal* reported:

> Apple Inc.'s efforts to line up a second supplier for its high-end smartphone screens—
> and reduce its dependence on Samsung Electronics Co.—have hit a hurdle, according to

people familiar with the matter. South Korea's LG Display Co. hopes to provide organic light-emitting diode displays for iPhones slated for release this fall, the people said. However, manufacturing problems have caused LG to fall behind the schedule that many suppliers follow to begin mass production for new iPhone models which usually starts around July, they said. …

LG Display is the leading maker of large-size OLED panels used in television sets. But the process for manufacturing such panels and that for smartphone displays involve different technologies, which LG has yet to nail down, one of the people said.

LG Display was recently ordered by Apple to undertake a third round of prototype production for the OLED smartphone screens, an extra step that most suppliers don't go through for many components, the people said.

The company is expected to supply as much as 20% of OLED displays for this year's new iPhones, according to supply-chain analysis by Susquehannan International Group.[80]

It took more than a year for LG Display to overcome these difficulties—a high price to pay for its negligence over the preceding ten years—and Samsung became the sole supplier of iPhone X's OLED screens.[81] As a result, Samsung absolutely dominated the market by the very end of the 2010s, with its market share dropping below 90% only in the fourth quarter of 2019, largely because of the rapid growth by that time of LG Display's OLED panel.

As head of the Mobile Development Group at LG Display, Ha played the key role in generating the company's OLED business for smartphones and small devices. Beginning in 2016, he led the development team to commercialize LG Display's own high-end OLED—Plastic OLED (or pOLED)—which uses plastic substrates that are more flexible and effective than the glass ones. LG Display had developed pOLED since 2012, when the Ministry of Economics elected it as the management company to develop a "flexible transparent display."[82] It was, however, only after Ha became responsible for the development of pOLED in 2016 that its development and commercialization for smartphones gathered speed. For example, LG Display's 2017 annual report notes that the company had increased the supply of smartphones "equipped with pOLED, owing to mass-production of pOLED in the second half of the year."[83] Thereafter, pOLED became a major topic in the company's annual reports, and LG Display began to supply its pOLED to Apple in 2019.[84] In 2018, Ha's team also developed a new pOLED for the smartwatch, in which both LTPS and oxide TFT are driven simultaneously.[85] The most memorable achievement was made in 2019, when his team succeeded in developing the multidisplay with pOLED for automotive applications and commercialized it "for the first time in the world."[86] LG Display soon dominated the market, accounting "for 92.5% of the market for vehicle OLED panels in terms of sales in 2020. Samsung Display accounted for 6.9% and China's BOE the remaining 0.6%."[87]

This dramatic turn in LG's OLED business became more apparent from 2020 on, as a South Korean business news portal described in March 2020:

LG Display's share of the global market for smartphone organic light emitting diode (OLED) panels has surpassed 10 percent for the first time. LG Display recorded a 10.8 percent share in the fourth quarter of 2019, the global market research company

IHS Markit said on March 12. The figure represents more than a five-fold increase from 2.1 percent in the previous quarter. In 2017, its share was in the 1 percent range.

The rapid growth of LG Display's market share is attributable to the company's supply of plastic OLEDs for Apple's iPhone 11. According to LG Display, mobile display panels accounted for 36 percent of its sales in the fourth quarter of last year, surpassing TV panel sales for the first time in company history.

In the smartphone OLED market, LG Display and Samsung Display have an overwhelming share of 92 percent. Front-running Samsung Display occupied 81.2 percent in the quarter. Its share shrank by more than 9 percentage points compared to the third quarter.[88]

This was not a reporter's exaggeration but close to the truth. LG Display's share in OLED panels for the iPhone series increased steadily from 2019 on, and the company became the second largest supplier, exceeded only by Samsung: for example, LG supplied 25 million OLED panels for iPhones in 2020 and expected to supply 50 million in 2021, while Samsung supplied 89.6 million in 2020 and planned to supply 110 million panels in 2021.[89] LG Display's annual reports for 2019 and 2020 confirm that the sales of "mobile" panels increased sharply while those of "TV" panels decreased: mobile panels (30.5%) finally surpassed TV panels (27.7%) in annual sales in 2020.[90]

As a result of aggressively expanding its OLED business for small and medium-sized panels by constructing new production lines, LG Display's connection to Apple seems to be stronger than ever in the 2020s.[91] South Korean media published more hopeful but speculative news. In February 2021, for example, *The Korea Times* reported that "As Apple is preparing to launch its first 'foldable iPhone' model—the first to be released in 2023, at the earliest—LG Display is involved in developing foldable display panels for the device" and that LG Display's Paju production lines could supply LTPO (low-temperature polycrystalline oxide)-based OLED panels for the planned foldable iPhones.[92] In a highly competitive and rapidly changing display market, nothing is quite certain. LG Display, however, is likely to remain one of the major players in that market as long as it can secure able engineers such as Ha (Figure 6.6).

NEW JOB AND NEW CHALLENGE

In December 2021, Ha was appointed director of the Research Lab for Future Technology at LG Display. The center is located at LG Sciencepark in Seoul, "the R&D hub of LG Group," where eight LG affiliate companies operate large research centers to "build a community of innovators pioneering new futures for LG and our customers."[93] Ever since 1999, when he was transferred from LG Electronics' research center to LG LCD (the future LG Display), Ha Yong-Min had worked on the development of new LTPS TFT-LCD and OLED panels in Gumi and Paju, where LG Display operated two large production compounds. Thus in 2021, after almost 23 years, Ha moved from a production site to a pure, "preceding technology" research center, where he is supposed to develop future technologies, often five or ten years ahead of time.

FIGURE 6.6 Executive Vice President Ha Yong-Min (in blue pullover sweater) listens to a presentation by a LG Display researcher in 2019. Ha has worked on TFT-LCD and OLED at LG for almost 30 years and is recognized as the master of these fields within the company.

Source: Courtesy of Ha Yong-Min.

This "sudden" transfer of Ha from the Paju production compound to Seoul's research center seems to result from LG Display's acute sense of crisis in the 2020s: LG Display no longer has a competitive edge in the TFT-LCD market since Chinese companies have already overtaken it in production capacity as well as in technology, and they are also challenging both LG Display and its rival Samsung Display in the OLED market. As the most experienced engineer and manager in OLED panels, Ha is certainly the best person for LG Display to task with developing future technologies for the company's survival. He is now working both on increasing the company's technological competitiveness in OLED panels for TVs, smartphones, and other information technology devices and on developing microdisplays and micro-LED displays for augmented reality (AR) or virtual reality (VR).

What might Ha's long-term future be at LG Display? As senior executive vice president, there is only one post left for Ha in the company—president (and there are only two president positions at LG Display). On the one hand, he has been a very successful engineer and manager, supervising various projects since the beginning of the company, but this does not guarantee his advancement to the top executive position. On the other hand, if Ha should leave LG Display after completing his missions, many possible jobs await him since he has a great amount of experience in R&D, production, and management. He could go to a reputable university and train the next generation there; be invited to become CEO or CTO by other companies, especially medium-sized semiconductor companies; or he could start his own company. In any case, some fascinating challenge awaits him that he has never experienced before.

NOTES

1 Ministry of Trade, Industry and Energy, "The Fifth Display Day" (October 6, 2014), 5.
2 Samsung Display Devices Inc. (SDI) developed and manufactured displays until 2012, when its display business was separated under the name Samsung Display.
3 Both Samsung Display and LG Display shifted their focus from TFT-LCD to OLED in the late 2010s, and this tendency accelerated in the beginning of the 2020s. For example, in the OLED panels for smartphones, Samsung Display's market share in 2021 was 49%, followed by BOE (16%) and LG Display (8%), https://optics.org/news/13/3/42. In the ever growing market for OLED panels for automobiles, LG Display's market share was more than 90% in 2020, https://m.pulsenews.co.kr/view.php?year=2021&no=144599.
4 Interview with Ha Yong-Min by Kim Dong-Won (May 28, 2021).
5 The Department of Electronic Engineering at Seoul National University became a separate department from the Department of Electrical Engineering in 1959.
6 Interview with Ha Yong-Min by Kim Dong-Won (May 28, 2021).
7 Ibid.
8 Ibid. Kim Choong-Ki himself graduated from SNU (Department of Electrical Engineering) in 1965.
9 Park Sang-In et al., *Uri Kim Choong-Ki Seonsaengnim* [*Our Teacher, Kim Choong-Ki*] (Daejeon: Privately printed, 2002), 89.
10 Yong-Min Ha, "Design and Implementation of Lamp-Heated LPCVD System and Poly-Si Deposition" (master's thesis, Seoul: KAIST, 1990).
11 LNF-Wiki, "Low Pressure Chemical Vapor Deposition," https://lnf-wiki.eecs.umich.edu/wiki/Low_pressure_chemical_vapor_deposition (searched on October 12, 2021).
12 Ha Yong-Min, Kim Tae-Sung, and Kim Choong-Ki, "Design and Implementation of Lamp-Heated LPCVD System," *Daehanjeonkihakhoe Haksuldaehoe Nonmunjip* (November 1991), 299–303.
13 Interview with Ha Yong-Min by Kim Dong-Won (May 28, 2021).
14 Park Sang-In et al., *Uri Kim Choong-Ki Seonsaengnim*, 87.
15 Yong-Min Ha, "Device Structure and Fabrication Process for High Performance Polysilicon Thin Film Transistors" (doctoral thesis, Daejeon: KAIST, 1994).
16 For more information about TFT and LCD, see Joseph A. Castellano, *Liquid Gold: The Story of Liquid Crystal Displays and the Creation of an Industry* (Singapore: World Scientific Publishing, 2005), and Yue Kuo, "Thin Film Transistor Technology—Past, Present, and Future," *The Electrochemical Society Interface, 22: 1* (2013), 55–61.
17 Nam-Deog Kim, "Fabrication of Hydrogenated Amorphous Silicon Thin-Film Transistor for Flat Panel Display" (master's thesis, Seoul: KAIST, 1986).
18 Interview with Ha Yong-Min by Kim Dong-Won (May 28, 2021).
19 Yong-Min Ha, "Device Structure and Fabrication Process for High Performance Polysilicon Thin Film Transistors," 1.
20 Ha Yong-Min, Han Chul-Hee, and Kim Choong-Ki, "Hydrogenation Mechanism of Top-Gated Poly-Si TFT," *Daehanjeonjagonahakhoe Haksuldaehoe* (January 1994), 431–432.
21 Interview with Ha Yong-Min by Kim Dong-Won (December 2, 2021).
22 Interview with Ha Yong-Min by Kim Dong-Won (May 28, 2021).
23 Sung-Kye Park, "Device Design for Suppression of Floating Body Effect in Fully Depleted SOI MOSFET's" (doctoral thesis, Daejeon: KAIST, 1994), 131; interview with Ha Yong-Min by Kim Dong-Won (May 28, 2021).
24 LG Electronics, *LGjeonja 50nyeonsa* [*LG Electronics 50-Year History*], Vol. 4: English Edition (Seoul: LG Electronics, 2008), 26–27.
25 GoldStar's (or LG Electronics') semiconductor business was, however, barely mentioned in LG Electronics' official history, *LGjeonja 50nyeonsa*, largely because of LG Semiconductor's merger with Hyundai Electronics in 1999.
26 Hankuk Kyungje, "Samsung/Geumseong Bandoche Segyejeokin Hoesaro Seongjang," *Hankuk Kyungje* (January 30, 1989), www.hankyung.com/news/article/1989013000921 (searched on October 18, 2021).
27 Yonhap News, "Koo Bon-Moohoejang Pyeongsaengui Han, Bandoche," *Yonhap News* (May 20, 2018), www.yna.co.kr/view/AKR20180518082300003 (searched on October 18, 2021).

28　LG Electronics, *LGjeonja 50nyeonsa, Vol. 1*, 150–151, 162–163.

29　Hankuk Kyungje, "LG Philips LCD, Koo Bon-Joonsajang interview," *Hankuk Kyungje* (July 11, 2000), www.hankyung.com/news/article/200007108154; Jeonja Shinmun, "IMD 2001: Daegieop," *Jeonja Shinmun* (August 28, 2001), https://m.etnews.com/200108240271?SNS=00004; and, Jeonja Shinmun, "Jeo-on poli-silicon TFT-LCD Seolbituja Gyeongjaeng Sijakdwaetda," *Jeonja Shinmun* (November 12, 2001), www.etnews.com/200111090199?m=1 (searched on August 17, 2022).

30　Moon Buhm-Jin and Ha Yong-Min, "IPS Type LCD." Ha continued working on the subject in the 2000s, and filed several patents, including "In Plane Switching Mode Liquid Crystal Display Device and Method of Fabricating the Same (2003)." For more information on these patents, see KIPRIS, "Ha Yong-Min, IPS," http://kpat.kipris.or.kr/kpat/searchLogina.do?next=MainSearch#page1 (searched on October 21, 2021).

31　LG Electronics, *LGjeonja 50nyeonsa, Vol. 1*, 182.

32　For more information about Samsung SDI's (and Samsung Display's) development of TFT-LCD and OLED panels, see Samsung Display, "History," www.samsungdisplay.com/eng/intro/history/2000s.jsp#anchor (searched on August 17, 2022).

33　LG Electronics, *LGjeonja 50nyeonsa, Vol. 1*, 182.

34　Ibid., 183.

35　All these successes were recorded as major achievements of LG Philips LCD in its annual reports. For example, see LG Philips LCD, *2002 Annual Report* (March 31, 2003), 8, and idem, *2006 Annual Report* (March 30, 2007), 9.

36　LG Philips LCD, *2005 Annual Report* (March 31, 2006), 9.

37　LG Philips LCD, *2006 Annual Report* (March 30, 2007), 16.

38　OLED-Info, "An Introduction to OLED Displays," www.oled-info.com/oled-introduction (searched on October 19. 2021).

39　Homer Antoniadis, "Overview of OLED Display Technology," www.ewh.ieee.org/soc/cpmt/presentations/cpmt0401a.pdf (searched on October 19, 2021).

40　They filed a few patents, such as "Bias-Aging Method and the Circuit Structure for AMOLED (2003)" and "Driving Circuit of Organic Light Emitting Diode Display (2004)." For more information about these patents, see Korea Intellectual Property Rights Information Service (KIPRIS), "OLED, Ha Yong-Min," http://link.kipris.or.kr/link/AJAX/CTOTAL.jsp (searched on October 20, 2021).

41　Ministry of Industry and Commerce, "Ten New Technologies in 2004" (January 20, 2004), 1, 6, and 14. The quotations are from p. 14.

42　LG Philips LCD, *2004 Annual Report* (March 31, 2005), 8.

43　Interview with Ha Yong-Min by Kim Dong-Won (May 28, 2021).

44　Ibid.

45　Park Sang-In et al., *Uri Kim Choong-Ki Seonsaengnim*, 89.

46　Jae-Hoon Lee, Woo-Jin Nam, Hee-Sun Shin, Min-Koo Han, Yong-Min Ha, Chang-Hwan Lee, Hong-Seok Choi, and Soon-Kwang Hong, "Highly Efficient Current Scaling AMOLED Panel Employing a New Current Mirror Pixel Circuit Fabricated by Excimer Laser Annealed Poly-Si TFT," *IEEE International Electron Device Meeting (IEDM) Technical Digest* (2005), 931–934; Jae-Hoon Lee, Woo-Jin Nam, Byeong-Koo Kim, Hong-Seok Choi, Yong-Min Ha, and Min-Koo Han, "A New Poly-Si TFT Current-Mirror Pixel for Active Matrix Organic Light Emitting Diode," *IEEE Electron Device Letters, 27:10* (2006), 830–833.

47　Interview with Ha Yong-Min by Kim Dong-Won (May 28, 2021).

48　World Ranking Guide, "Helsinki School of Economics (HSE)," https://worldranking.blogspot.com/2010/02/helsinki-school-of-economics-hse.html (searched on October 20, 2021). The HSE was merged with the Helsinki University of Technology and the University of Art and Design in 2010 to form Aalto University.

49　aSSIST, "Executive MBA with Aalto University," www.assist.ac.kr/OverseasMBA/Aalto/curriculum_introduction.php (searched on October 20, 2021).

50　Hankuk Kyungje, "Pajue $10 Billion LCD Gongjang … LG Philips LCD," *Hankuk Kyungje* (April 3. 2003), www.hankyung.com/news/article/2003020390601; Hankuk Kyungje, "LCD Sijang Judogwon norinda … LG Philips 'Paju project' Uimi," *Hankuk Kyungje* (April, 3, 2003), www.hankyung.com/news/article/2003020391391 (searched on October 21, 2021).

51　LG Philips LCD, *2008 Annual Report* (March 31, 2009), 105–109.

52   Ibid., 20 and 108.

53   Ibid., 36. Also, interview with Ha Yong-Min by Kim Dong-Won (May 28, 2021).

54   Hankyoreh, "LG Display Will Supply LCD for iPod," *Hankyoreh* (April 3, 2008), www.hani.co.kr/arti/economy/economy_general/279774.html (searched on October 23, 2021).

55   Reuters, "LG Signs LCD Supply Deal with Apple," *Reuters* (January 11, 2009), www.reuters.com/article/us-lgdisplay-apple/lg-signs-lcd-supply-deal-with-apple-idUSTRE50B0FW20090112 (searched on October 22, 2021). It was also recorded in LG Display's 2009 annual report. See LG Display, *2009 Annual Report* (March 17, 2010), 15.

56   Apple Inc., "Apple Presents iPhone 4" (June 7, 2010), www.apple.com/newsroom/2010/06/07Apple-Presents-iPhone-4/ (searched on October 21, 2021).

57   Ibid. Apple's iconic CEO Steve Jobs explained and demonstrated these points for almost 10 minutes in his 52-minutes introduction. See "Apple WWDC 2010—iPhone 4 Introduction," www.youtube.com/watch?v=z__jxoczNWc (searched on October 21, 2021).

58   Hankuk Kyungje, "Apple Sinmugi 'iPhone 4G,' Partnerneun LG," *Hankuk Kyungje* (March 23, 2010), www.hankyung.com/it/article/2010032245751; Maeil Kyungje, "iPhone 4 tteuteoboni CPU-Samsung, LCD-LGjepum," *Maeil Kyungje* (June 9, 2010), www.mk.co.kr/news/business/view/2010/06/298328/; and Money Today, "iPhone 4 Bupumjung LG Displayga gajang bissa," *Money Today* (June 29, 2010), https://news.mt.co.kr/mtview.php?no=2010062910394529760 (searched on October 21, 2021).

59   LG Display Co., "Image Display Device Including Oxide Thin Film Transistor and Method of Driving the Same (2012)," and idem, "Thin Film Transistor Substrate Having Metal Oxide Semiconductor and Manufacturing Method Thereof (2012)," in KIPRIS, http://engpat.kipris.or.kr/engpat/searchLogina.do?next=MainSearch#page1 (searched on October 25, 2021); Hoon Jeong et al., "Temperature Sensor Made of Amorphous Indium-Gallium-Zinc Oxide TFTs," *IEEE Electron Device Letters*, *34:12* (2103), 1569–1571; Hoon Jeong et al., "Long Life-Time Amorphous-InGaZnO TFT-Based Shift Register Using a Reset Clock Signal," *IEEE Electron Device Letters*, *35:8* (2014), 844–846.

60   Hoon Jeong et al., "Temperature Sensor Made of Amorphous Indium-Gallium-Zinc Oxide TFTs," 1571.

61   LG Display, *2013 Annual Report* (March 21, 2014), 37.

62   LG Display, *2014 Annual Report* (March 30, 2015), 34

63   LG Display, *2014 Annual Report* (March 30, 2015), 241–242, and idem, *2018 Annual Report* (April 1, 2019), 314.

64   CNB News, "LG Display, 2015 Imwon Insa Silsi," *CNB News* (November 27, 2014), https://m.cnbews.com/m/m_article.html?no=273443 (searched on October 23, 2021).

65   Magazine Hankyung, "LG Display, 2018 Imwon Insa … 'Yeokdae choedae 26myeong Seungjin,'" *Magazine Hankyung* (November 30, 2017), https://magazine.hankyung.com/business/article/201711307543b (searched on October 23, 2021). The "strategic client" certainly refers to Apple.

66   For the market share of OLED panels, see OLED-info, "UBI: The OLED Market Grew Only 0.7% in 2020, Details Revenue by Company," *OLED-info* (March 10, 2021), www.oled-info.com/ubi-oled-market-grew-only-07-2020-details-revenue-company.

67   Asia Kyungje, "Palyeogo naenwatdadeoni … LG Flex 20,000 Panmae Gulyok," *Asia Kyungje* (January 20, 2014), www.asiae.co.kr/article/2014012009315418154.

68   Joong-Ang Ilbo, "Kwon Young-Soo LG Displaysajang, 'Steve Jobs Geukchan deoke Gogeup Panel Gongkeup dalyeoyo,'" *Joong-Ang Ilbo* (July 24, 2010), www.joongang.co.kr/article/4335135#home (searched on October 21, 2021).

69   I News 24, "LG Display's IPS Panel," *I News 24* (May 22, 2011), www.inews24.com/view/575415 (searched on October 23, 2021).

70   LG Display, *2008 Annual Report* (March 31, 2009), 19; idem, *2010 Annual Report* (March 28, 2011), 17; and idem, *2012 Annual Report* (March 21, 2013), 25.

71   For the rapid growth of OLED and decline of LCD panels for small-sized devices, see, for example, OLED-info, "DSCC: OLED to Overtake LCD Production Capacity for Mobile Applications in 2020," *OLED-info* (September 23, 2019), www.oled-info.com/dscc-oled-overtake-lcd-production-mobile-applications-2020 (searched on August 18, 2021).

72   OLED-info, "UBI: The OLED Market Grew Only 0.7% in 2020, Details Revenue by Company"; Statista, "Global LCD TV Panel Unit Shipments from H1 2016 to H1 2020, by Vendor," *Statista* (July

4, 2020 www.statista.com/statistics/760270/global-market-share-of-led-lcd-tv-vendors/#statisticCo ntainer (searched on August 18, 2022).

73 OLED-info, "DSCC Details its 2020 OLED Market Estimates," *OLED-info* (March 24, 2020), www. oled-info.com/dscc-details-their-2020-oled-market-estimates (searched on August 18, 2022).

74 Forbes, "Apple Eyes OLED for iPhone Displays: Samsung, LG Likely Suppliers," *Forbes* (November 26, 2015), www.forbes.com/sites/brookecrothers/2015/11/26/apple-will-switch-to-oled-for-iphone-displays-report/?sh=ba96a2768a8b (searched on November 3, 2021).

75 CNET, "Apple's iPhone 8 Could Get Curved OLED Screens—and Make Samsung Billions," *CNET* (February 13, 2017), www.cnet.com/tech/mobile/apple-might-buy-million-curved-oled-screen-from-samsung-rumor/ (searched on February 7, 2021).

76 Maeil Kyungje, "LG Display, 10nyeon Georaecheo Apple notchyeotda," *Maeil Kyungje* (January 26, 2017), www.mk.co.kr/news/business/view/2017/01/60988/ (searched on October 26, 2021).

77 Apple, "iPhone 8 and iPhone 8 Plus: A New Generation of iPhone" (September 12, 2017), www. apple.com/newsroom/2017/09/iphone-8-and-iphone-8-plus-a-new-generation-of-iphone/; and idem, "The Future Is Here: iPhone X" (September 12, 2017), www.apple.com/newsroom/2017/09/the-fut ure-is-here-iphone-x/ (searched on November 3, 2017).

78 LG Display, *2017 Annual Report*, 192–198, especially 197–198. Previous annual reports seldom mentioned the development or sales of small or medium-sized OLEDs.

79 Marshable,"Apple Is Trying to Rely Less on Samsung by Having LG Make Some of Its iPhone OLED Screens," *Marshable* (June 28, 2018), https://mashable.com/article/apple-lg-samsung-oled-screens; Business Insider, "Apple Is Reportedly Investing $2.7 Billion in LG Display to Build OLED Panels for Future iPhones," *Business Insider* (July 28, 2018), www.businessinsider.com/apple-is-reportedly-investing-27-billion-in-lg-display-to-build-oled-panels-for-future-iphones-2017-7; and The Verge, "Apple Reportedly Adds LG as Second OLED Display Supplier for iPhone XS and XS Max," *The Verge* (September 14, 2018), www.theverge.com/circuitbreaker/2018/9/14/17860688/apple-iphone-xs-max-lg-oled-display-samsung (searched on October 26, 2021).

80 The Wall Street Journal, "Apple Can't Cut Its Dependence on Rival Samsung's Screens," *The Wall Street Journal* (April 20, 2018), www.wsj.com/articles/apple-struggles-with-effort-to-diversify-scr een-suppliers-1524216606?mod=Searchresults_pos7&page=3 (searched on November 3, 2021).

81 As the sole supplier of OLED panels for iPhone X (and its succeeding models for some time), Samsung Display's sales and profits were heavily influenced by the ups and downs of iPhone's sales between 2017 and 2018. See Maeil Kyungje, "iPhone X Bujin Jikgyeoktan … Samsung Display Gadongyul Bantomak Uryeo," *Maeil Kyungje* (February 2, 2018), http://vip.mk.co.kr/news/view/21/20/1571953.html; and Apple Insider, "Samsung Reports Weak Demand for OLED Displays Used in iPhone X," *Apple Insider* (April 26, 2018), https://appleinsider.com/articles/18/04/26/samsung-repo rts-weak-demand-for-oled-displays-used-in-iphone-x (searched on November 3, 2021).

82 LG Display, *2012 Annual Report* (March 21, 2013), 20.

83 LG Display, *2017 Annual Report* (April 2, 2018), 192.

84 Yonhap News, "LG Olhaebuteo Apple-e OLED Display Gongkeun Gidae," *Yonhap News* (May 25, 2019), www.yna.co.kr/view/AKR20190525020800091 (searched on October 26, 2021).

85 LG Display, *2018 Annual Report* (April 1, 2019), 40.

86 LG Display, *2019 Annual Report* (March 30, 2020), 42.

87 Business Korea, "LG Display Dominating Global Market for High-End Vehicle OLEDs," *Business Korea* (February 18, 2021), www.businesskorea.co.kr/news/articleView.html?idxno=74331 (searched on November 3, 2021).

88 Business Korea, "LG Display's Share of Global Smartphone OLED Panel Market Tops 10 Percent," *Business Korea* (March 13, 2020), www.businesskorea.co.kr/news/articleView.html?idxno=42672 (searched on October 27, 2021).

89 Gizmo China, "LG Display to Supply More OLED Panels to Apple for iPhones This Year," *Gizmo China* (April 22, 2021), www.gizmochina.com/2021/04/22/lg-display-supply-oled-panel-apple-iph one/ (searched on November 3, 2021).

90 LG Display, *2019 Annual Report*, 281; and idem, *2020 Annual Report* (March 15, 2021), 258.

91 iMore, "LG Displays Preparing Another OLED line for Apple, Says Report," *iMore* (July 29, 2021), www.imore.com/lg-display-preparing-another-oled-line-apple-says-report; The Korea Times, "LG

Getting Closer with Apple," *The Korea Times* (August 17, 2021), www.koreatimes.co.kr/www/tech/2021/08/133_314046.html; and Business Korea, "LG Display to Invest 3.3 Tri. Won in Small and Medium-Sized OLED Panels for 3 Years," *Business Korea* (August 18, 2021), www.businesskorea.co.kr/news/articleView.html?idxno=7433 (searched on November 3, 2021).

92 The Korea Times, "LG to Supply 'Foldable Panels' for Apple," *The Korea Times* (February 20, 2021), www.koreatimes.co.kr/www/tech/2021/02/133_304312.html (searched on November 3, 2021).

93 LG Sciencepark, "President Message," www.lgsciencepark.com/EN/about.php (searched on August 8, 2022).

7 Treasure of SK Hynix
Park Sung-Kye

Park Sung-Kye is one of the best known semiconductor engineers at SK Hynix, where he has developed highly sophisticated and efficient memory chips since 1994. Park's Korea Advanced Institute of Science and Technology (KAIST) classmate, Ha Yong-Min of LG Display (profiled in Chapter 6), has praised him as the "treasure of Hynix."[1] In November 2012, the South Korean government awarded Park the Order of Industrial Service Merit (Iron Tower Class) for his contributions to

> the development of the NAND flash memory with the world's smallest cell, which can be installed in either smartphones or tablets. With this new technology, SK Hynix could reduce the manufacturing cost (31%) and increase the sales (186%) and profit (20%) in 2011. Apple Inc. certified this new type of NAND flash memory and employed it in its smartphones and tablets.[2]

Park has a unique career history: since receiving his doctoral degree from KAIST in the summer of 1994, he has worked at four semiconductor companies, three of which belonged to three different large South Korean conglomerates (LG, Hyundai, and SK). Here's how that happened: (1) Park began work at GoldStar Electron in 1994, which changed its name to LG Semiconductor in 1995; (2) LG Semiconductor was then sold to the Hyundai conglomerate in 1999, changed its name to Hyundai Semiconductor, and was merged with Hyundai Electronics later that year; (3) in 2001, Hyundai Electronics was bankrupted, and its semiconductor division became independent of the Hyundai conglomerate, changing its name to Hynix Semiconductor; and (4) in March 2012, the SK conglomerate completed the purchase of Hynix and changed the company's name to SK Hynix. Park's singular career trajectory reflects the wild fluctuations of the semiconductor industry in South Korea during the late twentieth and early twenty-first centuries. As a key memory specialist in these "four" companies over the last three decades, first in dynamic random-access memory (DRAM) and then in NAND ("not and") flash memory, Park contributed greatly to the rise of Hynix as the number two memory manufacturer in the world.

EARLY YEARS

Park Sung-Kye was born on August 8, 1966, on a small farm near Pohang,[3] where the South Korean government would soon construct the nation's largest steel mill,

DOI: 10.1201/9781003353911-10

Pohang Iron and Steel Mill (better known as POSCO). As the youngest of five children, he was the darling of the family: his eldest brother is almost twenty years his senior, and his second brother and two sisters are also much older than he is. His father was a typical farmer, and Park played with several farm machines in his youth. He remembers that he tried hard to fix broken farm machines and electronics by himself and that he liked mathematics and physics during his middle and high school years but didn't much like studying Korean.[4] Since his family was not affluent enough to send him a college in Seoul, Park chose to study electronic engineering at nearby Kyungpook National University in Daegu. One of his elder brothers lived in Daegu and provided him with room and board. Kyungpook National University had been "selected by the government as Specialized University [sic] in the area of Electronics" in 1973, and its Department of Electronic Engineering soon emerged as one of the best departments in the field.[5] During the 1970s and 1980s, for example, the department sent many of its graduates to KAIST, leading to Kyungpook's becoming the fifth or sixth most prolific South Korean university in term of numbers of graduates accepted to KAIST.[6] At college, Park found physical electronics the most interesting subject and hoped to study semiconductor engineering in the future. Like most college graduates applying to KAIST during the 1970s and 1980s, two major reasons for Park to apply there were KAIST's offers of exemption from military service and its full financial support. He passed the entrance examination without difficulty and entered the Department of Electrical Engineering in the spring of 1988.

Since Park aspired to major in semiconductor engineering, he wanted to enter Kim Choong-Ki's laboratory in his second semester at KAIST. He remembers that the competition was intense that year because an unusually large number of students desperately hoped to join Kim's group: Park therefore had to visit Kim's office three times.[7] Kim finally allowed seven new students to enter his laboratory, and Park was one of those seven. Over the next year and a half, he faced severe "physical hardship" but overcame it without much difficulty, largely because he had become accustomed to physical labor on his father's farm during his youth.[8] Park recalls that Kim and senior students in the laboratory "deliberately" gave the physically demanding jobs to junior students in order to teach them that engineering can only be learned with the hands as well as with the brain:

> Professor Kim always emphasizes that "You must use something malleable within the hard nut on your neck" (i.e., your brain within the skull on your neck). We were often puzzled about whether we had entered KAIST to do miscellaneous chores (for example, cleaning rooms and tables, fixing broken machines, or going on errands to buy necessary parts on the street) or to study semiconductors. Our laboratory was notorious for being one of the physically hardest labs within KAIST, and we actually always moved our hands and feet. Whenever we moved busily, Kim scolded us, "Are you slaves? Why do you just carry your hard nut, without using something malleable within that hard nut?" However, when we sat and read something only, he scolded us again. I really didn't know what to do at Kim's laboratory in my first year. However, I now guess that he then wanted to train us to become familiar with theory and practice together. This was a great lesson for my life after graduation, and I often use his dictum for my juniors in the company.[9]

Through this hardship, Park experienced and developed the ability of problem solving (Figure 7.1).

Park's master's thesis is on the metal gate complementary metal-oxide semiconductor (CMOS) using the rapid thermal process.[10] His work aims to manage the impurity doping in order to improve the uniformity of the thickness of oxide on the wafer and to obtain good contact resistance by using three characteristics of the rapid thermal process (rapid thermal diffusion, rapid thermal oxidation, and rapid thermal alloy).[11] Park designed and manufactured CMOS devices using the rapid thermal process, tested them carefully to compare the results with those from the standard CMOS process, and concluded that both produced the same results.[12] He has particularly mentioned that he was grateful to several seniors, including Seo Kang-Duk, who gave him much valuable advice on his thesis.[13]

Like many of his classmates, Park advanced to the doctoral program in 1990, and he also became an industry-sponsored scholarship student funded by GoldStar Electron (which was renamed LG Semiconductor in 1995). He recalls that when Kim recommended that he become a GoldStar scholarship student, he immediately accepted the position without question. When asked why he accepted this offer rather than choosing Samsung Electronics, Park said that he trusted Kim Choong-Ki's advice because, as Park's mentor, Kim must have carefully considered which company would be most suitable for his student.[14] Although he had been busy moving KAIST's facilities from Seoul to the new Daejeon campus in the early 1990s, Park nonetheless started his research on "a new method for the extraction of device parameters and a solution for the suppression of the floating body problem" in fully depleted silicon-on-insulator (SOI) metal-oxide-semiconductor field-effect transistors (MOSFETs).[15] He proposed a new design based on maintaining a high body potential as much as possible and found that the new method "has less than 10% errors both for the Si film doping concentration and for the fixed interface charge densities, for a wide range of the device parameters."[16] Some of Park's results were published before his

FIGURE 7.1 Park Sung-Kye was drilled strictly under Kim Choong-Ki for six years and became an efficient problem solver when encountering any difficulties. **Left**: Park and a colleague conducting an experiment, wearing antistatic dust-free coveralls. **Right**: Park poses amid equipment in Kim's laboratory at KAIST.

Source: Courtesy of Park Sung-Kye.

FIGURE 7.2   Park Sung-Kye (left) and Ha Yong-Min (right) at their doctoral graduation ceremony at KAIST in August 1994. Park and Ha were close friends during their KAIST years and have respected each other ever since.

Source: Courtesy of Ha Yong-Min.

graduation in August (Figure 7.2).[17] In his doctoral thesis, Park thanks Choi Jin-Ho and Cho Byung-Jin (profiled in Chapter 5) for teaching him "how to survive in the laboratory and how to operate facilities."[18] Although Park was extremely busy doing research between 1990 and 1994, he also followed another of Kim's golden rules: he got married before finishing his doctorate, in order to "fulfill the most important mission as a human."[19]

In the summer of 1994, Park entered GoldStar Electron, where one of Kim Choong-Ki's former doctoral students, Kim Jeong-Kyu, already worked. He was immediately assigned to the development of static random-access memory (SRAM) and DRAM at the company's semiconductor's research center in Cheongju. The new chairman of the LG conglomerate from 1995, Koo Bon-Moo, was deeply interested in the growth of LG Semiconductor and sought to challenge Samsung's dominance in the semiconductor industry. It was Koo who had separated the semiconductor division from GoldStar Semiconductor and renamed it GoldStar Electron in 1989 in order to invigorate the semiconductor business.[20] By the mid-1990s LG Semiconductor and Hyundai Electronics were competing with each other to become number two in semiconductor manufacturing after Samsung Electronics in South Korea.

Park soon attracted his superiors' attention by solving several key problems in SRAM, DRAM, and other semiconductor developments. He remembers that his superiors asked him to make major presentations to high management because in many cases only he knew the details of the projects or of the developing products and could therefore answer any questions from them.[21] It is not easy to trace Park's activities at LG Semiconductor because few records were retained after the company was merged with Hyundai in 1999. Twelve team papers that he published between 1995 and 2000 fill the gap, and three of them are particularly important: "Moisture Induced Hump Characteristics of Shallow Trench-Isolated Sub 1/4μm nMOSFET" (1999), "CMOSFET Characteristics Induced by Moisture Diffusion from Inter-Layer Dielectric in 0.23 μm DRAM Technology with Shallow Trench Isolation" (2000), and "A New Extraction Method of Retention from the Leakage Current in 0.23 μm DRAM Memory Cell" (2000).[22] The handful of patents that Park filed between 1995 and 2000 also show his diverse interests during this early period: "A Method of Fabricating a Thin Film Transistor" (1995), "Structure of Transistor and Method for Manufacturing the Same" (1996), "CMOS and Method for Fabricating the Same" (1996), "Manufacturing Method of Thin-Film Transistor" (1997), "MOSFET and Method for Manufacturing the Same" (1998), "Method for Fabricating Semiconductor Device" (1998), "Fabrication Method of DRAM Cell Transistor" (1998), "Manufacturing Method for Cell Transistor in DRAM" (1999), and "Dynamic Random Access Memory Cell Forming Method" (2000).[23]

In 1999, LG Semiconductor was absorbed by the Hyundai conglomerate and changed its name to Hyundai Semiconductor, which was soon merged with Hyundai Electronics. Most of LG Semiconductor's employees, including Park, were soon transferred to Hyundai Electronics' research facility. A very different company culture awaited him. Meanwhile, Park's KAIST classmate Ha Yong-Min (profiled in Chapter 6), who had also received an industry-sponsored scholarship from GoldStar (LG Electronics from 1995 on), was sent to the Anyang Research Center after graduation to work on poly-Si TFT. Ha therefore remained at the LG conglomerate and would soon emerge as the key thin-film-transistor liquid-crystal display (TFT-LCD) and organic light-emitting diode (OLED) specialist at LG Philips LCD (LG Display from 2008) for the next two decades.

HYNIX SEMICONDUCTOR (2001–2011)

On February 23, 1983, Chung Ju-Yung, the founder and head of the Hyundai conglomerate, established Hyundai Electronics to manufacture home appliances, industrial electronics, and semiconductors.[24] Until the early 1980s Hyundai's major business areas had not included any electronics-related ones but focused instead on construction and heavy industry, including shipbuilding and automobiles. To bolster this new adventure, Chung himself became president of Hyundai Electronics for its first year and then appointed one of his sons, Chung Mong-Hun, as the next president. Chung Mong-Hun paid special attention to and had affection for the development of Hyundai Electronics and strongly supported the development of semiconductors.[25] Hyundai Electronics grew rapidly during the 1980s but, as a latecomer, could not rival either Samsung Electronics or GoldStar in home appliances or, later, mobile phones. It therefore focused on developing and manufacturing semiconductors and industrial electronics instead.

Semiconductors were, in fact, the core business and the hallmark of Hyundai Electronics from the very beginning. At a company ceremony in 1984, Chung

Ju-Yung answered a reporter's question as to why he had entered the electronics business: "South Korean youngsters would like to work on electronics, *such as semiconductors*. So, we must do it" (emphasis added).[26] To catch up with semiconductor forerunners such as Samsung Electronics, Hyundai Electronics depended heavily on foreign technology through licensing or technology partnerships in the 1980s and rapidly advanced its own research and development capacities. Nevertheless, the share of the company's semiconductor products in the world market remained very small, and it was even behind GoldStar in the domestic market. It was only in the early 1990s that Hyundai Electronics began to emerge as a major manufacturer of semiconductors, especially DRAM and SRAM, in the world market. It succeeded in developing and mass-producing 1Mb, 4Mb, 16Mb, and 64Mb DRAM, which brought much profit to the company: in 1992, for example, Hyundai Electronics' share of DRAM in the world market temporarily climbed to number nine.[27] By 1995, its market share had grown steadily, and the company was included within the top twenty in the world, defeating GoldStar to become the number two South Korean semiconductor manufacturer, next to Samsung Electronics.[28] Chung Mong-Hun even tried to develop CPU chips (Intel 80386 and 80486 compatible chips) with Kyung Chong-Min's team at KAIST in the mid-1990s (see Chapter 3). However, he gave up this ambitious plan in order to continue Hyundai Electronics' good relationship with Intel.

The "1995 Semiconductor Bubble" burst at the beginning of 1996, badly injuring the top three South Korean semiconductor manufacturers for the next few years. Samsung Electronics was the least hurt because of its diverse semiconductor products, and its market share even improved during and after the crisis. Neither Hyundai Electronics nor LG Semiconductor was as lucky, and their sales and profits plunged sharply. Nonetheless, neither the Hyundai nor the LG conglomerate had any intention of abandoning this promising business, despite frequent astronomical losses when prices of semiconductor memory plummeted. In the fall of 1997, the economic crisis hit South Korea, and the country asked the International Monetary Fund (IMF) for financial aid for the first time in its history. The new South Korean government, which seized power through the presidential election in December 1997, decided to reorganize the five big conglomerates, and the semiconductor business became one of its major targets.[29] Since Samsung Electronics' world market share had already reached the top ten in the early 1990s, Hyundai Electronics and LG Semiconductor became the targets of a merger, or the so-called Big Deal. In late 1998, the government decided that Hyundai Electronics would absorb LG Semiconductor. LG's chairman, Koo Bon-Moo, resisted to the end, but to no avail. In January 1999, Koo informed the South Korean president in a private meeting that LG would withdraw from the semiconductor business and sell all its shares of LG Semiconductor to the Hyundai conglomerate. In July 1999, the Hyundai conglomerate bought LG Semiconductor's shares and changed LG Semiconductor's name to Hyundai Semiconductor. In October 1999, Hyundai Semiconductor was then merged with Hyundai Electronics.[30]

Nevertheless, the enlarged Hyundai Electronics—now the second-largest semiconductor manufacturer in South Korea as well as a rising dark horse in the world semiconductor market—soon encountered the "winner's curse."[31] Too much spending to acquire LG Semiconductor's shares, a serious quarrel among Chung Ju-Yung's sons over inheriting their father's crown, and a sharp drop in the price of DRAM at the beginning of 2001 drove the company into a serious financial crisis. On March 7, 2001, Hyundai Electronics changed its name to Hynix Semiconductor and began to sell nonmemory divisions to resolve the crisis.[32] In June of that year, the Hyundai conglomerate gave up all its rights to Hynix Semiconductor in order for the troubled company to receive financial help from Hynix's creditor banks, so Hynix Semiconductor became independent of the Hyundai conglomerate.

Beginning in early 2002, there was serious talk about selling Hynix Semiconductor to a foreign company, and the South Korean government tried to sell it to Micron, an American semiconductor company. This effort, however, met with strong opposition from many sides, and ultimately failed.[33] Although the direct control of Hynix Semiconductor by the syndicate of creditor banks formally ended in July 2005, the company remained under the control of both the South Korean government and the syndicate of creditor banks until the end of 2011. Park recalls that some regional supermarkets refused to issue membership cards to the employees of Hynix in the early 2000s because they believed the company would soon become bankrupt.[34] Both the efforts of Hynix's employees and the occasional booms in the memory market during the 2000s saved the company, regardless of relatively little new investment from the creditor banks.

Park had begun to work at the Memory R&D Division of Hyundai Electronics in Icheon in early 2000. He encountered some difficulties adjusting to the new environment because Hyundai's engineers and staff were not always cordial to the "intruders" from LG. Park remembers that some Hyundai men tried not to give him proper work to do or avoided working with him and that he seriously considered quitting his job.[35] However, two senior Hyundai engineers, Oh Choon-Sik and Koh Yo-Hwan, both of whom had been trained under Kim Choong-Ki at KAIST, rescued Park from this difficult situation. Oh Choon-Sik had written his doctoral thesis on "MOSFET Source and Drain Structures for High-Density CMOS Integrated Circuits" and became Kim's first student to enter Hyundai Electronics in 1986, three years after Hyundai Electronics was established.[36] Oh was responsible for several developments of memory, especially DRAMs and SRAMs, during the 1990s, and quickly rose to high management posts.[37] When Park moved to Hyundai in 2000, Oh was the head of production technology, and in 2002 he was promoted to head of the memory research center. Koh Yo-Hwan had written his doctoral thesis on the "Latch-Free Self-Aligned Power MOSFET and IGBT Structure Utilizing Silicide Contact Technology."[38] After graduating in 1989, Koh entered Hyundai Electronics and participated in various memory projects in the 1990s and 2000s, especially NAND flash memory in the 2000s. Ko's work was closely related to Park's research, and he was eager to hear Park's ideas. Once these two key "Hyundai" engineers recognized Park's ability, others in the Icheon compound began to cooperate with him.

In the beginning, Park had concentrated on developing DRAMs, as he had done at LG. After Hynix's inception in 2001, the company focused on developing and producing DRAMs, which accounted for more than three-fourths of its total annual sales: its 2003 annual report, for example, indicates that 78% of its total sales were from DRAMs, 18% from system integrated circuits (ICs), and 5% from SRAMs and flash memories; the 2007 annual report, in contrast, shows that the company had slimmed its products further to focus on DRAMs and flash memories.[39] The major task of Park's research team was to improve the refresh time of these products. DRAM cells lose charge over time, so periodic refresh operations are required to avoid data loss.[40] Since the characteristics of the refresh time of DRAM are closely connected with the yield and the quality of the product, all memory manufacturing companies did their best to be the first to develop the new technology and then to keep it a secret. Park focused on developing new cell and peripheral transistors to improve the refresh time of DRAM and then applied the new transistors to the DRAMs that Hynix manufactured. He also developed a new technology of testing the sense amplifier offset and applied it to improve the sense margin, which contributed greatly to improving the yield.

Successful results followed. Park developed the ultra-high-speed 256Mb double-data-rate (DDR) synchronous DRAM (SDRAM); 0.1 micron 512Mb DDR; ultra-high-speed 256Mb DDR500; 512Mb graphics DDR (GDDR) (11.6 Gbps) DRAM; 200MHz 512Mb mobile DRAM; 185MHz 512Mb mobile DRAM; 200MHz 1Gb mobile DRAM; and 1Gb GDDR5.[41] His most notable achievement during this early period was the development of a new technology, the "Step gaTed AsymmetRic (STAR) Cell Transistor," to lower power consumption and to improve the speed in DRAM. This technology was soon adopted by Hynix to manufacture both the 512Mb GDDR DRAM and the 200MHz 512Mb mobile DRAM, which attracted special attention in the world market: 512Mb GDDR DRAM was considered the "DRAM industry's fastest and highest density graphic memory," while 200MHz 512Mb mobile DRAM was "the world's fastest and smallest 512Mb mobile DRAM, … [which] is about 1.5 times faster than existing mobile DRAM products."[42] In 2005 Park's team published two papers on this new technology, "Enhancement of Date Retention Time in DRAM Using Step gaTed AsymmetRic (STAR) Cell Transistors" and "Numerical Analysis of Deep-Trap Behaviors on Retention Time Distribution of DRAMs with Negative Worldline Bias."[43] Three more papers on DRAMs, based on Park's research before his move to another research team at Hynix, were published later: "Fully Integrated and Functioned 44nm DRAM Technology for 1Gb DRAM" (2008);"Characterization of Polymer Gate Transistors with Low-Temperature Atomic-Layer-Deposition-Grown Oxide Spacer" (2009); and "Fully Integrated 54nm STT-RAM with the Smallest Bit Cell Dimension for High Density Memory Application" (2010).[44] Park remembers that, as the chief researcher or team leader of various DRAM projects, he participated in the development of almost all DRAMs that Hynix developed between 2002 and 2007.[45]

In February 2005, the Korea Industrial Technology Association named Park "Engineer of the Month" (Figure 7.3):

> Park Sung-Kye of Hynix Semiconductor is a chief researcher who has developed semi-conductor devices for the last ten years. He is elected "Engineer of the Month" for his contributions to the development of the technology to lower power consumption and to improve the speed in DRAM, which greatly improves the wafer yield and lowers the defect rate. The low-consumption cell that Chief Engineer Park developed is based

FIGURE 7.3   Park Sung-Kye has received two awards for his contributions to the development of memory chips. **Left**: In February 2005, the Korea Industrial Technology Association named Park "Engineer of the Month" for his contributions to the development of the technology to lower power consumption and to improve speed in DRAM. **Right**: In November 2012, Park received the Order of Industrial Service Merit (Iron Tower Class) from the South Korean government for his development of the smallest cell for NAND flash memory. He posed with his wife at the ceremony.

Source: Courtesy of Park Sung-Kye.

on a design that enables keeping data safely in DRAM at low voltage when rewriting [data]. It also enables finding and controlling the cause of very small (1/10,000,000) defects. This technology is the core that is directly related to the wafer yield and the quality of the final products but is very difficult to acquire as DRAM's level of integration increases.

Hynix Semiconductor has applied Park's new technology to the development and production of Blue Chip (0.15µm), Prime Chip (0.13µm), Golden Chip (0.11µm), and Diamond Chip (0.09µm) DRAMs. In 2004, Hynix Semiconductor's total sales were over $5.4 billion, and the products that adopted Park's new technology made up 92% of this. Now Park is developing the next-generation DRAMs—Nova (0.07µm) and Post Nova (0.06µm).[46]

Park was the first Hynix engineer who received this honor. In May 2007, he was promoted to the position of research fellow (equivalent to being a candidate for director).[47]

In early 2008, Hynix suddenly moved Park from the DRAM development team to the NAND flash memory team. Park remembers: "One day, my superior asked me whether I was interested in NAND flash memory. I had little experience on NAND flash until then, but I replied, 'Yes, it sounds interesting and challenging.' My superior transferred me to the NAND flash development team the next morning."[48] So he started working on NAND flash memory, which became his major research area until 2019.

NAND flash memory, first developed by Toshiba in the 1980s, had been a fast-growing and popular form of semiconductor memory since the beginning of the twenty-first century and was widely adopted in notebooks, mobile phones, USB drives, digital cameras, and other small electronics. Samsung Electronics, Hynix's domestic rival in the memory business, had developed this new form of memory in the early 1990s and began to mass-produce it in the late 1990s. By the mid-2000s, Samsung was not only the number one manufacturer in the world of this type of memory but also the leader of the next generations of NAND flash memories.[49] Hynix was therefore a latecomer in the NAND flash memory market, and its sales were almost negligible in the beginning. In April 2003, however, Hynix signed a contract to develop flash memories and supply them to the American company, STMicroelectronics. This was widely considered a win–win deal. As Hynix's annual report that year noted, "We expect the maximum synergy from this alliance that results from our DRAM technology and production ability and ST's system application know-how."[50] Thanks to this new stimulus, Hynix began to actively develop and manufacture NAND flash memories from 2004 on: it succeeded in developing 512Mb NAND flash memory and began to mass-produce it in February 2004, and its sales in the fourth quarter of the year reached almost 10% of the total sales of the company and 3.3% of the world market.[51] In November 2005, Apple announced that it had made a long-term contract with "Hynix, Intel, Micron, Samsung Electronics, and Toshiba to secure the supply of NAND flash memory through 2010" for Apple's popular iPod.[52] By the end of 2007, NAND flash memory had become Hynix's second most important product, only behind DRAM, and Hynix's share of the world market of NAND reached 17%.[53] Nonetheless, it was still far behind Samsung, not only from the point of view of market share but also from that of the advance of technology. The decision to move Park to this fast-growing field in early 2008 was therefore indicative of Hynix's firm resolution to strengthen this area further.

As he had done in the development of DRAM, Park participated in developing almost all the NAND flash memories that the company has designed and manufactured since 2008. A few examples between 2008 and 2011 are, in 2008, Hynix began to manufacture the 48-nano 16Gb and the 16Gb triple-level-cell (TLC) NAND flash memories for the first time in the world; in 2009, Park's team developed the 41nm 32Gb multi-level-cell (MLC) flash and improved the 4/8/16Gb MLC flash; in 2010, his team succeeded in developing the 30nm 32Gb MLC flash, the 32Gb TLC NAND, faster 16Gb MLC and single-level-cell (SLC) NAND, the 40nm 4Gb SLC for mobile phones, the 20nm 64Gb MLC NAND, and 32Gb MLC high-speed (HS) NAND; and in 2011, Park's team developed the 26nm 16/32/64Gb MLC NAND and the 30nm 1Gb SLC NAND.[54] Many of these products were an improvement of the existing technology for the market, but some were technology breakthroughs: for example, 64Gb MLC NAND adopted the 20nm process that promises to deliver 60% more productivity than the existing 30nm 32Gb NAND, and 32Gb MLC HS NAND targets the fast-growing smartphone and tablet markets, being 24% smaller in size but including 30% more dies compared to the existing 30nm 32Gb NAND.[55] Owing to this series of achievements, Hynix strengthened its position as the number four (or number five) manufacturer of NAND flash memory in the world market by the end of 2011.[56] It is therefore no wonder that, in early 2012, Hynix was selected to supply Apple with its NAND flash memory for Apple's iPad series, along with Toshiba and Micron Technology.[57] In November 2012, Park himself received the Order of Industrial Service Merit (Iron Tower Class) from the government for these achievements (see Figure 7.3).

Beginning in early 2010, Park also participated in developing three-dimensional (3D) NAND flash memory, which is

> a type of non-volatile flash memory in which the memory cells are stacked vertically in multiple layers. The design and fabrication of 3D NAND memory is radically different than traditional 2D—or planer—NAND in which the memory cells are arranged in a simple two-dimensional matrix. ... This effectively multiplies the available memory cells for a given area, enabling far greater storage capacities while using smaller footprint areas. In addition, stacking enables shorter overall connections for each memory cell, which supports faster memory performance.[58]

For the next four years, his major efforts would concentrate on this emerging technology.

The number of high-quality papers that Park and his colleagues published in the early 2010s proves how enthusiastic they were to develop NAND flash memory after 2008. Some examples are: "3-D Vertical FG NAND Flash Memory with a Novel Electrical S/D Technique Using the Extended Sidewall Control Gate" (2010), "A Highly Manufacturable Integration Technology of 20nm Generation 64Gb Multi-Level NAND Flash Memory" (2011), "A Novel 3D Cell Array Architecture for Terra-Bit NAND Flash Memory" (2011), "Highly Reliable 26nm 64Gb MLC E2NAND (Embedded-ECC & Enhanced-Efficiency) Flash Memory with MSP (Memory Signal Processing) Controller" (2011), "Novel Negative Vt Shift Program Disturb Phenomena in 2X~3X nm NAND Flash Memory Cells" (2011), and "A Middle-1X nm NAND Flash Memory Cell (M1X-NAND) with Highly Manufacturable Integration Technology" (2011).[59]

Although Hynix grew rapidly as the number two DRAM and number four NAND flash memory manufacturer in the world, the company was still unstable in the first decade of the twenty-first century. It was a mammoth company that constantly required several-hundred-million-dollar or even some billion-dollar investments to update its research facilities and production lines. Yet Hynix had had no "owner" since 2001 to make such decisions quickly and decisively, as its domestic rival Samsung Electronics had done. The syndicate of creditor banks and the government-controlled board of directors reluctantly agreed to make necessary but as small as possible investments. Worse, their decisions were often influenced by either the government or politicians. The election of a new president of the company in 2007 is a good example: the government-controlled board of directors elected a former deputy minister of Industry and Energy over more qualified and engineering-oriented candidates such as Oh Choon-Sik.[60]

Hynix's other serious problem during this period was its heavy dependence on DRAM and, from 2004 on, partly on NAND flash memory. As the prices of these two memories, especially that of DRAM, fluctuated radically, the company's finances swung between heaven and hell. For example, in November 2006, *The New York Times* published an article about Hynix in which it praised the company for escaping its troubles in the early 2000s to become the second largest DRAM manufacturer in the world.[61] A little more than two years later, the same newspaper published another article about Hynix, this time with a very pessimistic tone: "Hynix Semiconductor, the world's No. 2 computer memory-chip maker, logged its fifth straight quarterly loss Thursday as chip prices plunged and output shrank amid the industry's worst downturn."[62] A year later, the company celebrated acquiring the highest profits in its history in the second quarter but faced another crisis within a year owing to the sharp drop in the price of DRAM.[63] The only way to end this cycle was to find a strong "owner" who was willing to invest aggressively for the future. The South Korean government recognized these problems and tried hard to find a "buyer" from late 2008 on. The company was officially put up for sale three times between 2009 and 2011 but elicited little interest. The government even approached the LG conglomerate, which had been forced to sell its semiconductor company to Hyundai ten years before, but LG refused to buy it back.[64] Finally, in November 2011, the SK conglomerate entered the bidding process "alone" and was immediately selected as the buyer of Hynix. Four months later, in March 2012, Hynix Semiconductor became SK Hynix.[65]

SK HYNIX (2012–2021)

The SK conglomerate's purchase of Hynix Semiconductor in November 2011 was big news in the world's semiconductor market. *The Wall Street Journal* reported it immediately:

> SK Telecom Co. agreed to purchase a 21% stake in Hynix Semiconductor Inc. for 3.427 trillion won ($3.04 billion). The deal allows nine banks to sell part of their 15% holding in Hynix, the world's second-largest memory-chip maker by revenue, after Samsung Electronics Co. The banks became major shareholders in debt-for-equity swaps after Hynix nearly collapsed in 2001 when chip prices plunged. The deal will let SK Telecom, South Korea's largest wireless operator, diversify into a new area. …
>
> Hynix last month reported a third-quarter net loss of 562.62 billion won, swinging from a record profit of 1,040 trillion won a year earlier. The company, whose customers include Apple Inc., said soft demand for personal computers damped chip sales and that a weaker won effectively increased Hynix's overseas debt.
>
> SK Telecom became the sole bidder in Hynix after South Korea's STX Corps. dropped its bid in September, citing market uncertainties and the financial burden.[66]

Although there were some worries that the SK conglomerate might become another victim of the "winner's curse," as had happened to the Hyundai conglomerate a decade earlier, SK's 2011 purchase soon proved to be a wise one. The SK conglomerate emerged as the third largest conglomerate in South Korea, after the Samsung conglomerate and the Hyundai Motor group, and succeeded in diversifying its business portfolio. SK Hynix not only contributed to this advance but also became a central company within the SK conglomerate.

It was indeed a great challenge for the SK conglomerate, which had little experience in the electronics business. Starting as Sunkyong (SK) Textile in 1953, the group had extended its business to the chemical and energy industries during the 1960s and 1970s.[67] In 1980, it succeeded in acquiring the Korea Oil Corporation, the largest oil refinery at that time, from the government, thereby becoming the fifth largest conglomerate in South Korea. In 1994, the SK conglomerate entered the mobile telephone service business and soon became the most popular service provider in South Korea. Until 2011, therefore, SK's major business areas were "energy & chemicals, information & communication technology."[68]

To bolster its newly acquired but unfamiliar semiconductor company, the SK conglomerate's chairman, Chey Tae-Won, himself served as chairman of SK Hynix from 2012 to 2019. He also invited the retired Samsung Electronics semiconductor specialist-entrepreneur Lim Hyung-Kyu (profiled in Chapter 4) to come to SK as vice chairman of the SK SUPEX Council, responsible for "the Information, Communication and Technology Committee," and also as a vice chairman of SK Hynix. Though Lim served at SK Hynix for only three years (2014–2016), he contributed greatly to the speedy stabilization and development of the company. Many SK Hynix engineers remember that Lim's most important contribution during his tenure was to effectively block the unnecessary interference by nonengineer managers or financiers who came to SK Hynix from other SK companies after the takeover.[69] However, Lim could not succeed in changing the company culture completely. At Samsung Electronics, engineers would not only supervise R&D and production but also manage other activities, including sales, while managers with a nonengineering background did supporting jobs; at SK Hynix, in contrast, nonengineers checked engineers' R&D and production with vigilant eyes.

Park remembers that Hynix encountered many dramatic changes after the SK conglomerate bought the company in 2012: salaries and benefits for employees were much improved, the amounts invested in SK Hynix were enlarged greatly, every process was expedited, and the new management of the company began to ask, "What else do you need?"[70] He also recalls that more people with nonengineering backgrounds occupied top management positions after the takeover and that these "finance/strategy" specialists behave differently compared with those who have engineering backgrounds: for example, whereas engineers such as Park had considered Intel's NAND as a rival product to compete with or overcome, these "finance/strategy" specialists decided to simply buy it.[71] It will be interesting to observe which business model proves better for SK Hynix in the future—Samsung Electronics' model of having engineers in top management positions vs. SK Hynix's approach of having a mixture of top managers with engineering and nonengineering backgrounds.

Park's activities did not change much after Hynix became SK Hynix in 2012 (Figure 7.4). In December 2013, he was promoted from research fellow to junior director responsible for management. He also moved back and forth between the R&D and production divisions in order to develop top-notch NAND flash memories, DRAM and NAND cores, and next-generation device technology, with different titles such as "leader of the flash technology development team," "head of the device and process integration technology group," and "head of the NAND device technology group." Some notable achievements between 2012 and 2017 are, in 2012, Park's team further developed 20nm 64Gb/32Gb/16Gb MLC NAND and 20nm 4/8/

FIGURE 7.4   Park Sung-Kye's activities accelerated after the SK conglomerate bought Hynix in 2012 and changed its name to SK Hynix. Shortly thereafter, in January 2013, *Media SK*, the SK conglomerate's web journal, uploaded an article about him entitled "Park Sung-Kye at SK Hynix, Who Lives in the Nanometer World."

Source: *Media SK*, January 18, 2013.

16/32Gb e-NAND for smartphones and tablets, as well as 20nm 64/128/256Gb STD 2.5-inch SSD for desktop and notebook computers; in 2013, his team applied "1× nm process" technology to existing NANDs to increase productivity greatly, which was expected to improve profitability; in 2014, they developed 1× nm 128Gb TLC and 3D 128Gb MLC for the first time at SK Hynix; in 2015, they applied 1× nm technology to various NAND memories and also developed a 36-layer 128Gb MLC that adopts the 3D NAND technology; in 2016, Park's team earnestly applied its 3D NAND technology to eMMC, PCIe SSD, UFC (universal flash controller), and SATA client SSD; and in 2017, they succeeded in developing a 72-layer 3D 256Gb TLC for the first time in the world.[72]

The years between 2012 and 2017 were also the period when Park was most prolific in publishing papers. Some examples published in IEEE journals are "A New Metal Control Gate Last Process (MCGL Process) for High Performance DC-SF (Dual Control Gate with Surrounding Floating Gate) 3D NAND Flash Memory" (2012), "Device Considerations for High Density and Highly Reliable 3D NAND Flash Cell in Near Future" (2012), "New Read Scheme Using Boosted Channel Potential of Adjacent Bit-Line Strings in NAND Flash Memory" (2012), "Novel Concept of the Three-Dimensional Vertical FG NAND Flash Memory Using the Separated-Sidewall Control Gate" (2012), "Highly Reliable MIX MLC NAND Flash Memory Cell with Novel Active Air-gap and p+ Poly Process Integration Technologies" (2013), "Comprehensive Analysis of Retention Characteristics in 3-D NAND Flash Memory Cells with Tube-Type Poly-Si Channel Structure" (2015),

FIGURE 7.5 In December 2017, SK Hynix launched a special task force (Parthenon TF) to develop 96-layer NAND flash memory and appointed Park Sung-Kye to lead it. Many top researchers from different research teams joined this project, and together they succeeded in developing a new technology, 4D NAND, which became a new platform for the coming generations of NAND flash memories. This picture shows the more than one hundred researchers who participated in the Parthenon project. Park is tenth from the left in the front tow.

Source: Courtesy of Park Sung-Kye.

"Technology Scaling Challenge and Future Prospects of DRAM and NAND Flash Memory" (2015), and "Influence of Intercell Trapped Charge on Vertical NAND Flash Memory" (2017).[73]

On December 7, 2017, Park was promoted to senior director and put in charge of a newly organized research and development team, the "Parthenon Task Force," that aimed to develop 96-layer 3D NAND flash (Figure 7.5).[74] Hynix's rivals in NAND flash memory, Toshiba (with Western Digital), Micron, and Samsung, were also developing 96-layer NAND flash, adopting different technologies from one another.[75] Park's appointment to the newly organized team therefore indicated SK Hynix's strong desire to enter this hotly competitive race. Although the development process was not always smooth, Park's team overcame all the difficulties they encountered. Park remembers:

> Even before the company officially launched the Parthenon project in December 2017, we [Park's research team at SK Hynix's Center for Future Technology] had discussed 96-layer NAND flash memory for almost two years. The first question was whether we would adopt (in fact, copy) other competitors' technologies for it, or develop our

own as we had done when developing 24-, 36-, 48- and 72-[layer] 3D NAND flash. There were several "hot" debates on the issue, but the conclusion was almost always the same: to develop a new platform by adopting the periphery under cell (PUC) and "modified pipeless" technologies. We named this new platform "4D" since we put the peripheral electronics under the cell, adding another dimension to 3D. Even after we decided the principal direction, there were at least five major technological barriers compared with the 72-layer NAND. After Parthenon launched, many talented leading researchers in the company volunteered to participate in the project, and we together overcame at least seven critical crises between December 2017, when the project started, and July 2019, when we transferred the technology for the mass production of 4D NAND flash memory. The support from the top management who had decided to adopt the new platform for 96-layer NAND flash was also critical. This decision enabled SK Hynix to possess strong competitiveness in the next generations of NAND by maintaining the same platform.[76]

In August 2018, at the Flash Memory Summit, SK's representative proudly announced that the company was developing a new "4D" NAND flash memory with the "introduction of charge trap (CTF) rather than [its] earlier floating gate NAND memory cell technology, which [the company] combined with the flash memory cells over the peripheral chip electronics (PUC) to achieve higher performance and to reduce real estate on their chips."[77] Two months later, in October, SK Hynix announced that it had succeeded in developing the "world's first 96-layer 512Gb 'CTF-based 4-D NAND Flash' based on its TLC (Triple-Level Cell) arrays, using 3-D CTF (charge trap flash) design paired with the PUC (peri. under cell)."[78] Park's task force also developed the mass production process technology for this 4D NAND flash and transferred it to the production line. Based on this success, SK Hynix also succeeded in developing 128-layer 1 terabit TLC 4D NAND flash memory in June 2019 and a 176-layer 512Gb TLC 4D NAND flash memory in December 2020.[79] In August 2022, SK Hynix announced that it had succeeded in developing a 238-layer NAND flash memory, which has "the smallest NAND flash chip in size, [and] boasts a 50% improvement in data transfer speed over previous generation chips and power efficiency as well, as it cuts the volume of energy consumed for data reading by 21%."[80]

After completing his mission successfully, Park returned to the Center for Future Technology in December 2019. His new job there has been to manage the Design Input Center (DIC), whose major tasks are (1) to develop design rules that can improve the reliability and process margin further; (2) to develop and provide technology computer-aided design (TCAD) solutions for both the prediction and the solution of the defect mechanism; and (3) to build the process design kit (PDK) for maximizing the quality of products.[81] Park initially estimated that he would need three years to build these new standards but soon realized that he would need more time (Figure 7.6).

THE FUTURE OF SUCCESSFUL ENGINEERS IN THE COMPANY

On January 18, 2013, less than a year after Hynix Semiconductor became SK Hynix, *Media SK*, the SK conglomerate's web journal, uploaded an article about Park

FIGURE 7.6 In March 2022, Park Sung-Kye's colleagues at SK Hynix who had worked with him during the Parthenon project celebrated his 30th anniversary at the company. Park had become GoldStar Electron's employee (as a company-sponsored scholarship student at KAIST) on March 1, 1992. Over the next 30 years Park's affiliation changed four times while he worked solely on memory chips and won the respect of his fellow engineers. Those pictured here often gather around their former boss even though the Parthenon task force was officially disbanded years ago.

Source: Courtesy of Park Sung-Kye.

entitled "Park Sung-Kye at SK Hynix, Who Lives in the Nanometer World" (see Figure 7.4).[82] After introducing his major achievements over the last fifteen years, the article declares that "From DRAM to NAND flash memory, most of SK Hynix's products have come to the world after his touch." It also delivers Park's own view of life as engineer:

> My dream has always been a moving target. If you climb the mountain only aiming to reach the peak, you may become soon tired or bored. An expert climber once advised me that I had better look down and just follow the route: if you do so for some hours, you may then realize that you have already reached some point in the middle. This way—paying attention to following the route—must offer climbers a better chance to reach the peak. Besides, it also gives you the opportunity to enjoy the process. I believe that a dream is the same: just continue working and concentrate on the present process.

The article concludes as follows: "Park became an engineer because he likes solving problems and applying technology to the real world much better than delivering already made knowledge to students. He is still interested in semiconductors although he has worked on them since his twenties."

In January 2022, several South Korean newspapers celebrated SK Hynix's very successful first ten years with the following sensational titles: "Chey Tae-Won Made a Decision Despite Many Objections of SK Directors ... Hynix Becomes the Company to Earn $12 Billion a Year" and "Hynix with SK Wing Has Been Transformed from a Headache (with $170 million Loss) into a Filial Son."[83] These articles pointed to the strong leadership of "Chairman Chey Tae-Won," timely investments, and effective management under the SK conglomerate as the secrets of the company's success. None, however, mentioned the endless efforts of SK Hynix's engineers, such as Park, as a major driver of the company's success over the decade. This may explain why many successful semiconductor engineers, once their prime years as an engineer are over, try to transform themselves into senior managers or high-ranking officers in their companies. Even in the twenty-first century, South Koreans seldom pay attention to engineers or their achievements but focus instead on titles such "president" or "vice chairman."

Park Sung-Kye, however, seems more interested in transferring his experience and visions to the next generation of engineers. When asked what he wishes to do in the future at SK Hynix, Park replied, "If the opportunity is given, I would like to become the director of the Center for Future Technology in order to build a system developing SK Hynix's unique products, which will eventually make the company number one in the world's memory market."[84] Park has been a semiconductor engineer through and through ever since he entered Kim Choong-Ki's laboratory in the summer of 1988, and he seems likely to continue being so until his retirement from SK Hynix and beyond.

The successful career of Park Sung-Kye as well as that of his KAIST friend, Ha Yong-Min of LG Display, thus raise an important question: What is the ultimate goal of a successful engineer in the South Korean semiconductor industry? Both Park and Ha have worked on one subject—Park on memories (DRAM and NAND flash) and Ha on TFT-LCD and OLED—for their entire careers. They were very successful as engineers in their companies and received coveted awards from the South Korean government for their contributions to the development of their respective fields. They also climbed the promotion ladder quickly in their rather conservative companies, both reaching the level of vice president in their early fifties. The question is whether they still have any future at their companies, where there are few positions to which they can advance from their present positions as vice presidents: unlike Samsung Electronics and Samsung Display, which have more positions of this kind, both LG Display and SK Hynix remain dominated at the top by officers who have no background in engineering. Even though both Park and Ha became managers during their forties, this is not necessarily enough to guarantee their promotion to the top management position of their companies. It will be very interesting to see whether Park and/or Ha are promoted to the president/CEO level in their companies in the future. If not, no doubt that other, totally different challenges and lives await them.

NOTES

1   Interview with Ha Yong-Min by Kim Dong-Won (May 28, 2021).
2   Ministry of Knowledge and Economy, "Soje-Bupum Seongkwabogo Daehoe" (November 1, 2012), 7 and 9. This is a summary for reporters.

3   Park's birthday is officially registered as June 21, 1967, about a year later than the true birthday. This
    kind of mismatch was not uncommon in South Korea during the 1960s if the parent registered a birth
    late for some reason.

4   Interview with Park Sung-Kye by Kim Dong-Won (June 1, 2021).

5   Kyungpook National University, "School of Electronics Engineering," https://see.knu.ac.kr/eng/cont
    ent/faculty/greeting.html (searched on November 9, 2021).

6   Kim Dong-Won et al., *Hankukgwahakgisulwon Sabansegi: Miraereul hyanghan kkeunimeopneun
    Dojeon* (Daejeon: KAIST, 1996), 127. The top four universities—Seoul National University, Yonsei
    University, Hanyang University, and Korea University—are all located in Seoul.

7   Interview with Park Sung-Kye by Kim Dong-Won (June 1, 2021).

8   Ibid.

9   Park Sang-In et al., *Uri Kim Choong-Ki Seonsaengnim* [*Our Teacher, Kim Choong-Ki*]
    (Daejeon: Privately printed, 2002), 86–87.

10 Sung-Kye Park, "Design of Metal Gate CMOS Process Using Rapid Thermal Process" (master's
    thesis, Daejeon: KAIST, 1990).

11 Ibid., 1–2.

12 Ibid., 64.

13 Ibid., 67.

14 Interview with Park Sung-Kye by Kim Dong-Won (June 1, 2021).

15 Sung-Kye Park, "Device Design for Suppression of Floating Body Effect in Fully-Depleted SOI
    MOSFETs" (doctoral thesis, Daejeon: KAIST, 1994), i.

16 Ibid., 112.

17 Park Sung-Kye and Kim Choong-Ki, "A Device Parameter Extraction Method for Thin Film SOI
    MOSFET's," *Daehanjeonkihakhoe Haksuldaehoe Nonmunjip, 7* (1992), 820–824; Park Jae-Woo, Park
    Sung-Kye, and Kim Choong-Ki, "LOCOS Process to Reduce the Edge Effect on SOI NMOSFET,"
    Daehanjeonjagonghakhoe Haksuldaehoe, 17:1 (1994), 209–210.

18 Sung-Kye Park, "Device Design for Suppression of Floating Body Effect in Fully-Depleted SOI
    MOSFETs," 131.

19 Park Sang-In et al., *Uri Kim Choong-Ki Seongsaengnim*, 86.

20 GoldStar Semiconductor changed its name to GoldStar Information & Communication in 1990. Since
    GoldStar Electron (LG Semiconductor from 1995) was forced to merge with Hyundai's counterpart in
    1999, LG Electronics' *50-Year History* deleted "GoldStar Electron" (and also "LG Semiconductor")
    from its organization chart. See LG Electronics, *LGjeonja 50nyeonsa* [*LG Electronics 50-Year
    History*], *Vol. 1* (Seoul: LG Electronics, 2008), Appendix.

21 Interview with Park Sung-Kye by Kim Dong-Won (June 1, 2021).

22 Sung-Kye Park, Young-Chul Lee, Moon-Sik Suh, Sang-Ho Lee, Hyung-Jae Lee, and Gyu-Han Yoon,
    "Moisture Induced Hump Characteristics of Shallow Trench-Isolated Sub 1/4 μm nMOSFET," *IEICE
    Technical Report, 99:234* (1999), 21–25; Sung-Kye Park, Moon-Sik Suh, Jae-Young Kim, and Sung-
    Ho Jang, "CMOSFET Characteristics Induced by Moisture Diffusion from Inter-Layer Dielectric in
    0.23 μm DRAM Technology with Shallow Trench Isolation," *2000 IEEE International Reliability
    Physics Symposium Proceedings* (2000), 164–168; and Choong-Mo Nam, Sung-Kye Park, Sang-
    Ho Lee, Jai-Bum Suh, Gyu-Han Yoon, and Sung-Ho Jang, "A New Extraction Method of Retention
    from the Leakage Current in 0.23 μm DRAM Memory Cell," *Proceedings of the 2000 International
    Conference on Microelectronic Test Structures* (2000), 102–105.

23 Korea Intellectual Property Rights Information Service (KIPRIS), "Park Sung-Kye (in Korean): No.
    21~No. 38," http://kportal.kipris.or.kr/kportal/search/search_patent.do (searched on January 28, 2022).

24 For the history of Hyundai Electronics, see Hyundai Electronics, *Hyundaijeonja 10nyeonsa* [*Ten-Year
    History of Hyundai Electronics*] (Icheon, Kyeongkido: Hyundai Electronics, 1994).

25 Chung Mong-Hun's special affection for Hyundai Electronics can be found in his long interview in
    Hyundai Electronics, *Hyundaijeonja 10nyeonsa*, 39–50. He was particularly proud of the company's
    achievements in semiconductors.

26 Hyundai Electronics, *Hyundaijeonja 10nyeonsa*, 39. This *Hyundaijeonja 10nyeonsa* devotes more
    pages to the development of semiconductors than to other products.

27 Ibid., 184.

28 Opinion News, "1995nyeon cheot Bandoche Hohwang eottaetna," *Opinion News* (December 22, 2020), https://www.opinionnews.co.kr/news/articleView.html?idxno=44364 (searched on November 15, 2021). The gap between LG Semiconductor and Hyundai Electronics was, however, very small, and often reversed.

29 For the "Big Deal" between LG and Hyundai in 1998–1999, see Sisa Journal, "Bandochesaeop Big Deal mueoteul namgyeotna," *Sisa Journal* (January 7, 1999), https://www.sisajournal.com/news/articleView.html?idxno=81583; Kyunghyang Shinmun, "Big Deal 6nyeon 'Gwanchi Akmong …' jukeotda salatda," *Kyunghyang Shinmun* (December 23, 2004), https://m.khan.co.kr/economy/indus try-trade/article/200412231747141; and Jeonja Shinmun, "Bandoche Big Deal Nonran," *Jeonja Shinmun* (September 17, 2012), https://www.etnews.com/201209110681 (searched on August 26, 2022).

30 SK Hynix Newsroom, "Hyundai Electronics-Hyundai Semiconductor, Tonghapbeopineuro Gongsikchulbeom" (October 15, 1999), https://news.skhynix.co.kr/presscenter/officially-launched-as-an-integrated-corporation (searched on January 13, 2022).

31 For more details, see Kim Soo-Yeon, Baek Yu-Jin, and Park Young-Ryul, "Hankuk Bandochesaneopui Seongjangsa: Memory Bandochereul jungsimeuro [The Historical Review of the Semiconductor Industry]," *Gyeongyeongsahak, 30:3* (2015), 145–166 on 159–160.

32 Hankuk Kyungje, "Hyundai Electronics, 'Hynix Semiconductor'ro Hoesamyeong Byeonkyeong," *Hankuk Kyungje* (March 8, 2001), https://www.hankyung.com/news/article/2001030872956 (searched on August 26, 2022).

33 For the opposition to the sale of Hynix to Micron, see MBC, "Jeongbuchaegwondan Hynix Heolgapmaegak Nonran," *MBC News* (April 19, 2002), https://imnews.imbc.com/replay/2002/nwd esk/article/1889228_30761.html; Dong-A Ilbo, "Hynix Heolgapmaegak Nonran," *Dong-A Ilbo* (April 22, 2002), https://www.donga.com/news/article/all/20020422/7810916/9; and Hankyoreh, "Hynix Heolgapmaegak Nonran beonjeo," *Hankyoreh* (April 23, 2002), http://legacy.www.hani.co.kr/section-004100021/2002/04/004100021200204231918035.html (searched on August 26, 2022).

34 Interview with Park Sung-Kye by Kim Dong-Won (June 1, 2021).

35 Ibid.

36 Choon-Sik Oh, "MOSFET Source and Drain Structures for High-Density CMOS Integrated Circuits" (doctoral thesis, Daejeon: KAIST, 1986).

37 Oh Choon-Sik's contributions to the development of semiconductor industry were duly noted by the South Korean government and by academia: the South Korean government awarded him the Order of Industrial Service Merit in 2002, and the College of Engineering at Seoul National University chose him as one of 60 engineers who had contributed most to the development of South Korean industry. Oh was promoted to vice president of Hynix in 2004 and was considered a serious candidate for president of the company in 2007.

38 Yo-Hwan Koh, "Latch-Free Self-Aligned Power MOSFET and IGBT Structure Utilizing Silicide Contact Technology" (doctoral thesis, Seoul: KAIST, 1989).

39 Hynix Semiconductor, *2003 Annual Report* (March 30, 2003), 40; idem, *2007 Annual Report* (March 28, 2008), 18–20.

40 For more information about refresh time, see The Utah Arch, "A DRAM Refresh Tutorial" (November 27, 2013), http://utaharch.blogspot.com/2013/11/a-dram-refresh-tutorial.html (searched on August 23, 2022).

41 Hynix Semiconductor, *2004 Annual Report* (March 31, 2005), 9; idem, *2005 Annual Report* (March 31, 2006), 9–10; and idem, *2007 Annual Report* (March 28, 2007), 9.

42 Phys Org, "Hynix Develops World's Fastest and Highest Density Graphic Memory," *Phys Org* (December 5, 2005), https://phys.org/news/2005-12-hynix-worlds-fastest-highest-density.html; and Firstpost, "Hynix Announces 512Mbit Mobile DRAM," *Firstpost* (December 4, 2006), https://www.firstpost.com/tech/news-analysis/hynix-announces-512mbit-mobile-dram-3551063.html (searched on January 10, 2022).

43 M. W. Jang et al., "Enhancement of Date Retention Time in DRAM Using $\underline{S}$tep ga$\underline{T}$ed $\underline{A}$symmet$\underline{R}$ic (STAR) Cell Transistors," *Proceedings of ESSDERC* (2005), 189–192; and Jeong-Hyong Yi, Sung-Kye Park, Young-June Park, and Hong Shick Min, "Numerical Analysis of Deep-Trap Behaviors

on Retention Time Distribution of DRAMs with Negative Worldline Bias," *IEEE Transactions on Electron Devices*, *52:4* (2005), 554–560.

44 Hyunjin Lee et al., "Fully Integrated and Functioned 44nm DRAM Technology for 1Gb DRAM," *2008 Symposium on VLSI Technology Digest of Technical Papers* (2008), 86–87; Ga-Won Lee et al., "Characterization of Polymer Gate Transistors with Low-Temperature Atomic-Layer-Deposition-Grown Oxide Spacer," *IEEE Electron Device Letters*, *30:2* (2009), 181–184; and Suock Chung et al., "Fully Integrated 54nm STT-RAM with the Smallest Bit Cell Dimension for High Density Memory Application," *2010 International Electron Devices Meeting* (2010), 12.7.1–12.7.4.

45 Interview with Park Sung-Kye by Kim Dong-Won (June 1, 2021).

46 Korea Industrial Technology Association, "Engineer of Korea (Formerly Engineer of the Month)," https://m.koita.or.kr/m/mobile/award/kor_ng_search_view.aspx?no=483 (searched on January 6, 2022). See also Dong-A Science, "Idaleu Engineersang 2wol Susangja, Park Sung-Kye, Jeong Jong-Seok," *Dong-A Science* (February 14, 2005), https://www.dongascience.com/news.php?idx=-44553 (searched on January 6, 2022). Two more Hynix engineers followed in Park's footsteps in 2006 and 2007, before the company became SK Hynix in 2012. SK Hynix has produced five more winners of this honor since 2012.

47 Joong-Ang Ilbo, "Hynix Bandoche Imwoninsa," *Joong-Ang Ilbo* (May 15, 2007), https://www.joong ang.co.kr/article/2727160#home (searched on January 14, 2022).

48 Interview with Park Sung-Kye by Kim Dong-Won (June 1, 2021).

49 Samsung Electronics, *Samsungjeonja 40nyeon* [*40-Year History of Samsung Electronics*] (Suwon, Kyeongkido: Samsung Electronics, 2010), 230–242.

50 Hynix Semiconductor, *2003 Annual Report* (March 30, 2004), 34. The contract also included the codevelopment of SRAM (p. 37).

51 Hynix Semiconductor, *2004 Annual Report* (March 31, 2005), 22–23.

52 Apple Inc., "Apple Announces Long-Term Supply Agreements for Flash Memory," (November 21, 2005), https://www.apple.com/newsroom/2005/11/21Apple-Announces-Long-Term-Supply-Agr eements-for-Flash-Memory/ (searched on January 5, 2022).

53 Hynix Semiconductor, *2008 Annual Report* (April 1, 2009), 29.

54 Hynix Semiconductor, *2008 Annual Report*, 46–48; idem, *2009 Annual Report* (March 31, 2010), 39–41; idem, *2010 Annual Report* (March 31, 2011), 44–46; idem, *2011 Annual Report* (March 30, 2012), 48.

55 Hynix Semiconductor, *2010 Annual Report* (March 31, 2011), 45–46.

56 Statista, "NAND flash Manufacturers Revenue Share Worldwide from 2010 to 2021 by Quarter," *Statista* (August 31, 2021), https://www.statista.com/statistics/275886/market-share-held-by-leading-nand-flash-memory-manufacturers-worldwide/ (searched on January 10, 2022).

57 Hankuk Kyungje, "Apple, juyo Bupumnappumeopche Dabyeonhwa," *Hankuk Kyungje* (March 19, 2012), https://www.hankyung.com/it/article/201203193802i (searched on January 11, 2022).

58 TechTarget, "3D NAND Flash," https://www.techtarget.com/searchstorage/definition/3D-NAND-flash (searched on January 25, 2022).

59 Except in cases where they are quoted or otherwise highlighted, these and subsequent team works coauthored by Park Sung-Kye and his students and colleagues are not listed in full in the References or in a note.

60 Maeil Kyungje, "Hynixsajang Kim Jong-Gabssi Naejeong," *Maeil Kyungje* (February 27, 2007), https://www.mk.co.kr/news/home/view/2007/02/102236/ (searched on January 19, 2022). Oh Choon-Sik quit Hynix and began a startup company.

61 The New York Times, "With Turnaround, Hynix Comes Full Circle," *The New York Times* (November 28, 2006), https://www.nytimes.com/2006/11/28/business/worldbusiness/28iht-hynix.3696087.html (searched on January 10, 2022).

62 The New York Times, "Losses Mount at Hynix as Chip Prices Slump," *The New York Times* (February 5, 2009), https://www.nytimes.com/2009/02/05/business/worldbusiness/05iht-chip.1.19949158.html (searched on January 10, 2022).

63 Hankuk Kyungje, "10nyeon Jeolchibusim ... Hynix Memory choegang kkumkkunda," *Hankuk Kyungje* (August 19, 2010), https://www.hankyung.com/news/article/2010081749681; and idem,

"Hynix Siljeok 1bunki Badak," *Hankuk Kyeongje* (January 7, 2011), https://www.hankyung.com/finance/article/2011010670241 (searched on January 11, 2022).

64   eDaily, "Hynix Insu 'Love-Call' ... LG Electronics sikundeunghan Iyuneun," *eDaily* (October 4, 2010), https://www.edaily.co.kr/news/read?newsId=02263206593129640&mediaCodeNo=257; and Newspim, "LGga Hynix Insue Chamyeo anhaetdeon Iyu," *Newspim* (July 13, 2011), https://www.newspim.com/news/view/20110713000206 (searched on January 10, 2022).

65   Yonhap News, "Hynix Bandoche Maegak Ilji," *Yonhap News* (November 11, 2011), https://www.yna.co.kr/view/AKR20111111149800002 (searched on January 11, 2022).

66   The Wall Street Journal, "SK Telecom to Buy Hynix Stake for $3.04 Billion," *The Wall Street Journal* (November 14, 2011), https://www.wsj.com/articles/SB10001424052970204190504577037864059713458 (searched on January 11, 2022).

67   For more details on the SK conglomerate, see the history of the group on its website (https://eng.sk.com/history).

68   SK, "Our Companies," https://eng.sk.com/companies#industry (searched on January 11, 2022).

69   Interviews with SK Hynix engineers who prefer to remain anonymous.

70   Interview with Park Sung-Kye by Kim Dong-Won (June 1, 2021).

71   Intel, "Intel Sells SSD Business and Dalian Facility to SK Hynix" (December 29, 2021), https://www.intc.com/news-events/press-releases/detail/1513/intel-sells-ssd-business-and-dalian-facility-to-sk-hynix (searched on August 25, 2022).

72   SK Hynix, *2012 Annual Report* (March 28, 2013), 34–35; idem, *2013 Annual Report* (March 31, 2014), 33–35; idem, *2014 Annual Report* (March 31, 2015), 33–35; idem, *2015 Annual Report* (March 20, 2016), 54–55; idem, *2016 Annual Report* (March 31, 2017), 35–37; and idem, *2017 Annual Report* (April 2, 2018), 47.

73   IEEE Xplore, "Sung-Kye Park," https://ieeexplore.ieee.org/search/searchresult.jsp?newsearch=true&queryText=Sung-Kye%20Park (searched on January 23).

74   Newspim, "SK Hynix, Segye Choego 96dan NAND Flash Gaebal bonkyeokhwa," *Newspim* (December 7, 2017), https://www.newspim.com/news/view/20171207000187; and ZD Net Korea, "SK Hynix, 96dan 3D NAND Gaebal Sidong," *ZD Net Korea* (December 18, 2017), https://zdnet.co.kr/view/?no=20171218140855 (searched on January 15, 2022).

75   Toshiba and Western Digital were the first to develop 96-layer NAND flash memory. See Western Digital, "Western Digital Announces Industry's First 96-Layer 3-D NAND Technology" (June 27, 2017), https://www.westerndigital.com/company/newsroom/press-releases/2017/2017-06-27-western-digital-announces-industrys-first-96-layer-3d-nand-technology (searched on January 14, 2022).

76   Interview with Park Sung-Kye by Kim Dong-Won (June 9, 2022).

77   Forbes, "Some Flash Memory Keynotes," *Forbes* (August 13, 2018), https://www.forbes.com/sites/tomcoughlin/2018/08/13/some-flash-memory-keynotes/?sh=3bbb2e943202 (searched on August 22, 2022). Both Toshiba and Micron also mentioned the development of 96-layer NAND flash memory in the same keynote speeches.

78   SK Hynix, "SK Hynix Inc. Launches the World's First 'CTF-based 4D NAND Flash' (96-Layer 512Gb TLC)," *SK Hynix News* (November 4, 2018), https://news.skhynix.com/sk-hynix-inc-launches-the-worlds-first-ctf-based-4d-nand-flash-96-layer-512gb-tlc/ (searched on January 23, 2022). See also Techspot, "SK Hynix Launches 96-layer 4D NAND Flash," *Techspot* (November 6, 2018), https://www.techspot.com/news/77269-sk-hynix-launches-96-layer-4d-nand-ssd.html (searched on August 23, 2022).

79   SK Hynix, "SK Hynix Starts Mass-Producing World's First 128-Layer 4-D NAND," *SK Hynix Newsroom* (June 26, 2019), https://news.skhynix.com/sk-hynix-starts-mass-producing-worlds-first-128-layer-4d-nand/; and idem, "SK Hynix Unveils the Industry's Most Multilayered 176-Layer 4-D NAND Flash," *SK Hynix Newsroom* (December 4, 2020), https://news.skhynix.com/sk-hynix-unveils-the-industrys-highest-layer-176-layer-4d-nand-flash/ (searched on January 23, 2022).

80   Reuters, "SK Hynix Says Has Developed Its Most Advanced 238-Layer Storage Chip," *Reuters* (August 2, 2022), https://www.reuters.com/technology/sk-hynix-says-has-developed-its-most-advanced-238-layer-storage-chip-2022-08-02/ (searched on August 22, 2022).

81   Park Sung-Kye's curriculum vitae (2021).

82   Media SK, "Park Sung-Kye at SK Hynix, Who Lives in the Nanometer World," *Media SK* (January 18, 2013), http://mediask.co.kr/1115 (searched on January 18, 2022).

83 Dong-A Ilbo, "SK Nalgaedan Hynix, 2000eok Jeokjaseo 'Hyoja'ro talbakkumhaetda," *Dong-A Ilbo* (January 19, 2022), https://www.donga.com/news/Economy/article/all/20220119/111328304/1?ref=main; and Maeil Kyungje, "Chey Tae-Won, Imwonbandae-edo Gyeoldan ... Hynix 12jo beoleodeulineun Hoesaro Daebyeonsin," *Maeil Kyungje* (January 20, 2022), https://www.mk.co.kr/news/business/view/2022/01/57791/ (searched on January 25, 2022).

84 Interview with Park Sung-Kye by Kim Dong-Won (June 9, 2022).

8 Engineer-Entrepreneur
Chung Han

Chung Han is a successful engineer-turned-startup-entrepreneur, whose i3system develops sophisticated infrared image sensors for both the defense and the commercial markets. His unceasing efforts since the late 1980s to develop better infrared devices began to be noted in the early 2010s: in 2013, for example, *Dong-A Ilbo*, an influential South Koran newspaper, chose him as among the one hundred persons who would glorify South Korea in the 2020s.[1] In 2021, he was awarded the prestigious POSCO TJ Park Technology Prize for playing

> a key role in making [South] Korea the seventh nation in the world to mass-produce infrared imaging sensors. ... [Among the many infrared sensors Chung Han has developed,] the 12μm ultra-small infrared imaging sensor, which was recently developed with its own technology, is an essential part of self-driving cars that has been commercialized only in the U.S. and it is evaluated as a steppingstone for Korea to become a technology leader in the field of infrared image sensors.[2]

In a 2022 paper by two Polish researchers reviewing the performance of two different materials for infrared sensors, i3system's data were chosen to compare with those from more famous institutes or companies, such as the Jet Propulsion Laboratory, Naval Research Laboratory, Raytheon Vision System, Teledyne Imaging Sensors, and IRnova AB (Sweden).[3]

Chung Han is unique among Kim Choong-Ki's former students in several respects: his background and entire career path differ greatly from those of most of Kim's former students. He was the only one who entered Kim's laboratory after finishing military service, and he is a rare figure who began devoting himself to startups soon after graduation. Although Chung is neither a professor like his mentor Kim nor an industrial engineer like most of Kim's former students but rather an entrepreneur (or more specifically engineer-entrepreneur), he has made every effort to emulate Kim his whole life.

EARLY YEARS

Chung Han was born on November 12, 1960, in Daegu, a politically and economically important city in the southeastern part of South Korea. When he was just two years old, his father passed away. Chung recalls that his mother had a tough time

DOI: 10.1201/9781003353911-11

raising her three children and that he himself had to work to support the family when he was a middle school and high school student.[4] Nonetheless he studied hard and in 1980 succeeded in entering the College of Engineering at Yonsei University, one of the top three universities in South Korea. When he became a sophomore, he chose electrical engineering as his major. His college years were not, however, as smooth as those of most of Kim Choong-Ki's other students. He had to take several part-time jobs to meet his living expenses in Seoul. He also became deeply involved in the prodemocracy movement that swept South Korean universities during the 1980s. Between his financial circumstances and political activism, as well as his two-year compulsory military service, Chung left the campus for several years. It was during his military service that he realized that studying hard was the only way for him to overcome his present difficulties (Figure 8.1). So, when he finally returned to Yonsei in 1985, he was a totally different man: he paid less attention to political activities and devoted himself to studying electrical engineering. Chung remembers that he was usually one of the first students to arrive at the university library in the morning and the last to leave at night.[5] Since he couldn't earn enough money to pay his tuition and living expenses while studying hard, Chung often asked for a leave of absence, and his years in college eventually extended to nine years. His hard work and earnestness paid off. He passed the entrance examination for KAIST and became a graduate student there in the spring of 1989, to study electrical engineering.

At the age of 28, Chung was much older than his classmates in the department. He was also very unusual for having already fulfilled his duty of military service, since most students entered KAIST to receive a military exemption. Chung wanted to study

FIGURE 8.1   Chung Han is a very rare figure among Kim Choong-Ki's former students, in that he finished his compulsory military service before entering Kim's laboratory in 1989. Here Chung poses on guard on a remote mountain near the demilitarized zone (DMZ).

Source: Courtesy of Chung Han.

semiconductors under Kim Choong-Ki and went to his office for an interview. He recalls his first meeting with Kim:

> I entered his office and told him that I would like to become his student to study semiconductors. He said, "Mr. Chung is not so young to carry out physically hard work. What do you think you can do best?" I replied, "I can do whatever you ask me to do. I am good at physically hard work." He laughed loudly and accepted me as his student.[6]

In fact, Chung was an enigma at Kim's laboratory, at least in the beginning, because of his age. Most "senior" members of the laboratory, whether master's or doctoral candidates, were actually two to seven years younger than Chung. Since Kim's laboratory dealt with many toxic chemicals and very delicate facilities, the discipline was very strict, and the authority of senior over junior members was absolute. Therefore, these younger "seniors" were supposed to give Chung instructions and, if necessary, reprimand him. This was quite an uncomfortable situation in Korean culture, where a younger person must pay some respect to an older one by adding the proper prefix and/or suffix to almost every word. Chung overcame this difficulty by carrying out unpleasant and hard work as a "junior" member of the laboratory without complaint. As a robust man whose fitness had been increased by his time in the South Korean Army, he volunteered to do any physically hard work needed in the laboratory or by other researchers. Chung's sincere attitude impressed many researchers in Kim's laboratory, some of whom later joined his startup company, although a few seniors mistakenly regarded him as just physically strong but not so intelligent.[7]

Kim Choong-Ki suggested that Chung study the infrared image sensor. The infrared image sensor is the detector that reacts to infrared radiation. Between the two main types—thermal and photonic (or photovoltaic)—the photonic-type sensor is more sensitive but requires a very low temperature to reduce thermal noise. Since the photonic-type infrared sensor offers clearer images, it was widely used in the defense industry, and its technology was strictly controlled by a handful of makers in a few advanced countries. In 1988, Kim received a research grant from the Agency for Defense Development (ADD) to develop the infrared image sensor using mercury cadmium telluride ($Hg_{1-x}Cd_xTe$): mercury cadmium telluride (MCT) is widely used as the material for the photonic-type infrared sensor because of its tunable bandgap spanning the short-wave infrared to the very long-wave infrared (LWIR) regions.[8] Kim asked two of his students, Han Pyong-Hee and Moon Chan, to work on the project, and Chung became the third member of the team. Han was the first South Korean engineer who developed the MCT photovoltaic infrared image detector in 1988, just before Chung entered KAIST, and Moon was working on the improvement of the detector when Chung started working on the subject.[9]

Chung's major task was to design and construct a truly working photovoltaic infrared image sensor. He recalls that he often returned to his dorm room at 5 a.m. and got up four hours later, with the wake-up call from the laboratory at 9 a.m.[10] There were two major difficulties for him to overcome: poor conditions of the facilities and difficulties in securing the necessary materials. He solved the first problem by fixing the facilities himself, and the second by searching the street markets and neighboring laboratories. Chung's master's thesis, "Fabrication and Characterization of Large Area $Hg_{0.7}Cd_{0.3}Te$ Photovoltaic Infrared Detector," clearly shows how he solved various problems step by step to achieve the planned goal (i.e., $R_0A = 200$ Ω–cm^2 with the detectivity greater than 1.0×10^{11} cm Hz$^{1/2}$/W).[11] In the conclusion, he confesses that he had to rely on his own trial-and-error efforts to get the result because most of the necessary technologies were secret: he even had to change the photomask five times to get the targeted photovoltaic sensor.[12]

In his second year as a master's candidate, Chung changed his advisor from Kim to Lee Hee-Chul. The reason was not because either Chung or Kim was not satisfied with the relationship but because Kim had decided to transfer many of his graduate students to Lee. Lee was a professor of semiconductor devices at the Korea Institute of Technology (KIT), a sister college of KAIST established in 1985 in Daejeon.[13] The institute, which aimed to provide KAIST with well trained college graduates in science and technology, was formally merged with KAIST in 1989, when KAIST began to move its campus to the Daejeon area. One delicate problem of the merger was how to unite two independent departments in the same area: until 1989, KAIST was a graduate-only institute, whereas KIT was an undergraduate-only college. Sharing graduate students was a most significant issue that involved the vital interests of both professors and graduate students. Kim became a model by transferring some of his students to his new colleagues, which was certainly not an easy decision but a goodwill gesture to the new members of the Department of Electrical Engineering. Chung's new advisor, Lee Hee-Chul, was a specialist in semiconductor devices but hadn't worked on the infrared image sensors until then. Since all semiconductor professors shared the same facilities in the same building and Kim did not stop coaching Chung, Chung actually had two advisors from 1990 on. This unusual but beneficial relationship continued until Chung received his doctoral degree in 1996.

In the fall of 1990, as his master's work was coming to an end, Chung began to look for a suitable job. He first sent his application to the Agency for National Security Planning (the South Korean version of the CIA) because he was very interested in this institute.[14] He received a rejection notice in March 1991 because of his past record in the prodemocracy movement.[15] He tried other electronics companies, but their regular hiring seasons were already over. He then visited Kim Choong-Ki for advice, and Kim immediately hired him as an operator in his laboratory. Since Chung had finished his military service before entering KAIST, it was no problem for Kim to hire him. (For other KAIST graduates with either master's or doctoral degrees, some restrictions applied when seeking jobs at KAIST or other places to compensate for their earlier exemption from military service.) Within a few months, Chung changed his status to that of a doctoral candidate. Chung later confessed that he had wanted to remain Kim's student but had faithfully followed Kim's advice to become Lee Hee-Chul's instead.[16] Kim had continued paying attention to Chung's progress, and Chung considered Kim to be his second advisor.

Like most of Kim's other students, professional know-how was not the only thing that Chung learned from Kim at KAIST. He remembers:

While I was a doctoral candidate at KAIST, I had learned from Kim how to behave as a senior person. One year, when the Korean Harvest Festival [the Korean version of the Thanksgiving holiday] approached, he received a bottle of expensive whisky from a company. He kept it for a while and donated it when the laboratory had a group dinner. That incident gave me a new insight into the world, as well as an example of how to behave as a senior person. Because of that memory, I always donate all gifts that I received to the employees of my company.[17]

The subject of Chung's doctoral thesis was same as that of his master's: the photovoltaic infrared image sensor. Entitled "The Fabrication of HgCdTe Photovoltaic Infrared Detector and Enhancement of Its Performance by ECR Plasma Hydrogenation," his doctoral thesis aims to develop a photovoltaic infrared image sensor with "good characteristics." For this, Chung first focused on finding the sources of leakage current and then suggesting effective solutions. He also proposed a new method for measurement of the steady-state effective minority carrier diffusion length in a HgCdTe photodiode, whose results were later published in the *Japanese Journal of Applied Physics*.[18] The highlight of his thesis was Chung's development of the hydrogenated photodiodes created by electron cyclotron resonance (ECR) that "showed increased photocurrents, improved ideality factor, and higher RoA."[19] He later boasted that this was the first commercial-grade photovoltaic infrared image sensor developed in South Korea.[20]

The opening of the Center for Electro-Optics at KAIST in December 1994 was important to Chung's future career. Funded by the ADD, the center aimed to develop three key technologies for South Korea's defense industry—thermal (infrared) imaging, fiber optics, and lasers.[21] For the first of these, the thermal infrared image research team intended to develop the focal plane array (FPA) device using HgCdTe, whose resolution is 256×256 pixels. The team also worked on developing a small TFT-LCD (thin-film-transistor liquid-crystal display) screen, as well as an automatic recognition algorithm. As *the* specialist in the infrared image sensor, Chung participated in the project and played an important role in laying the foundation of the center during his last year at KAIST. He also met many other specialists on the subject from other universities and research institutes. Some of those who worked on infrared image sensors at the center—Kim Byung-Hyuk, Kim Young-Ho, Bae Soo-Ho, Yoon Nan-Young, and Yang Ki-Dong—would later join Chung's startup company. His experience at the Center for Electro-Optics gave Chung two valuable opportunities: for the first time, he experienced the entire process of making the infrared image sensor, including the screen and algorithm, and he began to build a network for his future startup company.

In 1996, Chung joined Hyundai Electronics. Chung was an industry-sponsored student between 1992 and 1996, and Hyundai Electronics paid all his tuition, fees, and living expenses until his graduation. Some of Kim's former students, including Oh Choon-Sik, Koh Yo-Hwan, and Cho Byung-Jin (profiled in Chapter 5), had already worked in the company. Chung's major task was to do research on the development of DRAM chips. He remembers that he attended many conferences both in South Korea and abroad during this period, which not only broadened his perspective but also strengthened his interest in the infrared image sensor.

FIRST STARTUPS

In January 1998, Chung quit his job at Hyundai Electronics to launch his first startup. The idea of the startup first sprang from his friend and colleague at Hyundai Electronics, Lee Ho-Joon, who had also been educated at KAIST under Kim Choong-Ki. Chung remembers, "When we both were graduate students, we promised each other that if one of us began a startup, the other would join it without reservation."[22] The division at Hyundai Electronics in which Lee had worked was scheduled to be abolished in early 1998 because of the restructuring of the company during the economic crisis. Since neither Chung nor Lee had enough capital to start a company, an investor was added to make a three-person partnership. These three launched Seju Engineering to develop a sophisticated miniaturized reserve battery for bombs. The partnership ended very quickly, however, largely because of their very different business perspectives, especially between Chung and the investor. Chung sold his shares and departed the company within a few months. Lee soon left the company also and established his own startup, EMC Microsystem, to produce various electronic sensors, such as breath-analyzing sensors.[23]

On July 11, 1998, Chung established another startup, Hankum Engineering, at the Technology Business Incubator Center (TBIC) on the KAIST campus (Figure 8.2). The TBIC had first opened in 1994 to "support researchers and students in the Daejeon area to begin startups, utilize technologies developed by research institutes, and encourage small- and medium-sized engineering companies in the area."[24] The center provided the startups with the necessary space, initial matches with venture capital, legal and marketing services, workshops, and advertisements through its website

FIGURE 8.2   Chung Han started his own startup, Hankum Engineering, in a small room on the KAIST campus. KAIST's incubator program (TBIC) provided him with the space, initial funding, and necessary services.

Source: Courtesy of Chung Han.

and newsletter. Three years after they first enter, all tenant startups are examined for "graduation": about 500 startups have graduated from TBIC as of 2021. Chung rented a small space at the newly opened LG Hall, where his mentor Kim Choong-Ki had an office, and started to develop the infrared image sensor. He paid much attention to securing talented engineers, and several KAIST graduates joined the company. Chung assigned them to the research and development section.

Chung, of course, wanted to concentrate on developing his beloved infrared image sensors, but the overall situation in the next few years was not ideal for him to advance this ambitious plan. First, the South Korean defense industry believed that the infrared image sensor technology was too advanced for any South Korean company to develop. From the point of view of the South Korean defense industry, it could not trust the quality of the infrared image sensors that a South Korean startup developed and manufactured. This prejudice lasted even after the i3system (which Hankum Engineering adopted as its new name in 2003) manufactured them. Second, a spectacular failure by KC Tech in the mid-1990s seriously discouraged any hope for the development of the infrared image sensor for some time. Established in 1987, KC Tech manufactured key components and parts for the semiconductor industry.[25] The company started to develop infrared image sensors for the military in the mid-1990s with a massive grant from the ADD but soon gave it up. The ADD was so disappointed that it sharply cut the budget for the subject for the next few years. Third, the economic crisis that had begun in the fall of 1997 was still badly affecting the South Korean economy: the interest rate for loans was 20% or higher, and even that rate was not easy to secure for a startup company like Hankum. Under these circumstances, Chung could not concentrate solely on the development of the infrared image sensor but had to search out and produce other semiconductor devices for the company's survival. He remembers, "I finally realized that the real world is quite different from the world that I had expected from the laboratory."[26]

For the next four years, Hankum developed various semiconductor parts—"anything profitable," as Chung recalls—for larger semiconductor companies or research institutes.[27] KC Tech became Hankum's primary buyer during this early period, but many other customers also gave the company small orders. An interesting device that Hankum developed in this period was the cryptograph that the Agency for National Security Planning ordered: the machine was so successful that the developer resigned from Hankum to begin his own startup.[28] Meanwhile, KAIST's semiconductor laboratory generously allowed the company to use its facility during its off-hours.

The first commercially successful product Hankum developed was therefore not an infrared image sensor but the capsule endoscope, a small device that "visually examines the midsection of the gastrointestinal tract, which includes portions of the small intestine. ... Inside the capsule is a tiny wireless camera that takes pictures as it passes through the small intestine. Images are transmitted to a recording device worn on a belt around the patient's waist."[29] First developed by Gabbi Iddan and Paul Swain in the mid-1990s, it soon became widely used in the medical community. The size of the first-generation capsule endoscope was, however, too large to easily swallow, and the images from the capsule were not clear enough.

In 1999, the South Korean Ministry of Science and Technology launched its ambitious Frontier Technology in the 21st Century program and chose development of the intelligent microsystem as its first project.[30] The capsule endoscope was included in this first project, and a consortium was organized for it: a microsystem research team from the Korea Institute of Science and Technology (KIST), a medical team from Yonsei University, and Hankum were selected as the three major participants. The original mission of Hankum was to develop a more efficient and smaller communication device for the capsule. This device was technically the most difficult part of the microsystem, largely because the limited choice of wavelengths affects the life of the battery.[31] The company, however, soon extended its role to developing other core parts and assembling them. In January 2003, the Ministry of Science and Technology proudly announced that the intelligent microsystem development team had succeeded in developing "the MIRO [or Miro], the world's smallest capsule endoscope with much clearer images, ... whose size is just 10 mm (width) × 25 mm (length)."[32] It also emphasized that the price of MIRO was about one-third that of the recently developed Israeli endoscope and that MIRO had passed all medical experiments successfully. South Korean media reported extensively on this success, and Kim Tae-Song of KIST, who had been the project director, was highly praised.[33]

The real winners, however, were Chung and his company, Hankum (Figure 8.3). Hankum's engineers learned all the details of the capsule endoscope and doubled their efforts to commercialize it. As the popularity of and demand for the capsule endoscope rose quickly, the profit and fame of Hankum did likewise. For the first time since the beginning of the company, the name "Hankum" began to appear in South Korea's nationally circulated newspapers. Chung poured the money that the company earned from this first success into hiring more talented researchers, often offering much higher salaries than usual or other incentives to attract them.[34]

Chung, however, still desperately wanted to concentrate on the development of the infrared image sensor. Between 1998 and 2003, Hankum experimented with and developed several infrared image sensors, such as a 256×256 pixel infrared image sensor (1999), the first infrared cooled package (1999), and the second infrared cooled package (2000).[35] The company also obtained a few official certificates from military agencies. But Hankum had not yet developed any real commercial-grade infrared image sensors that could compete with other existing products manufactured in advanced countries, nor did it possess the necessary facilities and space to manufacture them. Chung finally decided to sacrifice the golden opportunity to become one of the leading manufacturers of the capsule endoscope in the world. He sold some key patents and rights to the capsule endoscope to a new startup company, IntroMedic, in order to raise the capital needed to develop the infrared sensor.[36] This was not an easy decision for Chung or for Hankum, but he pushed ahead in his chosen direction. To make the change more permanent, Chung decided to change the company's name and asked its employees to suggest suitable ones. After some discussion and voting, the new name "i3system, Inc." was elected in September 2003: "i3" means the entirety of "intelligence, image, and information."[37]

The capsule endoscope, however, continued to be a trademark of i3system for the next few years. Despite selling its key technology to IntroMedic, i3system remained the best company to develop and manufacture the capsule endoscope. i3system filed several new patents for capsule-endoscope-related technology up until 2013, and the capsule endoscope continued earning hard cash for the

매일경제(03.1.29)

세계최소 캡슐 내시경 개발

과기부 개발사업단 영상 선명도등 성능도 뛰어나

김태송 박사

경 한 박사

송시몰 교수

FIGURE 8.3 The capsule endoscope was the first hit item of Chung Han's company Hankum Engineering. **Upper left**: When the capsule endoscope was first developed by the joint research team of KIST–Yonsei University–Hankum, it was widely publicized by the South Korean media. **Upper right**: The small size of the capsule endoscope was often compared to the size of a US penny. **Bottom**: The development of the capsule endoscope continued after Hankum changed its name to i3system and concentrated on the development of infrared sensors. On October 25, 2007, Chung had the opportunity to explain the most advanced capsule endoscope to South Korean President Roh Moo-Hyun and his wife.

Sources: *Maeil Kyungje*, January 29, 2003, A42; *Topclass*, March 2008; and The President Roh Moo-Hyun Archive, registration number 54243.

company. It also received large research grants from the Ministry of Trade, Industry and Energy to develop communication technology for the capsule endoscope until 2009: for example, the company received about $2.3 million between 2003 and 2006 and another $3.8 million between 2007 and 2010.[38] Chung and his research team even presented a paper, "Micro Capsule Endoscope for Gastro Intestinal Tract," at the Twenty-Ninth Annual International Conference of the IEEE Engineering in Medicine and Biology Society in 2007.[39] When South Korean President Roh Moo-Hyun and his wife attended the 2007 Future Growth Engine Exhibition on October 25, 2007, Chung had the opportunity to explain i3system's capsule endoscope to them (see Figure 8.3). In 2009, several newspapers reported that the company had succeeded in developing a special capsule endoscope to examine the throat (esophagus) that is technically more complicated than ordinary capsule endoscopes.[40] It was only in the early 2010s that i3system's infrared sensors and X-ray detectors finally replaced the capsule endoscope as the company's major products and icons.

I3SYSTEM

The renamed company, i3system, met many difficulties in the beginning, especially during its first six years. However, this was also the period when the company expanded rapidly, not only adding a new factory building (Figure 8.4) but also recruiting more researchers for research and development of the infrared image sensor.

The beginning seemed rosy. As soon as the new company started, Chung succeeded in securing a $2 million research fund from the ADD to develop the infrared image sensor. To this sum, he added all the money Hankum had earned and then purchased the necessary facilities at a bargain price from KC Tech, which had given up on the development of the infrared image sensor and desperately wanted to sell its facilities. In October 2003, just one month after the inception of the new company, Chung

FIGURE 8.4 Left: Chung Han and his guests, including Kim Choong-Ki (second from right in the left photo), celebrating the opening of the company's own building in September 2003. With the opening of the building, Chung changed the company's name from Hankum to i3system, Inc. and began to develop infrared sensors in earnest. **Right**: Researchers at i3system working in a cleanroom.

Source: Courtesy of Chung Han.

and his researchers started developing i3system's first "real" infrared image sensor, a two-dimensional "cooled" infrared detector. Since the research funds came from the ADD, the choice of the "cooled" type of sensor was natural. The cooled-type infrared sensor not only provides much clearer and more accurate images but also reacts more quickly than an "uncooled" one. For these better results, however, the cooled-type sensor requires cryocoolers that lower the sensor temperature into the cryogenic level (below −150°C, −238°F) to reduce thermal noise. It is more sophisticated and expensive than the "uncooled" type infrared sensor, and only the military has widely used this "cooled" infrared sensor, regardless of its much higher price tag. In other words, this was one of the most tightly controlled defense technologies. The South Korean military had depended entirely on the import of cooled infrared sensors from one of six countries that had manufactured this type of sensor. Chung and his i3system entered this uncharted marketing territory that no South Korean company had ever broken into or had posed a challenge to before.

The progress of i3system was painfully slow, however, which invited severe criticism from the ADD. Chung remembers that some of those who visited the company to monitor its progress said, "A small startup company like yours cannot develop the cooled infrared sensor," "I told you that it is impossible," or "Why don't you stop now?"[41] The agency then pressed him to contact an Israeli company that had supplied cooled infrared sensors to the South Korean military to find out whether or not that company would employ i3system as a subcontractor. Chung traveled to Israel in early 2004 to visit the company, but the negotiations ended quickly: the Israeli company demanded an extravagant price for the transfer of technology, with many restrictions.[42] On the flight back home, Chung resolved again to develop the cooled infrared image sensor himself. The researchers at i3system redoubled their efforts to speed up the development, and Chung himself almost lived at the company facilities for several months.

Chung and his researchers had developed a prototype of the cooled infrared image sensor in May 2004, when the deadline approached for the company's midway report to the ADD. However, the prototype did not work properly, and the researchers found that the error might have resulted from the connection between the sensor and the signal system when the sensor was cooled at −190°C. They tried everything to fix it but failed. Chung then visited the agency on May 29, two days before the deadline, to ask for an extension of the deadline for a few more weeks, but the ADD flatly refused his request. He remembers: "I was so depressed at that moment that I did not return to the company from the agency but drove my car to the beach on the East Sea [about 150 miles from the company]."[43] If i3system could not solve the problem within two days, the Agency would stop providing funding, and all efforts would become nil instantly. Chung stared at the sea for a few hours, considering several options, and finally returned to the company. When he returned, a miracle awaited him. The researchers reported that the problem had been "mysteriously" solved and that the prototype detector had produced the expected results. On May 31, i3system proudly submitted its midway report, along with its prototype. This success in 2004 not only provided researchers at i3system with much needed confidence but also promoted a positive image of i3system at the ADD and within the defense industry.

More successes followed. In January 2005, the i3system succeeded in acquiring its first 320×240 pixel image using the focal plane array (FPA) or staring array. The FPA is "a special type of bolometer and is used as a thermal detector in infrared cameras. The FPA is based on a series of small thin-film bolometers arranged in a matrix in the focal plane of the detector. This results in a planar pixel sensor, allowing an improved price-performance ratio in various applications."[44] Its applications include missile or related weapon-guidance sensors, medical imaging, infrared astronomy, and manufacturing inspection. The development of FPA and the following elaboration of it indicated that i3system was now approaching being able to manufacture and supply commercial-grade cooled infrared sensors for the defense industry. One serious problem for both Chung and i3system was that this success would not produce any profits until the company could actually manufacture and supply the cooled infrared image sensors—and how could the company maintain its balance sheet until then?

To increase the company's profit margin while developing the cooled infrared image sensors, Chung decided to develop two less-sophisticated devices simultaneously. The first was "uncooled" infrared image sensors, which can operate at normal temperatures and are much cheaper than the cooled ones, but offer poorer-quality images in almost every case. For example, an uncooled infrared image sensor can be used for night vision or for surveillance, whereas a cooled infrared image sensor is required for missile guidance or for the gunner's primary sightings in an armored tank. In April 2006, i3system started developing high-quality uncooled infrared image sensors (quarter video graphics array, or QVGA, level), and successfully produced a series of uncooled infrared image sensors for the market.[45] They soon became popular because they were especially "robust" and therefore "have high immunity for shock environment."[46] The second device was the line-scanning X-ray detector for dental panoramic or cephalometric applications, which "ensures the sharpest images among the current products in the market" but is also "compact in size and easy to install."[47] The company began to manufacture two different types in the spring of 2007, and they soon dominated the domestic market. These two lines have been the major products of i3system, Inc. ever since. The capsule endoscope added much-needed profit to the company, too. Revenues from all these less-sophisticated devices allowed Chung to continue developing his cooled infrared image sensors without interruption.

In May 2009, the South Korean Defense Acquisition Program Administration determined that i3system's two-dimensional cooled infrared image sensors met the standard for specific military uses. Through Samsung Thales (a munitions company that is part of the Samsung conglomerate), i3system then received its first order to supply the military with cooled infrared image sensors that would be installed in the newest tanks, at a cost of more than $20,000 each, by the end of September.

Everything seemed to be going well, but a final hurdle was awaiting Chung. In early September 2009, he and his researchers discovered that the defect rate was unusually high: only two out of twenty sensors (10%) operated correctly. They tried hard to fix the errors but couldn't find the source of the problem. Chung begged both the Defense Acquisition Program Administration and Samsung Thales to extend the deadline, which was moved slightly, to October 24. He remembers that there was nothing more he could do except pray.[48] Just like five years before, in May 2004, the problem was mysteriously solved: one day in October, eighteen out of twenty sensors (90%) operated perfectly. Chung and his researchers later discovered that they had failed to adequately control both temperature and humidity in the newly constructed air-room, which was not tightly insulated. The hot and very humid weather in August and September may have influenced the sensitive facilities, which were then able to operate normally as the weather became cooler and much drier in October, allowing the defect rate to drop sharply. Chung later recalled that he learned a lesson from this last trouble, which might have shaken the company's future seriously: "If I become comfortable, heaven will impose a severe punishment."[49]

Although this success of i3system in 2009 was not widely publicized because of its connection with hi-tech weapons, it was truly a great leap for the South Korean defense industry. Until then, South Korea had been completely dependent on imports of the cooled infrared sensors needed for its tanks, guided missiles, fighter planes, and other highly sophisticated weapons. Only a select few companies in six countries (the US, Germany, Great Britain, France, Israel, and Japan) manufactured this core sensor, and they also controlled the market as they wished. If there were any problems with the imported sensors, there were only two ways to fix them: either they were sent back to the manufacturer for repair and the military had to wait until they were returned; or the manufacturer sent their specialists to fix them. As a result, both the initial price of the sensors and the maintenance costs were very high. For a nation like South Korea, which has been technically on "ceasefire" status with North Korea since 1953, this was not just an inconvenience but a matter of life and death. The success of i3system in 2009 made South Korea the seventh country in the world to manufacture the coveted cooled infrared image sensors, not only freeing it from the shackles of foreign companies but enabling it to export this profitable item to other countries. Chung humbly recognized that i3system's success owed a great deal to the government research institute (ADD), academy (KAIST), industry (Samsung Thales and, later, Hanwha Systems), and the South Korean government.[50] However, it was Chung and his i3system that combined these players together, and actually developed and manufactured the cooled infrared image sensors (Figure 8.5). The Ministry of Economy duly appointed i3system a "defense contractor" in July 2010.

The successful development of the cooled infrared image sensors for military use in 2009 not only brought fame to the company but also helped improve its finances. First of all, several government agencies began to present awards to Chung's i3system from 2009 on. The earliest was an award by the head of the ADD on December 31, 2009, for the company's contribution to "the development of two-dimensional infrared detectors." On June 28, 2011, i3system received a presidential citation for its contribution to "the national security." The company received six more citations from government agencies in 2011 alone. More honors followed over the next ten years. Some notable examples are: the Ministry of Defense selected i3system as the "Outstanding Defense Contract Company" in 2012, and presented another award for its contribution to the development of the localization of defense-related parts in 2016; and LIG Nex1, South Korea's major defense company, awarded it the "Best Partner Prize" in 2013.[51] Although the detailed reasons for these awards were not revealed or explained in the mass media, they nonetheless helped spread the company's name in wider industry circles. The South Korean public began to connect i3system with its (cooled) infrared image sensor in early 2013, when the sensor was included in a South Korean satellite that was launched by the nation's first rocket. *Dong-A Ilbo*, for example, reported that "The infrared camera developed by i3system will take photos of the Earth's surface according to the temperature differences, which will be used for weather forecasting, catastrophe detection, and [detecting] changes of ocean temperature. i3system modified the infrared camera for the military for this specific purpose"; *Jeonja Shinmun* interviewed the head researcher of i3system, who had developed the infrared camera supplied to the satellite; and *Asia Kyeongje* pinpointed i3system as representative of the small or medium-sized companies that had participated in the satellite project.[52]

The company's improved finances, however, were a more important result of its 2009 success. Between 2010 and 2015, i3system's sales and profits increased

InSb 1280×1024 (10um)

Related images

FIGURE 8.5 One of the most sophisticated cooled infrared sensors, the InSb 1280x1024 (10μm) type. The bottom photos show sample clear images taken in the dark.

Source: i3system's website.

steadily, largely because of its sale of cooled infrared sensors for military use: the total sales in 2009 were under $5 million but rose to over $10.5 million in 2010, $18 million in 2011, $21.5 million in 2013, and $32.5 million in 2015; its profits also rose steadily, from just over $750,000 in 2009 to $1.1 million in 2010, $1.25 million in 2011, $2.66 million in 2013, and $5.4 million in 2015.[53] The capital of the company therefore increased from just under $500,000 in 2010 to $1.4 million in 2014. The number of permanent employees also rose, from 124 in 2009 to 239 in 2015. Since about 70 to 80% of its sales during the period went to a handful of larger defense contractors (Samsung Thales, LIG Nex1, and the ADD were the three largest clients), the items sold must have been mostly cooled infrared image sensors for military use. Non-military devices—uncooled infrared sensors, X-ray devices, and capsule endoscopes—contributed only slightly to the growth of sales and profits during this period. The improved finances also resulted in upgrading the credit rating of the company, from BB+ in 2010 to BBB- in 2011, BBB in 2013, and BBB+ in 2015, which enabled i3system to attract more investment and to borrow money from the bank.

Chung invested both the profits that the cooled infrared sensors had brought to the company and the new investments into the expansion of i3system's production lines and the development of new products.[54] In October 2010, the company began to construct a production line for the cooled (QVGA level) infrared image sensor, and a year later, it built another line for the uncooled (QVGA level) infrared image sensor. In 2012, it finished the development of both cooled and uncooled infrared image sensors with much sharper resolutions (VGA level). The next year, the company launched the development of more sophisticated infrared image sensors (both cooled and uncooled types) and also a new type of sensor—the short-wave infrared (SWIR) sensor. These three types of sensors—cooled and uncooled infrared image sensors, and SWIR image sensors—along with X-ray sensors would become the four major products of the company after it went public (Figure 8.6).

Based on its rapid and successful growth between 2010 and 2015, Chung finally decided to open i3system to the public in July 2015. For the first time in its history, the company was in the public spotlight. Several mass media outlets reported extensively on the company's IPO, often including interviews with Chung. *Hankuk Kyungje*, an influential financial newspaper, reported as follows:

i3system, the only domestic manufacturer of infrared image sensors, is going public at the Korean Securities Dealers Automated Quotations (KOSDAQ) on July 30. ... Established in July 1998, i3system successfully developed the infrared image sensors for the first time in South Korea, which made the country the seventh in the world to possess the technology. This technology has been applied solely to military uses until now but extends its application to civilian uses such as security cameras, smartphones, or automobile cameras. i3system plans to export its products and also to explore new domestic markets for civilian uses. ... Chung Han, CEO of the company, said, "Although company started the development of the infrared sensor about thirty years later than other countries, it now possesses the technology to compete with them with confidence."[55]

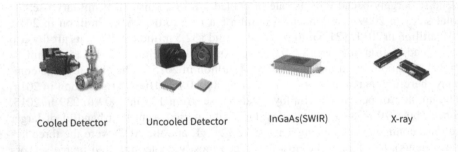

Cooled Detector   Uncooled Detector   InGaAs(SWIR)   X-ray

FIGURE 8.6 i3system's four major product lines in 2022. Although cooled infrared sensors for the defense industry remain the company's major product, it began diversifying its product lines in the late 2000s. Since 2015, when the company went public, it has accelerated its development of less sophisticated detectors for the commercial market.

Source: i3system's website.

The IPO was quite a success, and the initial rate of subscription reached 1,506:1. Chung secured about $1.7 million from the IPO and invested it in the development of mass production of the infrared image sensor that would be installed in the new guided weapons.[56] The company continued growing steadily over the next five years: its capital, sales, and profits, for example, almost doubled over the next two years (to $3 million in capital, $58 million in sales, and $7.7 million in profit), and a new factory building was added for the mass production in November 2016. In short, by the end of the 2010s, i3system was no longer just a promising startup but a medium-sized high-tech company.

The most notable change after the IPO in 2015 was the changing distribution of the company's products between military and non-military. From 2016 on, as the above-quoted article indicates, i3system accelerated its development of more products for non-military uses, while the portion of military-related products began to decline. The number of contractors or distributors also increased as the company manufactured and supplied more uncooled infrared, X-ray, and SWIR sensors. For example, in November 2015 i3system began to produce its first (uncooled) infrared camera, "Thermal Expert," for smartphones, and by 2021 it manufactured seven different versions of cameras for smartphones and stand-alone use.[57] Both X-ray and SWIR sensors have also continued to boost the company's efforts to diversify its sales since 2015: the company, for example, has supplied its direct conversion type X-ray sensors to major medical equipment companies, including Ray Medical, since 2013. As its non-military products increased, i3system's exports rose sharply beginning in 2015; in 2017, for instance, 22% of its total sales came from abroad.[58]

Maintaining an ideal balance between military and non-military products was, however, more difficult than Chung had expected. After its peak in both sales and profits in 2017 and 2018, i3system had hard time in 2019, when both its sales and its profits declined sharply (to $45 million and $155,000, respectively). The primary reason was the huge decrease in exports and in military demand that year.[59] Chung decided to pay more attention to the development of various uncooled types of infrared sensors that have more applicability to diverse areas and whose market was growing rapidly. For that purpose, a new series of sensors with higher resolution (LWIR 12μm, up to 1024x768), and also "with a reliable, small and lightweight, low input power focal plane array," was developed and added to the product lines.[60] These sensors aimed to supply surveillance/security or night vision when installed on an automobile, drone, Internet of Things (IoT), or smartphone.

Ironically, the outbreak of the COVID-19 pandemic in early 2020 gave i3system an opportunity to reverse its 2019 slump. The soaring demand for both uncooled infrared sensors and infrared cameras in South Korea and abroad meant that the company went into "full operation 24 hours a day."[61] Despite difficulties in transportation and the sharp drop in sales of other products during the pandemic, i3system recovered from the worst of its 2019 slump and recorded healthy sales and profits in 2020 ($57 million and $2.6 million, respectively).[62]

As the development of driverless automobiles began to accelerate at the beginning of the 2020s, the demand for excellent (uncooled) infrared sensors for night vision increased rapidly and the market's interest in i3system amplified accordingly. *Aju Kyungje* reported in January 2022 that

> The night vision device, which has already been installed in some high-end automobiles, such as Mercedes-Benz, BMWs, and Cadillacs, is an essential part for automatic driving. Since i3system has been the number-one supplier of the infrared sensor in the domestic market, the company may become the major beneficiary if automatic driving becomes commercialized.[63]

In his CEO's message on the company's website, Chung himself confirms that i3system will strengthen this direction: "We will keep working on not only enhancing untact [i.e., touchless] and autonomous driving technology but also competitive price[s] in future so we can contribute [to] both defense and commercial industries even more than now."[64]

The rapid and successful growth of i3system since 2010, and especially after 2015, not only made Chung Han famous as a successful startup entrepreneur but also assigned him a new mission—being a connector between civilian startups and the military. Until the early 2010s, Chung's name had occasionally, but not frequently, appeared in the mass media when i3system's capsule endoscopes or infrared image sensors were mentioned. This changed suddenly when, in 2013, *Dong-A Ilbo* selected him as one of the "100 Persons Who Will Glorify South Korea Ten Years Later": for the first time, his solo photo appeared in an influential newspaper, accompanied by an extensive interview.[65] From then on, Chung frequently appeared in influential newspapers and mass media (including on two TV shows) as the most successful example of the founder of a defense startup: for example, YTN (a South Korean 24-hour news channel) broadcast a special 30-minute program on i3system and Chung; and the state-run Korea Broadcasting System (KBS) selected i3system as one of the twelve most promising medium-sized companies, and broadcast a special show about i3system on December 22, 2019.[66]

However, Chung's increasing personal fame in the 2010s was less important than his new role as a bridge between startups and the military. Until very recently, only a select few large companies were involved in the defense industry in South Korea, and very few startups have contributed to it. The interaction between military and non-military technologies remained very low. The growth of i3system, which manufactured both military and non-military products, was, therefore, a perfect example for the future. Since the mid-2010s, Chung has been invited to speak at many defense-related committees as a representative of startups, where he has highlighted the important role of startups in the defense industry, as well as their difficulties.[67] He has also been a frequent guest at meetings where representatives of government research institutes, the military, and civilian companies discuss the cooperation among them. In the summer of 2019, for example, he attended such a meeting and remarked as follows:

> In order to raise the defense industry to a major national industry, it is compulsory to convert it into an export-oriented industry. To do so, the defense industry must strengthen its competitiveness in the domestic market and also prepare detailed strategies for individual items for export. In the defense industry, the high quality of materials and parts is more important than what they cost. So, we must concentrate our efforts on the development of advanced technologies in order to secure our competitiveness.[68]

Chung's tenacious devotion to the development of infrared image sensors for more than thirty years has thus made him a quintessential representative of startups.

SECRETS OF CHUNG'S SUCCESS AND A NEW CHALLENGE

In two interviews with Chung Han, he repeatedly emphasized that he owes a great deal to both his wife and his i3system employees.[69] He credits his wife's "disinterest based on trust in her husband" as a secret of his success that enabled him to concentrate all his efforts on developing infrared image sensors. He recalls that she didn't object when he decided to quit his stable job at Hyundai Electronics in order to begin his first startup. Nor did she complain when Chung was so busy he could not be with her when she delivered any of their three children. Instead, she has always supported him to realize his seemingly impossible dream. Chung equally thanks his employees at i3system. Unlike most other startups, i3system's employees have seldom left the company, and many original employees are still working there. Considering that about one-third of the company's employees have higher degrees in either science or engineering and that i3system was a relatively small company for its first fifteen years, less than 10% turnover is a quite remarkable result. Since the company has developed and manufactured many sensitive and even secretive devices, this low turnover was an important element in i3system's success.

It is, however, Chung's leadership that must be selected as the decisive factor in the success of i3system. Since 1998, Chung's whole life has been devoted to the growth of this startup company. His obsession to develop the infrared image sensors was the most important driving force of the company from the very beginning. If he had continued to focus on the capsule endoscope in the early 2000s, giving up the development of the infrared sensors, the company might have grown steadily, and Chung would have easily become richer. Instead, he chose the more difficult route, and he and his company experienced many hard times until 2010. Those few employees who didn't share Chung's dream or philosophy left the company, while those who remained supported his policy of investing all secured money either in the development of the infrared sensors or in hiring more R&D personnel. Employees of i3system know very well that their CEO neither evolved into an arrogant dictator nor behaved like an overnight millionaire but remained a researcher: he therefore prefers being called "Dr. Chung" rather than "CEO Chung," and pays more attention to research and development than to the management of the company. The researchers at i3system often complain that Chung must pay more attention to the management of the company and leave research and development to them.[70] In short, Chung, i3system, and infrared image sensors have been one since 1998.

It is very interesting to note that Chung continuously expresses his deepest gratitude to Kim Choong-Ki, who, he said, taught him how to behave as a leader, or senior. Although Chung had a different professor as his official advisor, he has, in fact, become one of the most faithful followers of Kim. For instance, he has tried hard to emulate Kim's hallmark disinterest in money and lack of greed. In his second interview with the author, he said that his wife often teases him for being a sincere "disciple" of the Kim Choong-Ki religion.[71] Chung even served for many years as the secretary of the informal gathering known as "Banjomo" (Those Who Like Semiconductors), whose members are former students of Kim Choong-Ki or of the late Han Chul-Hi.[72]

A serious new challenge awaits Chung. Just as Kim became a very successful engineering professor by training so many semiconductor engineers who grew to be "more" successful than Kim himself, so Chung needs to recruit an outstanding successor (or successors) who will further nurture i3system to become one of the top infrared image sensor companies in the world. Such growth is a very difficult task in South Korea because its business culture is not particularly friendly toward small and medium-sized companies. Moreover, i3system is not just another successful medium-sized company but a unique one with a specific technology— the cooled infrared sensor—that has already made South Korea the seventh country in the world able to produce and market it. There are two possibilities. The first is that i3system becomes a subsidiary of a big defense company, as in the case of other countries. If this occurs, the company may soon lose its identity as well as its unique enthusiasm to become one of the best manufacturers of infrared sensors in the world market. For the last twenty years, Chung has continuously reinvested what the company has earned so as to develop better infrared sensors, but there is no guarantee that a big defense company, once it had absorbed i3system as its subsidiary, would do so. In the worst-case scenario, even i3system's developed infrared sensor technology could disappear. The second, more preferable solution is to find a successor who possesses a good knowledge of semiconductors, enthusiasm to continue Chung's legacy, and proper leadership skills. It seems, however, almost impossible in South Korea to find such a person and persuade him or her to head this medium-sized company.

Perhaps Chung's most difficult challenge is just beginning.

NOTES

1   Dong-A Ilbo, "10nyeondui Hankukeul bitnael 100in," *Dong-A Ilbo* (April 2, 2013), www.donga.com/news/People/article/all/20130402/54131148/1 (searched on September 1, 2021).

2   POSCO TJ Park Foundation, "2021 POSCO TJ Park Prize Awardees," www.postf.org/en/page/award/history.do (searched on September 5, 2021).

3   Malgorzata Kopytko and Antoni Rogalski, "Performance Evaluation of Type-II Superlattice Devices Relative to HgCdTe Photodiode," *IEEE Transactions of Electron Devices*, 69:6 (2022), 2992–3002.

4   Interview with Chung Han by Kim Dong-Won (September 21, 2020).

5   Ibid.

6   Ibid.

7   Interview with Kim Choong-Ki's former students who asked to remain anonymous.

8   For the benefits of using mercury cadmium telluride (MCT), see Han Pyong-Hee, Kim Choong-Ki, Lee Sang-Don, and Kim Dong-Ho, "Fabrication of HgCdTe p-n Junction Diode," *Jeonkihakhoenonmunji*, 38:8 (1989), 593–599 on 593–594. The last two authors of the paper were researchers at the Agency for Defense Development.

9   Pyong-Hee Han, "Fabrication of $Hg_{1-x}Cd_xTe$ Photovoltaic Infrared Detector" (master's thesis, Seoul: KAIST, 1989); Chan Moon, "Fabrication of $Hg_{0.7}Cd_{0.3}Te$ IR Photodiode and Investigation on the Reverse Leakage Current" (master's thesis, Seoul: KAIST, 1990).

10 Interview with Chung Han by Kim Dong-Won (September 21, 2020).

11 Han Chung, "Fabrication and Characterization of Large Area $Hg_{0.7}Cd_{0.3}Te$ Photovoltaic Infrared Detector" (master's thesis, Seoul: KAIST, 1991).

12 Ibid., 57.

13 For more information about KIT, see Kim Dong-Won et al., *Hankukgwahakgisulwon Sabansegi: Miraeruel hyanghan kkeunimeopneun Dojeon* (Daejeon: KAIST, 1996), 90–96.

14 This spy agency was the number one enemy of the democratic movement during the 1980s since it arrested anyone it wished without due process and often tortured them.

15 Interview with Chung Han by Kim Dong-Won (May 21, 2021).

16 Ibid.

17 Park Sang-In et al., *Uri Kim Choong-Ki Seonsaengnim* [*Our Teacher, Kim Choong-Ki*] (Daejeon: Privately printed, 2002), 91.

18 Han Jung [Han Chung], Hee-Chul Lee, and Choong-Ki Kim, "Measurement of the Steady-State Minority Carrier Diffusion Length in a HgCdTe Photodiode," *Japanese Journal of Applied Physics*, 35:10B (1996), L1321–L1323.

19 Han Chung, "The Fabrication of HgCdTe Photovoltaic Infrared Detector and Enhancement of Its Performance by ECR Plasma Hydrogenation" (doctoral thesis, Daejeon: KAIST, 1996), 100.

20 Interview with Chung Han by Kim Dong-Won (September 21, 2020).

21 Kim Choong-Ki, Nam Chang-Hee, and Kim Byung-Yoon, "Jeonjakgwanghakteukhwayeonkusenteo (Center for Electro-Optics) Yeonkuhyeonhwang," *Kwanghakgwa Gisul, 2:1* (1998), 35–36; and Kim Choong-Ki, "Center for Electro-Optics," *Gukbangkwa Gisul* (January 2001), 36–41.

22 Interview with Chung Han by Kim Dong-Won (September 21, 2020).

23 EMC Microsystem closed in the mid-2000s.

24 For more information about the TBIC, see Kim Dong-Won et al., *Hankukgwahakgisulwon Sabansegi*, 137–143. See also TBIC's own website, https://tbic.kaist.ac.kr/#01_02_sect.

25 For more information about KC Tech, see www.kctech.com/index.php (searched on August 15, 2021).

26 Geumgang Ilbo, "Chung Han i3system-daepyo," *Geumgang Ilbo* (March 25, 2014), www.ggilbo.com/news/articleView.html?idxno=171588 (searched on September 1, 2021).

27 Interview with Chung Han by Kim Dong-Won (September 21, 2020).

28 Interview with Chung Han by Kim Dong-Won (May 21, 2021). *Maeil Kyuungje*, a financial newspaper, described Chung as a well-known figure in the South Korean intelligence community with his "security technology." See Maeil Kyungje, "Dr. Chung Han," *Maeil Kyungje* (February 4, 2003), www.mk.co.kr/news/home/view/2003/02/38971/ (searched on September 1, 2021).

29 Michigan Medicine, University of Michigan, "Capsule Endoscopy," www.uofmhealth.org/conditions-treatments/digestive-and-liver-health/capsule-endoscopy (searched on April 20, 2001).

30 The intelligent microsystem project, however, was transferred to the Ministry of Trade, Industry and Energy in 2004.

31 For technical difficulties of the capsule endoscope, see Park Jong-Man, "Technological Trend for Wireless Ingestible Capsule Design," *Hankuktongshinhakhoenonmunjip*, *34:12* (2009), 1524–1534.

32 Ministry of Science and Technology, "Chosohyeong Gogineung Capsulhyeong Naesigyeong 'MIRO' Gaebal" (January 28, 2003). This document was issued for the use of reporters.

33 Hankuk Kyungje, "Sege Choesohyeong Choegogineung Capsulhyeong Naesigyeong Gaebal," *Hankuk Kyungje* (January 28, 2003), www.hankyung.com/news/article/2003012802588; Maeil Kyungje, "Sege choeso Capsule Naesigyeong Gaebal," *Maeil Kyungje* (January 28, 2003), www.mk.co.kr/news/it/view/2003/01/31687/; and Top Class, "Segye Choesohyung Microcapsule Naesigyeong 'Miro' mandeun Kim Tae-Songbaksa," *Top Class* (March 2008), http://topclass.chosun.com/mobile/board/view.asp?catecode=K&tnu=200803100005 (searched on September 1, 2021).

34 One of Kim Choong-Ki's former students who joined Hankum remembers that Chung offered him/her a large sum of recruitment money, almost the equivalent of the price of a small condominium, as incentive to work for the company.

35 i3system, *Internal Report*, 2. This report summarizes the chronicles of the company and also records its development of the products.

36 IntroMedic actually started its business in September 2004, about a year after it purchased key technologies and patents from Hankum. The company grew rapidly to become one of the major manufacturers of the capsule endoscope in the world. For more information about the company, see www.intromedic.com:549/eng/main/ (searched on September 1, 2021).

37 For more information about the i3system, see http://i3system.com (searched on September 2, 2021).

38 Ministry of Trade, Industry and Energy, "21segi Frontier Gisulgaebalsaeop Project Sujuhyeonhwang (circa 2005, exact date unknown)"; and i3system, "Major Contract (2006–2022)," in *Internal Report*.

39 Tae-Song Kim, Si-Young Song, Han Jung [Chung], Jinseok Kim, and Eui-Sung Yoon, "Micro Capsule Endoscope for Gastro Intestinal Tract," *29th Annual International Conference of the IEEE Engineering in Medicine and Biology Society* (2007), 2823–2826. This paper has been cited ten times and viewed 271 times, https://ieeexplore.ieee.org/document/4352916 (searched on August 27, 2022).

40 Jeonja Shinmun, "i3system, Sikdogeomsa Capsulehyeong Naesigyeong Yeonnae Gaebal," *Jeonja Shinmun* (August 13, 2009), www.etnews.com/200908120125 (searched on September 9, 2021).

41 Interview with Chung Han by Kim Dong-Won (September 21, 2020).

42 Chung Han, "2021 POSCO Chungamsang Gisulsang Huboja Eopjeopseo (Achievement Report for the 2021 POSCO TJ Park Technology Prize)," 5.

43 Interview with Chung Han by Kim Dong-Won (September 21, 2020).

44 Optris, "Focal Plane Array (FPA)," www.optris.com/focal-plane-array (searched on September 6, 2021).

45 ETNews, "Jeokoeseon Yeongsangkamerayong Haeksimchip Guksanhwa Seongkong," *ETNews* (February 25, 2009), www.etnews.com/200902240266 (searched on September 9, 2021).

46 i3system, "Uncooled (Infrared) Detectors," http://i3system.com/uncooled-detector/?lang=en (searched on September 7, 2021).

47 i3system, "X-ray," http://i3system.com/x-ray/?lang=en (searched on September 7, 2021).

48 Interview with Chung Han by Kim Dong-Won (September 21, 2020).

49 Ibid.

50 Interview with Chung Han by Kim Dong-Won (May 21, 2021).

51 i3 System, *Internal Report*, 3–4.

52 Dong-A Ilbo, "Naroho 3cha Balsa D-3 … Tapjaedoeneun Gwahakwiseong museum il hana," *Dong-A Ilbo* (October 19, 2012), www.donga.com/news/It/article/all/20121019/50223421/1; Jeonja Shinmun, "Naroho Balsa Seongkong: sumeun Juyeok–Wang Hui-Jip i3system Suseokyeonkuwon," *Jeonja Shinmun* (January 31, 2013), https://m.etnews.com/201301310364?obj=Tzo4OiJzdGRDbGF zcyI6Mjp7czo3OiJyZWZlcmVyIjtOO3M6NzoiZm9yd2FyZCI7czoxMzoid2ViHRvlG1vYmlsZSI 7fQ%3D%3D; and Asia Kyungje, "Naroho Seongkong dwie algo boni Jungki iteotnae," *Asia Kyungje* (February 10, 2013), https://m.asiae.co.kr/article/2013020622254504955 (searched on September 10, 2021).

53 E Credible, *2010 Credit Report of i3system* (in Korean), *2011 Credit Report o2 i3system* (in Korean), *2012 Credit Report of i3system* (in Korean), *2013 Credit Report of i3system* (in Korean), *2014 Credit Report of i3system* (in Korean), and *2015 Credit Report of i3system* (in Korean), www.srms.co.kr/svc/ssr/mn/SSR01010100.do (searched on June 7, 2021).

54 i3system, *Internal Report*, 3–5.

55 Hankuk Kyungje, "KOSDAQ Ipseong i3system, 'Minsuyong Shijang Jinchulro Jaedoyakhanda,'" *Hankuk Kyungje* (July 15, 2015), www.hankyung.com/finance/article/2015071590836 (searched on September 15, 2021).

56 Consumer Times, "Chung Han, i3system Daepyoisa," *Consumer Times* (July 27, 2015), www.cstimes.com/news/articleView.html?idxno=183931 (searched on September 15, 2021).

57 i3system, "Product: Uncooled Detector," http://i3system.com/uncooled-detector/?lang=en (searched on September 15, 2021). The company's English website displays only two kinds of Thermal Experts, whereas its Korean website shows seven versions.

58 E Credible, *2017 Credit Report of i3system* (in Korean), www.srms.co.kr/svc/ssr/mn/SSR01010100.do (searched on September 15, 2021).

59 Two explosion accidents in 2018 and 2019 at a defense company seriously affected the South Korean defense industry in 2019, and i3system and other related enterprises were also affected.

60 i3system, "LWIR 1024x768 (12μm)," http://i3system.com/uncooled-detector/lwir1024/?lang=en (searched on September 15, 2021).

61 Herald Kyungje, "i3system, Corona 19 Global Hwaksane Yeolhwasangkamera Jumunswaedo–'Gongjang Pulgadongjung'," *Herald Kyungje* (March 12, 2020), http://biz.heraldcorp.com/view.php?ud=20200312000747&ACE_SEARCH=1; and Hello DD, "Corona Sajeonchadan Ildeungkongsin Jeokoeseon Senseo Jeoldaegangja," *Hello DD* (May 19, 2020), www.hellodd.com/news/articleView.html?idxno=71847 (searched on September 15, 2021).

62 E Credible, *2020 Credit Report of i3system* (in Korean), www.srms.co.kr/svc/ssr/mn/SSR01010100.do (searched on September 15, 2021).

63 Aju Kyungje, "i3system, Jayuljuhaeng Sangyonghwashi Suhye Gidae," *Aju Kyungje* (January 27, 2022), www.ajunews.com/view/20220127081409405 (searched on July 27, 2022).

64 i3system, "CEO's Message," http://i3system.com/ceos-message/?lang=en (searched on July 28, 2022).

65 Dong-A Ilbo, "10nyeondui Hankukeul bitnael 100in."

66 YTN, "Kangsogieopi Himida: Dokbojeokin Gisulro mandeun Cheomdan Mugiui Nun, i3system," *YTN* (March 26, 2017), www.youtube.com/watch?v=5Pub-BdT3ck; and KBS, "Gisulgangkuk Project Jungkyeongmanri: IDIS, i3system," *KBS* (December 22, 2019), www.youtube.com/watch?v=7snCkez-jcg (searched on September 17, 2021).

67 Seoul Kyungje, "Jeong Kyeong-Do gukbang, '4chsanseophyeokmyeong Haeshimgisul Hwalyonghae Bangsansuchul Gyeongjaengryeok Ganghwa,'" *Seoul Kyungje* (December 14, 2018), www.yna.co.kr/view/AKR20181214119400503; and Break News, "Bangwisaneopcheong–i3system, DAPA-GO Silsi," *Break News* (March 22, 2019), www.breaknews.com/642217 (searched on September 17, 2021).

68 Hello DD, "Anbowigi Hanbando, Gukbang R&D tonghae Bangwisaneop Sokdo naeya," *Hello DD* (July 20, 2019), www.hellodd.com/news/articleView.html?idxno=69230 (searched on September 17, 2021).

69 Interviews with Chung Han by Kim Dong-Won (September 21, 2020, and May 21, 2021).

70 Interview with Chung Han by Kim Dong-Won (May 21, 2021).

71 Ibid.

72 In 2022, 117 former students were registered as members. Han Chul-Hi was a student of Kim Choong-Ki who received his doctorate from KAIST in 1983. After working at GoldStar Semiconductor for three years, Han became a professor at his alma mater in 1987. He trained about three dozen semiconductor engineers before he died in 2001.

Epilogue

Who Will Become the Next Kim Choong-Ki?

Today, the South Korean semiconductor industry dominates the world market, but that dominance is now under serious threat. Although South Korea has competed furiously with Taiwan in recent decades, its most formidable challenger in the future will likely be China, whose ambitious Made in China 2025 plan prioritizes semiconductor development.[1] With plenty of capital, a huge number of highly educated engineers and scientists (including semiconductor specialists trained in the United States, Japan, and South Korea), and a government-controlled domestic market, Chinese semiconductor companies have already made inroads into the Chinese domestic market, which has been South Korean semiconductor companies' major market since the beginning of the twenty-first century.[2] Since the late 2010s, China has been the number one manufacturer of TFT-LCD panels, and it has also come closer to South Korea in DRAMs and NAND flash memories as well as in OLED panels. After SK Hynix announced the development of the "World's Highest 238-layer 4D NAND Flash" memory on August 2, 2022, for example, a Chinese chipmaker Yangtze Memory Technologies Company (YMTC) announced the next day that it had succeeded in developing a 232-layer NAND Flash memory.[3]

The other serious challenge comes from inside South Korea. The close triangular relationship among government, industry, and the academy, which contributed critically to the development and success of the country's semiconductor industry, has not worked well since the beginning of the twenty-first century. The major culprit is the South Korean government, which has neglected its role of supporting industry and academia with proper industrial and educational policies and financial support. Despite companies' pleas for more workers and universities' calls for policies that advance academic education and research, the government has done little in response. Nearly 50 years after Kim Choong-Ki began educating South Korea's first semiconductor engineers, the industry again faces a significant workforce shortage. Experts estimate that several thousand new engineering specialists are needed each year, but the country produces only a few hundred.[4] The Japanese government's 2018 decision to strictly control the export of core components and materials to South Korea and the serious shortage of semiconductors during the COVID-19 pandemic have finally pushed the South Korean government to pay more attention to improving the semiconductor environment, but its efforts have not yet been sufficient or fast

DOI: 10.1201/9781003353911-12

enough to restore a healthy balance among the three players. Even in the 2020s, it has been industry–academy cooperation, not the government's initiative, that has enabled the establishment of more semiconductor departments and programs at universities to increase the number of necessary chip specialists: for example, Korea University and SK Hynix established a Department of Semiconductor Engineering at Korea University in April 2020 to train the necessary specialists for SK Hynix; and KAIST and Samsung Electronics made a contract in November 2021 to establish a Department of Semiconductor System Engineering at KAIST whose graduates would be hired by Samsung Electronics after graduation.[5] Nevertheless, a much needed special law to support South Korea's semiconductor industry was hung up in Parliament for many years and only became active in August 2022, with a different title and different aims that support almost all high-tech industries.[6] The more specific Special Law for Semiconductors is still stalled in Parliament, despite pressure from government, industry, and the academy.[7]

Escalation of the tension between the United States and China at the beginning of the 2020s has also put the South Korean semiconductor industry in a very difficult position. The so-called Chip 4 Alliance—a supply-chain partnership among the United States, Japan, South Korea, and Taiwan proposed by the US government in March 2022—as well as the recently enacted CHIPS and Science Act 2022 have required that South Korea choose between its largest semiconductor market/major manufacturing site (China) and its closest ally/the source of its technology (the US).[8] Although most South Korean semiconductor specialists maintain that South Korea must side with the United States despite some damage, perhaps even serious damage to sales, the business community and politicians are very cautious about making a clear decision that might be irreversible. The development and possession of cutting-edge technology and manufacturing skill must be the solutions to this dilemma. But can or will South Korea acquire them in the future?

To overcome these grave new challenges and to prevail in this rapidly changing environment, South Korea needs new leading figures who, like Kim Choong-Ki, have both the courage to break with tradition and a bold, prescient vision for the future. Just as South Korea needed "engineer-minded" Kim Choong-Ki in the 1970s to train the first two generations of semiconductor specialists to serve South Korea's infant semiconductor industry, so it needs "creative-minded" new leaders today, capable of pioneering truly groundbreaking technologies that will secure the country's leadership on the world stage in the twenty-first century. A different time in a different environment requires a different leader. The future of South Korea depends on it.

NOTES

1 For the full report of "Made in China 2025" in English, see US Chamber of Commerce, "Made in China 2025" (March 16, 2017), www.uschamber.com/international/made-china-2025-global-ambitions-built-local-protections-0 (searched on September 9, 2022).
2 For the market share of TFT-LCD in the late 2010s and early 2020s, see Omdia, "BOE Becomes World's Largest Flat-Panel Display Manufacturer in 2019 as China Continues Rise to Global Market Dominance," *Omdia* (June 4, 2019), https://omdia.tech.informa.com/OM003804/BOE-Becomes-Worlds-Largest-Flat-Panel-Display-Manufacturer-in-2019-as-China-Continues-Rise-to-Global-Market-Dominance; and Statista, "Global LCD TV Panel Unit Shipments from H1 2016 to H1 2020, by

Vendor," *Statista* (March 9, 2022), www.statista.com/statistics/760270/global-market-share-of-led-lcd-tv-vendors/ (searched on September 9, 2022).

3   SK Hynix, "SK Hynix Develops World's Highest 238-Layer 4D NAND Flash" (August 2, 2022), https://news.skhynix.com/sk-hynix-develops-worlds-highest-238-layer-4d-nand-flash/; and Reuters, "China's Memory Upstart YMTC Edges Closer to Rivals with 232-layer Chip," *Reuters* (August 4, 2022), www.reuters.com/technology/chinas-memory-upstart-ymtc-edges-closer-rivals-with-232-layer-chip-2022-08-04/ (searched on September 9, 2022).

4   Hankuk Kyungje, "Bandoche Inryeok 10nyeonkan 3manmyeong Bujokhande ... Deahakseo Baechule nyeon 650myeongbbun," *Hankuk Kyungje* (May 5, 2022), www.hankyung.com/economy/article/2022050556851; Joong-Ang Ilbo, "'Pyeongtaekkkaji motga,' 'Gyosudo motguhaeseo' ... Hyeonjangseo naon 'Bandoche Inryeoknan," *Joong-Ang Ilbo* (June 10, 2022), www.joongang.co.kr/article/25078089#home; and KBS, "'Maenyeon 3cheonmyeong bujok,' Bandoche Inryeoknan ... Daechaekeun?" KBS News (June 21, 2022), https://news.kbs.co.kr/news/view.do?ncd=5489649 (searched on September 9, 2022).

5   Department of Semiconductor Engineering, https://se.korea.ac.kr/main/main.html; and Department of Electrical Engineering, KAIST, "KAIST-Samsungjeonja Bandochesystemgonghakgwa Seolip Hyeopyak Chegyeol" (November 25, 2021), https://ee.kaist.ac.kr/presses/22826/ (searched on October 1, 2022).

6   Hankuk Kyungje, "'Bandocheteukbyeolbeop' Naeil Sihaeng ... Teukhwadanjideung Gukgacheomdanjeonryaksanseop Yukseong," *Hankuk Kyungje* (August 3, 2022), www.hankyung.com/economy/article/202208033926Y (searched on October 1, 2022).

7   Maeil Kyungje, "Bandoche Wigigam keojineunde Teukbyeolbeop mung-gaegoitneun Gukhoe-eu Musa-anil," *Maeil Kyungje* (September 7, 2022), www.mk.co.kr/opinion/editorial/view/2022/09/792545/ (searched on October 1, 2022).

8   The Korea Herald, "Seoul Weighs 'Chip 4' Alliance as Ties with China Hang in Balance," *The Korea Herald* (August 8, 2022), www.koreaherald.com/view.php?ud=20220808000653; Focus Taiwan, "U.S.-led Chip Alliance Aimed at Curbing China Influence: Analyst," *Focus Taiwan* (August 21, 2022), https://focustaiwan.tw/business/202208210007; and The Diplomat, "The Chip 4 Alliance Might Work on Paper, But Problem Will Persist," *The Diplomat* (August 25, 2022), https://thediplomat.com/2022/08/the-chip4-alliance-might-work-on-paper-but-problems-will-persist/ (searched on September 9, 2022).

References

INTERVIEWS

Interviews with Kim Choong-Ki by Kim Dong-Won (September 21, 2020; May 22, 2021; November 27, 2021; May 28, 2022).

Interviews with Kyung Chong-Min by Kim Dong-Won (September 23, 2020; May 20, 2021).

Interviews with Lim Hyung-Kyu by Kim Dong-Won (February 14 and September 17, 2020; May 29, 2021).

Interviews with Cho Byung-Jin by Kim Dong-Won (September 28, 2020; May 21, 2021).

Interviews with Ha Yong-Min by Kim Dong-Won (May 28, 2021; December 2, 2021).

Interviews with Park Sung-Kyu by Kim Dong-Won (June 1, 2021; June 9, 2022).

Interviews with Chung-Han by Kim Dong-Won (September 21, 2020; May 21, 2021).

Interview with Jung Hee-Bum by Kim Dong-Won (May 23, 2021).

Interview with Yoon Nan-Young by Kim Dong-Won (May 22, 2021).

Interview with Kim Oh-Hyun by Kim Dong-Won (December 3, 2021).

Interviews with anonymous engineers by Kim Dong-Won (2020, 2021, 2022).

Interview with Kim Choong-Ki by Jeon Chihyung (March 7, 2016).

Interview with Lee Yong-Hoon by Jeon Chihyung (June 27, 2016).

Bibliography

Aju Kyungje. "i3system, Jayuljuhaeng Sangyonghwashi Suhye Gidae." *Aju Kyungje* (January 27, 2022). www.ajunews.com/view/20220127081409405.

Antoniadis, Homer. "Overview of OLED Display Technology." www.ewh.ieee.org/soc/cpmt/presentations/cpmt0401a.pdf.

Apple Inc. "Apple Announces Long-Term Supply Agreements for Flash Memory" (November 21, 2005). www.apple.com/newsroom/2005/11/21Apple-Announces-Long-Term-Supply-Agreements-for-Flash-Memory/.

———. "Apple Presents iPhone 4" (June 7, 2010). www.apple.com/newsroom/2010/06/07Apple-Presents-iPhone-4/.

———. "The Future is Here: iPhone X" (September 12, 2017). www.apple.com/newsroom/2017/09/the-future-is-here-iphone-x/.

———. "iPhone 8 and iPhone 8 Plus: A New Generation of iPhone" (September 12, 2017). www.apple.com/newsroom/2017/09/iphone-8-and-iphone-8-plus-a-new-generation-of-iphone/.

Apple Insider. "Samsung Reports Weak Demand for OLED Displays Used in iPhone X." *Apple Insider* (April 26, 2018). https://appleinsider.com/articles/18/04/26/samsung-reports-weak-demand-for-oled-displays-used-in-iphone-x.

Asia Kyungje. "Naroho Seongkong dwie algo boni Jungki iteotnae." *Asia Kyungje* (February 10, 2013). https://m.asiae.co.kr/article/2013020622254504955.

———. "Palyeogo naenwatdadeoni … LG Flex 20,000 Panmae Gulyok." *Asia Kyungje* (January 20, 2014). www.asiae.co.kr/article/2014012009315418154.

aSSIST. "Executive MBA with Aalto University." www.assist.ac.kr/OverseasMBA/Aalto/curriculum_introduction.php.

Berlin, Leslie. *Troublemakers: Silicon Valley's Coming of Age*. New York: Simon & Schuster, 2017.

Bloom, Sahil. "The Amazing Story of Morris Chang." *The Curiosity Chronicle* (January 25, 2021). https://sahilbloom.substack.com/p/the-amazing-story-of-morris-chang.

Boyle, Willard S., and George E. Smith. "Charge Coupled Semiconductor Devices." *The Bell System Technical Journal*, *49:4* (April 1970), 587–593.

Break News. "Bangwisaneopcheong–i3system, DAPA-GO Silsi." *Break News* (March 22, 2019). www.breaknews.com/642217.

Britannica. "Morris Chang: Chinese-Born Entrepreneur." www.britannica.com/biography/Morris-Chang

Business Insider. "Apple Is Reportedly Investing $2.7 Billion in LG Display to Build OLED Panels for Future iPhones." *Business Insider* (July 28, 2018). www.businessinsider.com/apple-is-reportedly-investing-27-billion-in-lg-display-to-build-oled-panels-for-future-iphones-2017-7.

Business Korea. "LG Display Dominating Global Market for High-End Vehicle OLEDs" (February 18, 2021). www.businesskorea.co.kr/news/articleView.html?idxno=74331.

———. "LG Display's Share of Global Smartphone OLED Panel Market Tops 10 Percent." *Business Korea* (March 13, 2020). www.businesskorea.co.kr/news/articleView.html?idxno=42672.

———. "LG Display to Invest 3.3 Tri. Won in Small and Medium-Sized OLED Panels for 3 Years." *Business Korea* (August 18, 2021). www.businesskorea.co.kr/news/articleView.html?idxno=7433.

Business Wire. "With Its Highest Growth Rate in 14 Years, the Global Semiconductor Industry Topped $429 Billion in 2017, HIS Markit Says." *Business Wire* (March 28, 2018). www.businesswire.com/news/home/20180328006092/en/With-its-Highest-Growth-Rate-in-14-Years-the-Global-Semiconductor-Industry-Topped-429-Billion-in-2017-IHS-Markit-Says

Castellano, Joseph A. *Liquid Gold: The Story of Liquid Crystal Displays and the Creation of an Industry.* Singapore: World Scientific Publishing, 2005.

Chang, Chun-Yen, and Po-Lung Yu, eds. *Made by Taiwan: Booming in the Information Technology Era.* Singapore: World Scientific, 2001.

Chang, Yang-Mo et al. *Uri Doja Iyagi.* Ichon, Kyeongkido: Eksupo, 2004.

Chin, Dae-Je. *Yeoljeong-eul Gyeongyeong hara.* Paju, Kyeongkido: Kimyeongsa, 2006.

Chinjukan Pottery. https://chinjukanpottery.com.

Cho Byung-Jin. *See also under* Jo Byung-Jin.

Cho, Byung-Jin. "Design Study of a Rapid Thermal Annealing System and Temperature Control." Master's thesis. Seoul: KAIST, 1988.

———. "Modeling of Rapid Thermal Diffusion of Phosphorous into Silicon and Its Application to VLSI Fabrication." Doctoral thesis. Seoul: KAIST, 1991.

Cho, Byung-Jin, and Choong-Ki Kim. "Elimination of Slips on Silicon Wafer Edge in Rapid Thermal Process by Using a Ring Oxide." *Journal of Applied Physics*, 67 (1990), 7583–7586.

Cho, Byung-Jin, Kim Kyeong-Tae, and Kim Choong-Ki. "Experimental Results of a Prototype Rapid Thermal Annealing System." *Daehanjeonjagonghakhoe Haksulbalpyohoe Nonmunjip*, 5:1 (1986), 71–73.

Cho, Byung-Jin, Sung-Kye Park, and Choong-Ki Kim. "Estimation of Effective Diffusion Time in a Rapid Thermal Diffusion Using a Solid Diffusion Source." *IEEE Transactions on Electron Devices*, 39:1 (1992), 111–117.

Cho, Byung-Jin, Peter Vandenabeele, and Karen Maex. "Development of Hexagonal-Shaped Rapid Thermal Processor Using a Vertical Tube." *IEEE Transactions on Semiconductor Manufacturing*, 7:3 (1994), 345–353.

Cho, Yong-Joon. *Ilbon Dojagi Yeohaeng: Kyushu 7dae Chosun Gama.* Seoul: Do Do Publication, 2016.

Chosun, Ilbo. "[Gongdaegyosudeul] Hynix Jolsokmaegak Bandochesaneop Akhwa." *Chosun Ilbo* (June 9, 2002). https://biz.chosun.com/site/data/html_dir/2002/06/09/2002060970253.html

———. "Gukmin boda Sujuni hwolssin najeon Saibi Bosujeongchi-ui Silpae." *Chosun Ilbo* (November 28, 2016). www.chosun.com/site/data/html_dir/2016/11/28/2016112800222.html.

———. "Guknaechoecho Intel 386hohwan CPU Gaebal: Hankukgwahakgisulwon Kyung Chong-Min-gyosu." *Chosun Ilbo* (September 7, 1995). https://biz.chosun.com/site/data/html_dir/1995/09/07/1995090772304.html.

———. "Hynix Maekak Bandaehaetdeon Kyung Chong-Min-gyosu." *Chosun Ilbo* (February 28, 2004). https://biz.chosun.com/site/data/html_dir/2004/02/28/2004022870039.html.

———. "KAIST Wearable Baljeonki, UNESCOga bbopeun Sesangeul bakkul Choegogisule Seonjeong." *Chosun Ilbo* (February 4, 2015). https://biz.chosun.com/site/data/html_dir/2015/02/04/2015020403318.html.

———. "SK, Samsung CTOchulsin Lim Hyung-Kyu ICT Gisul-Seongjang Chonggwalbuhoejang Yeongip." *Chosun Ilbo* (January 22, 2014). https://biz.chosun.com/site/data/html_dir/2014/01/22/2014012203125.html.

Christian Forum in Science and Engineering. "Cho Byung-Jin (1)." (July 19, 2015). www.sciengineer.or.kr/board_kVuz18/19005?ckattempt=1.

———. "Cho Byung-Jin (2)." (July 20, 2015). www.sciengineer.or.kr/board_kVuz18/18998.

Chun, Doo-Hwan. *Chun Doo-Hwan Hoegorok, Vol. 2*. Seoul: Jajaknamusup, 2017.

Chung, Han. *See also under* Jung Han.

Chung, Han. "2021 POSCO Chungamsang Gisulsang Huboja Eopjeopseo" [Achievement Report for the 2021 POSCO TJ Park Technology Prize].

———. "Fabrication and Characterization of Large Area $Hg_{0.7}Cd_{0.3}Te$ Photovoltaic Infrared Detector." Master's thesis. Seoul: KAIST, 1991.

———. "The Fabrication of HgCdTe Photovoltaic Infrared Detector and Enhancement of Its Performance by ECR Plasma Hydrogenation." Doctoral thesis. Daejeon: KAIST, 1996.

Chung, Suock, et al. "Fully Integrated 54nm STT-RAM with the Smallest Bit Cell Dimension for High Density Memory Application." *2010 International Electron Devices Meeting* (2010), 12.7.1–12.7.4.

Chung, Young-Iob. "The Impact of Chinese Culture on Korea's Economic Development," in Hung-Chao Tai, ed., *Confucianism and Economic Development: An Oriental Alternative?* Washington, D.C.: The Washington Institute Press, 1989, 149–165.

CISS. "Agreement." http://ciss.re.kr/english/agreement.

———. "Introduction." http://ciss.re.kr/english/introduction.

Clancey, Gregory. "Intelligent Island to Biopolis: Smart Minds, Sick Bodies and Millennial Turns in Singapore." *Science, Technology and Society, 17:1* (2012), 13–35.

CNB News. "LG Display, 2015 Imwon Insa Silsi." *CNB News* (November 27, 2014). https://m.cnbnews.com/m/m_article.html?no=273443.

CNET. "Apple's iPhone 8 Could Get Curved OLED Screens—and Make Samsung Billions" *CNET* (February 13, 2017). www.cnet.com/tech/mobile/apple-might-buy-million-curved-oled-screen-from-samsung-rumor/.

College of Engineering (SNU). "Distinguished Alumni Award." http://eng.snu.ac.kr/node/13.

Computer History Museum. "Fairchildren." https://computerhistory.org/fairchildren/.

———. "Oral History of Fujio Masuoka" (reference number: X6623.2013). http://archive.computerhistory.org/resources/access/text/2013/01/102746492-05-01-acc.pdf.

———. "Oral History of Sze, Simon." www.computerhistory.org/collections/catalog/102746858.

Consumer Technology Association. "CES 2020 Innovation Award Product, ThermoReal." www.ces.tech/Innovation-Awards/Honorees/2020/Honorees/T/ThermoReal-Made-with-Flexible-Thermoelectric-De.aspx.

Consumer Times. "Chung Han, i3system Daepyoisa." *Consumer Times* (July 27, 2015). www.cstimes.com/news/articleView.html?idxno=183931.

Daejeon Ilbo. "Insaeng-ui Vision boindamyeon miri Junbihaneun Jase Pilyo: Kim Choong-Ki KAST Teukhun-gyosu." *Daejeon Ilbo* (March 13, 2007). www.daejonilbo.com/news/articleView.html?idxno=675390.

Department of Electrical Engineering (KAIST). https://ee.kaist.ac.kr/en/node/10981.

———. https://ee.kaist.ac.kr/node/10278?language=ko.

———. "Brief History," in *2008/2009 Annual Report*. https://ee.kaist.ac.kr/en/node/21.

———. "KAIST-Samsungjeonja Bandochesystemgonghakgwa Seolip Hyeopyak Chegyeol" (November 25, 2021). https://ee.kaist.ac.kr/presses/22826/.

———. "Park Song-Bai." https://ee.kaist.ac.kr/en/professor_s6?language=en&combine=Park%2C%2BSong-Bae.

Department of Semiconductor Engineering (KAIST). https://se.korea.ac.kr/main/main.html.

Digital Times. "Smartsensore nameun Yeonkuinsaeng geolgeot. IDECsojang Satoeuisa balkin Kyung Chong-Min-gyosu." *Digital Times* (November 7, 2011). www.dt.co.kr/contents.html?article_no=2011110802012069758002.

The Diplomat. "The Chip 4 Alliance Might Work on Paper, But Problem Will Persist." *The Diplomat* (August 25, 2022). https://thediplomat.com/2022/08/the-chip4-alliance-might-work-on-paper-but-problems-will-persist/.

Do, Jae-Young, et al. "A 256K EEPROM with Enhanced Reliability and Testability." *1988 Symposium on VLSI Circuits*, held in Tokyo (1988), 83–84.

Dong-A Ilbo. "10nyeon dwi Hankukeul bitnael 100in: 'Hankyeneun eupda' ... Silpaedo jeulgimyeo dalyeo-on 100gaeui kkum." *Dong-A Ilbo* (April 2, 2013). www.donga.com/news/People/article/all/20130402/54131148/1.

———. "386hohwanchip Guknae cheot Gaebal." *Dong-A Ilbo* (September 3, 1995), 12.

———. "Hynix Heolgapmaegak Nonran." *Dong-A Ilbo* (April 22, 2002). www.donga.com/news/article/all/20020422/7810916/9.

———. "Microprocessorgisul, 10nyeonane Segejeongsang." *Dong-A Ilbo* (January 10, 1995), 17.

———. "Naroho 3cha Balsa D-3 ... Tapjaedoeneun Gwahakwiseong museum il hana." *Dong-A Ilbo* (October 19, 2012). www.donga.com/news/It/article/all/20121019/50223421/1.

———. "SK Nalgaedan Hynix, 2000eok Jeokjaseo 'Hyoja'ro talbakkumhaetda." *Dong-A Ilbo* (January 19, 2022). www.donga.com/news/Economy/article/all/20220119/111328304/1?ref=main.

Dong-A Science. "Idaleu Engineersang 2wol Susangja, Park Sung-Kye, Jeong Jong-Seok." *Dong-A Science* (February 14, 2005). www.dongascience.com/news.php?idx=-44553.

E Credible. *2010 Credit Report of i3system* (in Korean). www.srms.co.kr/svc/ssr/mn/SSR01010100.do.

———. *2011 Credit Report of i3system* (in Korean). www.srms.co.kr/svc/ssr/mn/SSR01010100.do.

———. *2012 Credit Report of i3system* (in Korean). www.srms.co.kr/svc/ssr/mn/SSR01010100.do.

———. *2013 Credit Report of i3system* (in Korean). www.srms.co.kr/svc/ssr/mn/SSR01010100.do.

———. *2014 Credit Report of i3system* (in Korean). www.srms.co.kr/svc/ssr/mn/SSR01010100.do.

———. *2015 Credit Report of i3system* (in Korean). www.srms.co.kr/svc/ssr/mn/SSR01010100.do.

———. *2017 Credit Report of i3system* (in Korean). www.srms.co.kr/svc/ssr/mn/SSR01010100.do.

———. *2020 Credit Report of i3system* (in Korean). www.srms.co.kr/svc/ssr/mn/SSR01010100.do.

Eckert, Carter. "Korea's Transition to Modernity: A Will to Greatness," in Merle Goldman and Andrew Gordon, eds., *Historical Perspectives on Contemporary East Asia*. Cambridge, Mass.: Harvard University Press, 2000, 119–154.

eDaily. "1/5ro julin Memory Chip ... Samsung, Ilboneul apdohada." *eDaily* (January 5, 2012). www.edaily.co.kr/news/realtime/realtime_NewsRead.asp?newsid=01088966599394768.

———. "Hynix Insu 'Love-Call' ... LG Electronics sikundeunghan Iyuneun." *eDaily* (October 4, 2010). www.edaily.co.kr/news/read?newsId=02263206593129640&mediaCodeNo=257.

eeNews Analogue. "TSMC, Taiwan to Increase Foundry Market Share in 2022." *eeNews Analogue* (April 25, 2022). www.eenewsanalog.com/en/tsmc-taiwan-to-increase-foundry-market-share-in-2022/.

Electronics and Telecommunications Research Institute (ETRI). *ETRI 30nyeonsa* [*Thirty-Year History of ETRI*]. Daejeon: ETRI, 2006.

————. *ETRI 35nyeonsa* [*Thirty-Five Year History of ETRI*]. Daejeon: ETRI, 2012. www.etri.
    re.kr/korcon/sub7/sub7_13.etri.

Emcrafts. www.emcrafts.com/en/products/se_main.php.

Engineering and Technology History Wiki. "Charge-Coupled Device." http://ethw.org/Charge-
    Coupled_Device.

ETNews. "Jeokoeseon Yeongsangkamerayong Haeksimchip Guksanhwa Seongkong." *ETNews*
    (February 25, 2009). www.etnews.com/200902240266.

etoday. "'Seungsengjangku' SK Hynix sumeun Joryeokja Lim Hyung-Kyu Buhoejang."
    etoday (November 6, 2014). www.etoday.co.kr/news/view/1013860.

ETRI. *See* Electronics and Telecommunications Research Institute.

The European Space Agency. "What is Concurrent Engineering?" www.esa.int/Enabling_Supp
    ort/Space_Engineering_Technology/CDF/What_is_concurrent_engineering.

Fairchild Camera and Instrument Corporation. www.fairchildimaging.com/our-history.

Firstpost. "Hynix Announces 512Mbit Mobile DRAM." *Firstpost* (December 4, 2006).
    www.firstpost.com/tech/news-analysis/hynix-announces-512mbit-mobile-dram-3551
    063.html.

Focus Taiwan. "U.S.-led Chip Alliance Aimed at Curbing China Influence: Analyst." *Focus
    Taiwan* (August 21, 2022). https://focustaiwan.tw/business/202208210007.

Forbes. "Apple Eyes OLED for iPhone Displays: Samsung, LG Likely Suppliers." *Forbes*
    (November 26, 2015). www.forbes.com/sites/brookecrothers/2015/11/26/apple-will-
    switch-to-oled-for-iphone-displays-report/?sh=ba96a2768a8b.

————. "Morris Chang." www.forbes.com/profile/morris-chang/?sh=1680cbbe5fc4.

————. "Some Flash Memory Keynotes." *Forbes* (August 13, 2018). www.forbes.com/sites/
    tomcoughlin/2018/08/13/some-flash-memory-keynotes/?sh=3bbb2e943202.

————. "Unsung Hero." *Forbes* (June 23, 2002). www.forbes.com/global/2002/0624/030.
    html#299c6b903da3.

Frost and Sullivan. "Samsung Surpasses Intel in 2017-Q2 Revenue, Grabs the Pole Position!
    But Will the Lead Continue?" www.frost.com/frost-perspectives/samsung-surpasses-
    intel-in-2017-q2-revenue-grabs-the-pole-position-but-will-the-lead-continue/.

Future Science Prize. "2021 The Mathematics and Computer Science Prize Laureate: Simon
    Sze." www.futureprize.org/en/laureates/detail/56.html.

Geumgang Ilbo. "Chung Han i3system-daepyo." *Geumgang Ilbo* (March 25, 2014). www.ggi
    lbo.com/news/articleView.html?idxno=171588.

Gizmo China. "LG Display to Supply More OLED Panels to Apple for iPhones This Year."
    Gizmo China (April 22, 2021). www.gizmochina.com/2021/04/22/lg-display-supply-
    oled-panel-apple-iphone/.

Government of the Republic of Korea. *Je4cha Gyeongjegaebal 5gaenyeon-gyehoek* [*The
    Fourth Five-Year Economic Development Plan (1977–1981)*]. Seoul: Government of
    the Republic of Korea, 1976.

————. *Jeonjagongeop Yukseong-gyehoek* [*Electronics Promotion Plan*]. Seoul: President
    Secretarial Office, September 25, 1976.

Gunsagar, K. C., C. K. Kim, and J. D. Phillips. "Performance and Operation of Buried Channel
    Charge Coupled Devices." *1973 International Electron Devices Meeting (IEDM)*,
    Washington, D.C. (December 3–5, 1973), 21–23.

Ha Yong-Min. "Design and Implementation of Lamp-heated LPCVD System and Poly-Si
    Deposition." Master's thesis. Seoul: KAIST, 1990.

————. "Device Structure and Fabrication Process for High Performance Polysilicon Thin
    Film Transistors." Doctoral thesis. Daejeon: KAIST, 1994.

Ha Yong-Min, Han Chul-Hee, and Kim Choong-Ki. "Hydrogenation Mechanism of Top-Gated
    Poly-Si TFT." *Daehanjeonjagonahakhoe Haksuldaehoe* (January 1994), 431–432.

Ha Yong-Min, Kim Tae-Sung, and Kim Choong-Ki. "Design and Implementation of Lamp-Heated LPCVD System." *Daehanjeonkihakhoe Haksuldaehoe Nonmunjip* (November 1991), 299–303.

Han Pyong-Hee, Kim Choong-Ki, Lee Sang-Don, and Kim Dong-Ho. "Fabrication of HgCdTe p-n Junction Diode." *Jeonkihakhoenonmunji, 38:8* (1989), 593–599.

Han Pyong-Hee. "Fabrication of $Hg_{1-x}Cd_xTe$ Photovoltaic Infrared Detector." Master's thesis. Seoul: KAIST, 1989.

Han Sang-Bok. *Oebaljajeongeoneun Neomeojiji anneunda.* Seoul: Haneulchulpansa, 1995.

Hankuk Kyungje. "10nyeon Jeolchibusim … Hynix Memory choegang kkumkkunda." *Hankuk Kyungje* (August 19, 2010). www.hankyung.com/news/article/201008 1749681.

———. "Apple, juyo Bupumnappumeopche Dabyeonhwa." *Hankuk Kyungje* (March 19, 2012). www.hankyung.com/it/article/201203193802i.

———. "Apple Sinmugi 'iPhone 4G,' Partnerneun LG." *Hankuk Kyungje* (March 23, 2010). www.hankyung.com/it/article/2010032245751.

———. "Bandoche Inryeok 10nyeonkan 3manmyeong Bujokhande … Deahakseo Baechule nyeon 650myeongbbun." *Hankuk Kyungje* (May 5, 2022). www.hankyung.com/econ omy/article/2022050556851.

———. "'Bandocheteukbyeolbeop' Naeil Sihaeng … Teukhwadanjideung Gukgacheomdan-jeonryaksanseop Yukseong." *Hankuk Kyungje* (August 3, 2022). www.hankyung.com/ economy/article/202208033926Y.

———. "Bimemori Ganghwa … Samsungjeonja, System LSIsaeop Yukseong Uimi." *Hankuk Kyungje* (August 27, 2002). www.hankyung.com/news/article/2002082731911.

———. "Dynalith System, kkomkkomhan Bandoche Seolgyegeomjeung." *Hankuk Kyungje* (June 18, 2014). www.hankyung.com/society/article/2014061887151.

———. "Hitech geu Juyeokdeul: KAIST Kyung Chong-Min-gyosutim." *Hankuk Kyungje* (September 18, 1995). www.hankyung.com/news/article/1995091800521.

———. "Hynix-Micron Maegakilji." *Hankuk Kyungje* (April 22, 2002). www.hankyung.com/ finance/article/2002042295058.

———. "Hynix Siljeok 1bunki Badak." *Hankuk Kyeongje* (January 7, 2011). www.hankyung. com/finance/article/2011010670241.

———. "Hyundai Electronics, 'Hynix Semiconductor'ro Hoesamyeong Byeonkyeong." *Hankuk Kyungje* (March 8, 2001). www.hankyung.com/news/article/2001030872956.

———. "KAIST Jeongnyeonbojangsimsa daegeo Talak, Gyosusahoe Cheolbaptong kkaejina?" *Hanguk Kyungje* (October 12, 2007). https://sgsg.hankyung.com/article/ 2007101143171.

———. "KOSDAQ Ipseong i3system, 'Minsuyong Shijang Jinchulro Jaedoyakhanda.'" *Hankuk Kyungje* (July 15, 2015). www.hankyung.com/finance/article/2015071590836.

———. "Kyeongjaengkuk Sahwalgeoneundae Jeongbuneun Dwitjim … Bandoche Saengtaegye, 40nyeonjeonboda Yeolak." *Hankuk Kyungje* (June 30, 2022). www.hanky ung.com/economy/article/2022062292911.

———. "LCD Sijang Judogwon norinda … LG Philips 'Paju project' Uimi." *Hankuk Kyungje* (February 3, 2003). www.hankyung.com/news/article/2003020391391.

———. "Lee Yoon-Woo 'Bupum', Choi Jisung 'Jepum' … '2gaeui Samsungjeongja' Chulbum." *Hankuk Kyungje* (January 17, 2009). httwww.hankyung.com/news/article/ 2009011640701.

———. "LG Philips LCD, Koo Bon-Joonsajang interview." *Hankuk Kyungje* (July 11, 2000). www.hankyung.com/news/article/200007108154.

———. "Pajue \$10 Billion LCD Gongjang…LG Philips LCD." *Hankuk Kyungje* (February 3. 2003). www.hankyung.com/news/article/2003020390601.

———. "Samsung, 4gae Bimemori Jipjung Yukseong … LDI, MCU Segye 1wi Mokpyo." *Hankuk Kyungje* (October 3, 2000). www.hankyung.com/news/article/2000100204661.

———. "Samsung, Bimemori Daeyakjin, … Olmaechul 80% Geupjeung 18eokbul." *Hankuk Kyungje* (December 11, 2000). www.hankyung.com/news/articles/2000121074891.

———. "Samsung/Geumseong Bandoche Segyejeokin Hoesaro Seongjang." *Hankuk Kyungje* (January 30, 1989). www.hankyung.com/news/article/1989013000921.

———. "Samsungchulsin Lim Hyung-Kyu, SK Miraesanup Chonggwalhanda." *Hankuk Kyungje* (January 23, 2014). www.hankyung.com/news/article/2014012228711.

———. "Samsungjeonja, Bimemori Daedaejeok Tuja … Gaebalbi 2000eok Chuga." *Hankuk Kyungje* (April 15, 2002). www.hankyung.com/news/article/2002041562761.

———. "Samsungjeonja Graphene Doip …Memory Gyeongjaengryeok Ganghwa." *Hankuk Kyungje* (January 4, 2012). www.hankyung.com/finance/article/2012010469456.

———. "Sege Choesohyeong Choegogineung Capsulhyeong Naesigyeong Gaebal." *Hankuk Kyungje* (January 28, 2003). www.hankyung.com/news/article/2003012802588.

Hankyoreh. "Hynix Heolgapmaegak Nonran beonjeo." *Hankyoreh* (April 23, 2022). http://leg acy.www.hani.co.kr/section-004100021/2002/04/004100021200204231918035.html.

———. "LG Display Will Supply LCD for iPod." *Hankyoreh* (April 3, 2008). www.hani.co.kr/arti/economy/economy_general/279774.html.

Hello DD. "Anbowigi Hanbando, Gukbang R&D tonghae Bangwisaneop Sokdo naeya." *Hello DD* (July 20, 2019). www.hellodd.com/news/articleView.html?idxno=69230.

———. "Bunjaeng Gajeonghaji anko sseun Teuheoneun Ssregi." *Hello DD* (March 18, 2015). www.hellodd.com/news/articleView.html?idxno=52495.

———. "Corona Sajeonchadan Ildeungkongsin Jeokoeseon Senseo Jeoldaegangja." *Hello DD* (May 19, 2020). www.hellodd.com/news/articleView.html?idxno=71847.

———. "Hulryunghan Gwahakjaga doeryeomyeon Yeonaesoseol mani ilgeoyajo." *Hello DD* (December 15, 2015). www.hellodd.com/news/articleView.html?idxno=56176.

———. " 'Jakge mandeuni mot mandeulge eopseo' … Yeoljeonsoja sae Yeoksa sseunda." *Hello DD* (September 22, 2020). www.hellodd.com/ncws/articleView.html?idxno= 72943.

Herald Kyungje. "i3system, Corona 19 Global Hwaksane Yeolhwasangkamera Jumunswaedo– 'Gongjang Pulgadongjung.' " *Herald Kyungje* (March 12, 2020). http://biz.heraldcorp. com/view.php?ud=20200312000747&ACE_SEARCH=1.

The Ho-Am Foundation. "Award." http://hoamprize.samsungfoundation.org/eng/award/part_view.asp?idx=10.

———. "Introduction." http://hoamprize.samsungfoundation.org/eng/foundation/intro.asp.

Hong, Sungook. "The Relationship between Science and Technology in Korea from the 1960s to the Present Day: A Historical and Reflective Perspective." *East Asian Science, Technology and Society*, 6 (2012), 259–265.

Hynix Semiconductor. *2003 Annual Report* (in Korean). March 30, 2004.

———. *2004 Annual Report* (in Korean). March 31, 2005.

———. *2005 Annual Report* (in Korean). March 31, 2006.

———. *2007 Annual Report* (in Korean). March 28, 2008.

———. *2008 Annual Report* (in Korean). April 1, 2009.

———. *2009 Annual Report* (in Korean). March 31, 2010.

———. *2010 Annual Report* (in Korean). March 31, 2011.

———. *2011 Annual Report* (in Korean). March 30, 2012.

Hyun Won-Bok. *Uri Gwahak, keu Baeknyeon eul Bitnaen Saramdeul: Hankukeu Gwahak Gisulin Baeknyeon, Vol. 1*. Seoul: Gwahak Sarang, 2009.

Hyundai Electronics. *Hyundaijeonja 10nyeonsa [Ten-Year History of Hyundai Electronics]*. Icheon, Kyeongkido: Hyundai Electronics, 1994.

I News 24. "LG Display's IPS Panel." *I News 24* (May 22, 2011). www.inews24.com/view/
575415.

i3system. http://i3system.com.

———. *Internal Report*.

———. "LWIR 1024x768 (12μm)." http://i3system.com/uncooled-detector/lwir1024/?lang=
en (searched on September 15, 2021).

———. "Uncooled (Infrared) Detectors." http://i3system.com/uncooled-detector/?lang=en.

IEEE Xplore. "Simon M. Sze." https://ieeexplore.ieee.org/author/37294788800.

———. "Sung-Kye Park." https://ieeexplore.ieee.org/search/searchresult.jsp?newsearch=
true&queryText=Sung-Kye%20Park.

iMore. "LG Displays Preparing Another OLED line for Apple, Says Report." *iMore* (July 29,
2021). www.imore.com/lg-display-preparing-another-oled-line-apple-says-report.

Insight Korea. "Samsungchulsin Lim Hyung-Kyu Buhoejang Yeongjp." *Insight Korea* (June
26, 2015). www.insightkorea.co.kr/news/articleView.html?idxno=13394.

Intel. "Intel Sells SSD Business and Dalian Facility to SK Hynix." (December 29, 2021). www.
intc.com/news-events/press-releases/detail/1513/intel-sells-ssd-business-and-dalian-
facility-to-sk-hynix.

Interference Technology. "New Research Suggests Monolayer Graphene is the Most Effective
Material for EMI Shielding." *Interference Technology* (October 25, 2012). https://int
erferencetechnology.com/new-research-suggests-monolayer-graphene-is-most-effect
ive-material-for-emi-shielding/.

Interuniversity Microelectronics Center (IMEC). www.imec-int.com/en/about-us#about.

IntroMedic. www.intromedic.com:549/eng/main/.

Jang, M. W., et al. "Enhancement of Date Retention Time in DRAM Using Step gaTed
AsymmetRic (STAR) Cell Transistors." *Proceedings of ESSDERC* (2005), 189–192.

Jeon, Sang-Woon. *A History of Korean Science and Technology*. Singapore: National University
of Singapore Press, 2011.

Jeong, Hoon, et al. "Long Life-Time Amorphous-InGaZnO TFT-Based Shift Register Using a
Reset Clock Signal." *IEEE Electron Device Letters*, *35:8* (2014), 844–846.

Jeong, Hoon, et al. "Temperature Sensor Made of Amorphous Indium-Gallium-Zinc Oxide
TFTs." *IEEE Electron Device Letters*, *34:12* (2103), 1569–1571.

Jeonja Shinmun. "Bandoche Big Deal Nonran." *Jeonja Shinmun* (September 17, 2012). www.
etnews.com/201209110681.

———. "Cho Byung-Jin KAISTgyosu 'Wearable Baljeonsoja', UNESCO Netexplo Award
Daesang." *Jeonja Shinmun* (February 4, 2015). www.etnews.com/20150204000275?m=1.

———. " 'Hynix Satae' Daeeung Yugam." *Jeonja Shinmun* (April 3, 2003). www.etnews.com/
200304020229.

———. "i3system, Sikdogeomsa Capsulehyeong Naesigyeong Yeonnae Gaebal." *Jeonja
Shinmun* (August 13, 2009). www.etnews.com/200908120125.

———. "IMD 2001: Daegieop." *Jeonja Shinmun* (August 28, 2001). https://m.etnews.com/
200108240271?SNS=00004.

———. "Intel 387 Hohwanchip Gaebal." *Jeonja Shinmun* (July 2, 1994). www.etnews.com/
199407020050.

———. "Jeo-on poli-silicon TFT-LCD Seolbituja Gyeongjaeng Sijakdwaetda." *Jeonja
Shinmun* (November 12, 2001). www.etnews.com/200111090199?m=1.

———. "KAIST Goseongneungjikjeopsistem Yeonkucenteo." *Jeonja Shinmun* (October 17,
2002). https://m.etnews.com/200210160056.

———. "KAIST Kim Choong-Ki Gyosutim, 3chawon jipjeop-inductor chut Gaebal." *Jeonja
Shinmun* (July 15, 1999). www.etnews.com/199907150074.

———. "Kyung Chong-Min–Naedal Munyeoneun 'Bandoche Seolgyegyoyukcenteo' Ssenteojang." *Jeonja Shinmun* (June 28, 1995). www.etnews.com/199506280039?m=1.

———. "Naroho Balsa Seongkong: sumeun Juyeok–Wang Hui-Jip i3system Suseokyeonkuwon." *Jeonja Shinmun* (January 31, 2013). https://m.etnews.com/201301310 364?obj=Tzo4OiJzdGRDbGFzcyI6Mjp7czo3OiJyZWZlcmVyIjtOO3M6NzoiZm9yd2 FyZCI7czoxMzoid2ViHRvIG1vYmlsZSI7fQ%3D%3D.

———. "Yoo Hoi-Jun and Cho Byung-Jinkyosu, 6hoe Kahng Dawonsang Susang ... 'AImit Goyujeonche Gisulbaljeon Giyeo'," *Jeonja Shinmun* (February 15, 2023), www.etnews. com/20230214000122.

Jesus de la Fuente. "Understanding Graphene." www.graphenea.com/pages/graphene#.YEqv TS2cYgp.

Jo [Cho], Byung-Jin, Choi Jin-Ho, and Kim Choong-Ki. "Slip Elimination in Rapid Thermal Processing." *Teukjeongyeonku Gyeolgwa Balpyohyoe Nonmunjip 1* (1989): 206–209.

Joong-Ang Ilbo. "3hoe Ho-Amsang Susangja Seonjeong." *Joong-Ang Ilbo* (February 23, 1993). www.joongang.co.kr/article/2788676.

———. "30nyeonjeon Eoryeopge Ppurin Ssiat, Yeon 300eok Dollar Yeolmaero." *San Francisco Joong-Ang Ilbo* (August 1, 2013). http://m.koreadaily.com/news/read.asp?art _id=1877832&referer=.

———. "Bandoche Misegongjeong 2-3nanoga Hangye, ijen 3D-soja Yeonku." *Joong-Ang Ilbo* (August 27, 2021). www.joongang.co.kr/article/24132626#home.

———. "'Bandoche Shinhwa Rival' Lim Hyung-Kyu, Hwang Chang-Gyu 2 Round." *Joong-Ang Ilbo* (January 23, 2014). https://news.joins.com/article/13712601.

———. "Bimemori Maechul 5nyeonhu 5wiro." *Joong-Ang Ibo* (August 28, 2002). https:// news.joins.com/article/4334076.

———. "Buhwalhan Hynix, nuga salryeotna." *Joong-Ang Ilbo* (November 26, 2014). https:// news.joins.com/article/16526233.

———. "Guknae Hakja 68%ga Gwahakagye Munjejeom 'Igongge Chabyeol' kkoba." *Joong-Ang Ilbo* (September 18, 2012). www.joongang.co.kr/article/9354007#home.

———. "Guknaeyeonkujin, Jeonjagiyong 'Graphene' Silyonghwa apdangkyeo." *Joong-Ang Ilbo* (February 29, 2012). https://news.joins.com/article/7497201.

———. "Hynix Bandoche Imwoninsa." *Joong-Ang Ilbo* (May 15, 2007). www.joongang. co.kr/article/2727160#home.

———. "Hynix Buhwal Dwitiyagi—500il." *Jung-Ang Ilbo* (March 14, 2007). www.joongang. co.kr/article/2661835#home.

———. "Jukgetda Shipdorok Gongbuhaetda ... Miss Yangseo Yang Hyang-Jassi Doetda." *Joong-Ang Ilbo* (September 17, 2018). https://news.joins.com/article/22975746.

———. "Kwon Young-Soo LG Displaysajang, 'Steve Jobs Geukchan deoke Gogeup Panel Gongkeup dalyeoyo.'" *Joong-Ang Ilbo* (July 24, 2010). www.joongang.co.kr/article/ 4335135#home.

———. "Lim Hyung-Kyu Samsung Gisulwonjangdeung 5myeong 'Olhae Gisul Gyeongyeonginsang.'" *Joong-Ang Ilbo* (February 22, 2007). https://news.joins.com/article/2641809.

———. "'Pyeongtaekkkaji motga,' 'Gyosudo motguhaeseo' ... Hyeonjangseo naon 'Bandoche Inryeoknan.'" *Joong-Ang Ilbo* (June 10, 2022). www.joongang.co.kr/article/ 25078089#home.

———. "Samsung 1ho Janghaksaeng 'Mr. NAND,' SK Hynix Deungkiisae." *Joong-Ang Ilbo* (March 5, 2014). https://news.joins.com/article/14075010.

———. "Samsung Jeonjareul Umjikineun Saramdeul." *Joong-Ang Ilbo* (February 23, 2002). https://news.joins.com/article/673226.

———. "Samsung LDI Bandoche Maechul 10eokbul Dolpa." *Joon-Ang Ilbo* (May 10, 2005). https://news.joins.com/article/44630.

————. "Samsungjeonja Bencheowa Sonjapgo Bimemori Bakcha." *Joong-Ang Ilbo* (July 30, 2001). https://news.joins.com/article/4110902.

————. "Samsungjeonja, 'Bimemori Seungbu' … 2001nyeonkkaji 12eokbul Tuip." *Joong-Ang Ilbo* (June 22, 1999). https://news.joins.com/article/3792150.

————. "System-on-chipdeung 11gae Chumdan Bimemory Bandoche, Samsungjeonja Guksanhwa Wanryo." *Joong-Ang Ilbo* (July 26, 2002). https://news.joins.com/article/4317696.

Jung [Chung], Han, Hee-Chul Lee, and Choong-Ki Kim. "Measurement of the Steady-State Minority Carrier Diffusion Length in a HgCdTe Photodiode." *Japanese Journal of Applied Physics, 35:10B* (1996), L1321–L1323.

KAIST. *See* Korea Advanced Institute of Science and Technology.

Kang, Byung-Jin, Jeong-Hun Mun, Chang-Yong Hwang, and Byung-Jin Cho. "Monolayer Graphene Growth on Sputtered Thin Film Platinum." *Journal of Applied Physics, 106* (2009), 104309. https://doi.org/10.1063/1.3254193.

Kang Jin-Ku. *Samsungjeonja Sinhwawa keu Bikyeol*. Seoul: Koryeowon, 1996.

Kang Ki-Dong. *Hankang-eu Gijeok, Hankuk Bandoche: Sijakeu Jinsil*. www.kdkelectronics.com/korea_semi_pdf/kdk-2014-0823mm.pdf.

Kang Kyung-Seok. *Hanguk Dojasa*. Seoul: Yegyeong, 2012.

KC Tech. www.kctech.com/index.php.

Kiln of 15th Chin Jukan. www.chin-jukan.co.jp/history.html.

Kim Byung-Woon. *Myeonbangjeok [Cotton Spinning]*. Seoul: Eulyumunhwasa, 1949.

Kim, Choong-Ki. "Carrier Transport in Charge-Coupled Devices." *1971 IEEE International Solid-State Circuits Conference*, University of Pennsylvania, Philadelphia (February 19, 1971), 158–159.

————. "Current Conduction in Junction-Gate Field-Effect." Ph.D. thesis, Columbia University, New York, 1970.

————. "Design and Operation of Buried Channel Charge-Coupled Devices," presented at CCD Applications Conference sponsored by the Naval Electronics Laboratory Center, San Diego (September 18–20, 1973).

————. "Guknae Bandoche Gongeop-eu Baljeon Hoego." *Jeonjagonghakhoeji, 13:5* (1986), 436–438.

————. "Jeonjagwanghak Yeonkuso" [Center for Electro-Optics]. *Gukbang-gwa Gisul, 263* (January 2001), 36–41.

————. "Rapid Thermal Processing for Submicron Devices." *Teukjeongyeonku Gyeolgwa Balpyohyoe Nonmunjip 1* (1988), 155–159.

————. "Two-Phase Charge-Coupled Linear Imaging Devices with Self-Aligned Implanted Barrier." *1974 International Electron Devices Meeting (IEDM)*, Washington, D.C. (December 9–11, 1974), 55–58.

Kim, Choong-Ki, J. M. Early, and G. F. Amelio. "Buried-Channel Charge-Coupled Devices," presented at Northeast Electronics Research and Engineering Meeting (NEREM), Boston (November 1–3, 1972).

Kim Choong-Ki, and Kim Jeong-Kyu. "Two-Step Rapid Thermal Diffusion of Phosphorous and Boron into Silicon from Solid Diffusion Sources." *Teukjeongyeonku Gyeolgwa Balpyohyoe Nonmunjip 1* (1989), 192–196.

Kim Choong-Ki, Lee Dong-Yup, and Jo [Cho] Byung-Jin. "Metal Alloy and Implantation Annealing by Rapid Thermal Processing." *Teukjeongyeonku Gyeolgwa Balpyohyoe Nonmunjip 1* (1989), 202–205.

Kim Choong-Ki, Nam Chang-Hee, and Kim Byung-Yoon. "Jeonjakwanghakteukhwayeonk usenteo (Center for Electro-Optics) Yeonkuhyeonhwang." *Gwanghakgwa Gisul, 2:1* (1998), 35–36.

Kim Choong-Ki, and E. H. Snow. "P-Channel Charge-Coupled Devices with Resistive Gate Structure." *Applied Physics Letters*, *20* (1972), 514.

Kim, Dong-Won. "The Godfather of South Korea's Chip Industry: How Kim Choong-Ki Helped the Nation Become a Semiconductor Superpower." *IEEE Spectrum* (August 28, 2022). https://spectrum.ieee.org/kim-choong-ki, and *IEEE Spectrum* (October 2022), 32–39.

———. "Transfer of 'Engineer's Mind': Kim Choong-Ki and the Semiconductor Industry in South Korea." *Engineering Studies*, *11:2* (2019), 83–108.

———. "Two Chemists in Two Koreas." *Ambix*, *52:1* (2005), 67–84.

———, ed. *Miraereul hyanghan kkeunimeopneun Dojeon: KAIST 35nyeon*. Daejeon: KAIST, 2005.

Kim Dong-Won, and Stuart W. Leslie. "Winning Markets or Winning Nobel Prizes? KAIST and the Challenge of Late Industrialization." *Osiris*, *13* (1998), 154–185.

Kim Dong-Won et al. *Hankukgwahakgisulwon Sabansegi: Miraereul hyanghan kkeunimeopneun Dojeon [The First Quarter Century of KAIST: An Endless Challenge to the Future]*. Daejeon: KAIST, 1996.

Kim Geun-Bae. *Hankuk Geundae Gwahakgisulinryeok-eu Chulhyeon*. Seoul: Munhakgwa Jiseongsa, 2005.

Kim Il-Gyeong. "Gyeongjaegaebalgyehoek." *Encyclopedia of Korean Culture*. http://encyko rea.aks.ac.kr/Contents/Item/E0002782.

Kim, Jeong-Gyoo, Byung-Jin Cho, and Choong-Ki Kim. "AES Study of Rapid Thermal Baron Diffusion into Silicon from a Solid Diffusion Source in Oxygen Ambient." *Journal of the Electrochemical Society, 137* (1990), 2857–2860.

Kim, Linsu. *Imitation to Innovation: The Dynamics of Korea's Technological Learning*. Boston: Harvard Business School Press, 1997.

Kim, Nam-Deog. "Fabrication of Hydrogenated Amorphous Silicon Thin-Film Transistor for Flat Panel Display." Master's thesis. Seoul: KAIST, 1986.

Kim, Su-Jin, Ju-Hyung We, and Byung-Jin Cho. "A Wearable Thermoelectric Generator Fabricated on a Glass Fabric." *Energy and Environmental Science*, *7* (2014), 1959–1965.

Kim Su-Yeon, Baik You-Jin, and Park Young-Ryeol. "Hankuk Bandochesaneopui Seongjangsa: Memory Bandochereul jungsimeuro (The Historical Review of the Semiconductor Industry)." *Gyeongyeongsahak*, *30:3* (2015), 145–166.

Kim, Tae-Song, Si-Young Song, Han Jung [Chung], Jinseok Kim, and Eui-Sung Yoon. "Micro Capsule Endoscope for Gastro-Intestinal Tract." *29th Annual International Conference of the IEEE Engineering in Medicine and Biology Society* (2007), 2823–2826.

Ko Dae-Seung. "The Establishment of the Office of Atomic Energy." Master's thesis, Seoul National University, 1991.

Koh, Yo-Hwan. "Latch-Free Self-Aligned Power MOSFET and IGBT Structure Utilizing Silicide Contact Technology." Doctoral thesis. Seoul: KAIST, 1989.

Kopytko, Malgorzata, and Antoni Rogalski. "Performance Evaluation of Type-II Superlattice Devices Relative to HgCdTe Photodiode." *IEEE Transactions of Electron Devices*, *69:6* (2022), 2992–3002.

Korea Advanced Institute of Science and Technology (KAIST). "A Ceremony in Honor of Professor Kim Choong-Ki (February 25, 2008)." KAIST Archive, video footage 2–10237.

Korea Broadcasting System (KBS). "Gisulgangkuk Project Jungkyeongmanri: IDIS, i3system." *KBS* (December 22, 2019). www.youtube.com/watch?v=7snCkez-jcg.

———. "Hankuk Wearable Baljeonjangchi, UNESCO Choegoui Gisul Seonjeong." *KBS News* (February 4, 2015). https://news.kbs.co.kr/news/view.do?ncd=3014902.

———. " 'Maenyeon 3000myeong Bujok' Bandoche Inryeoknan … Daechekeun?" *KBS News* (June 19, 2022). https://news.kbs.co.kr/news/view.do?ncd=5489649.

Korea Electronics Association. *Gijeokui Sigan, 50 (1959–2009): The Miraculous Time.* Seoul: Korea Electronics Association, 2009.

The Korea Herald. "Seoul Weighs 'Chip 4' Alliance as Ties with China Hang in Balance." *The Korea Herald* (August 8, 2022). www.koreaherald.com/view.php?ud=2022080 8000653.

Korea Industrial Technology Association. "Engineer of Korea (formerly Engineer of the Month)." https://m.koita.or.kr/m/mobile/award/kor_ng_search_view.aspx?no=483.

Korea Institute of Science and Technology (KIST). *KIST 25nyeonsa [Twenty-Five Year History of KIST]*. Seoul: KIST, 1994.

Korea Intellectual Property Rights Information Service (KIPRIS). "Ha Yong-Min, IPS." http://kpat.kipris.or.kr/kpat/searchLogina.do?next=MainSearch#page1.

———. "Ha Yong-Min, OLED." http://link.kipris.or.kr/link/AJAX/CTOTAL.jsp.

———. "Park Sung-Kye: No. 21~No. 38." http://kportal.kipris.or.kr/kportal/search/search_pat ent.do.

Korea National NanoFab Center. www.nnfc.re.kr.

Korea Net. "Charge Your Battery by Wearing It." *Korea Net* (April 11, 2014). www.korea.net/NewsFocus/Sci-Tech/view?articleId=118808.

The Korea Times. "LG Getting Closer with Apple," *The Korea Times* (August 17, 2021), www.koreatimes.co.kr/www/tech/2021/08/133_314046.html.

———. "LG to Supply 'Foldable Panels' for Apple." *The Korea Times* (February 20, 2021). www.koreatimes.co.kr/www/tech/2021/02/133_304312.html.

Kuo, Yue. "Thin Film Transistor Technology—Past, Present, and Future." *The Electrochemical Society Interface, 22: 1* (2013), 55–61.

Kyung Chong-Min. "Bandoche Seolgyegisului Hyeonhwangkwa Jeonmang," *Jeonjajinheung, 4:2* (1984), 12–15.

———. "Bandoche Seolgyegyoyukcenteo (IDEC) Saeopeul tonghan Bimemory mit System-seolgye Hwalseonghwa Bangan." *Hankuktongshinhakhoeji, 13:11* (1996), 1232–1240.

———. "Bandoche Seolgyegyoyukcenteo (IDEC)ui Saeopgyehoek." *Jeonjagonghakhoeji, 22:10* (1995), 1122–1130.

———. *Changeopga [Startup Entrepreneur]*. Seoul: Yulgok, 2020.

———. "Daeman Bandoche igiryeomyeon." *Jeonja Shinmun* (September 27, 2010), www.etnews.com/201009200041.

———. "Daemanui Jeongbuchulyeon Yeonkuso." *Jeonja Shinmun* (June 7, 1994). www.etnews.com/199406070035.

———. "Dojeonhajianneun Jeoleumeun eopda," in Park Yoon-Chang et al., *Nobelsangeul Gaseume pumgo*. Seoul: Dong-A Ilbo, 1994, 131–145.

———. "Gukchaek Inryeokyangseong Bichaek." *Jeonja Shinmum* (August 23, 2004), www.etnews.com/200408220004.

———. "Guknae Bandochesaneopui Miraewa Gisulinryeokui Yangseong munje." *Jeonjajinheung, 6:5* (1986), 2–5.

———. *Igongkyega salaya Naraga sanda*. Seoul: Yas Media, 2004.

———. "Jeolmeun Gongdaegyosu 7gyemyeong." *Jeonja Shinmun* (July 1, 1993), 4.

———. "Jeonmunkaga shinnaneun Sahoereul mandeulja." *Hankyoreh* (January 29, 1994). www.hani.co.kr/arti/legacy/legacy_general/L291691.html.

———. "Jikjeophoero Jejogongjeongui Model." *Jeonjagonghakhoejapji, 11:5* (1984), 1–4.

———. "KAISTbaljeon mit Kyeongyeongbanghyange daehan Sogyeon," unpublished statement (October 30, 2020).

———. *Keun Namuga jaraneun Ttang*. Seoul: Sigma Press, 1999.

———. "Low Level Currents in Buried Channel MOS Transistor." Master's thesis. Seoul: KAIS, 1977.

————. *Nano Devices and Circuit Techniques for Low-Energy Applications and Energy Harvesting: KAIST Research Series.* Dordrecht: Springer, 2016.

————. "A New Charge-Coupled Analog-to-Digital Converter." Doctoral thesis. Seoul: KAIST, 1981.

————. "Populisme daehan Gyeongkye." *Gukmin-Ilbo* (November 25, 2004), http://news.kmib.co.kr/article/viewDetail.asp?newsClusterNo=01100201.20041125000002301.

————. "Sangsikeul Jonjunghaneun Jidoja." *Gukmin-Ilbo* (April 21, 2005), http://news.kmib.co.kr/article/viewDetail.asp?newsClusterNo=01100201.20050421100000306.

————. "Uri Cheomdansaneopi kkot Piryeomyeon." *Computer World* (January 1992), 81.

————, ed. *Theory and Applications of Smart Cameras: KAIST Research Series.* Dordrecht: Springer, 2016.

Kyung, Chong-Min, and Choong-Ki Kim. "Charge-Coupled A/D Converter." *1981 Custom Integrated Circuit Conference*, Rochester, New York (May 1981), 621–626.

————. "Pipeline Analog-to-Digital Conversion with Charge-Coupled Devices (correspondence)." *IEEE Journal of Solid-State Circuits, 15* (April 1980), 255–257.

————. "A Two-Dimensional Analysis of the Low-Level Currents in Buried Channel MOS Transistors." *Jeonjagonghakhoeji, 15:6* (1978), 35–38.

Kyung, Chong-Min, et al. "HK386: an X86-compatible 32-bit CISC Microprocessor." *Proceedings of the ASP-DAC 1997* (January 28–31, 1997), 661–662.

Kyung, Chong-Min, et al. *The Structure and Design of the High-Performance Microprocessor* (in Korean). Seoul: Daeyoungsa, 2000.

Kyunghyang Shinmun. "Big Deal 6nyeon 'Gwanchi Akmong …' jukeotda salatda." *Kyunghyang Shinmun* (December 23, 2004). https://m.khan.co.kr/economy/industry-trade/article/200412231747141.

Kyungpook National University. "School of Electronics Engineering." https://see.knu.ac.kr/eng/content/faculty/greeting.html.

Kyungseong Bangjik. *Kyungseong Bangjik 50nyeon (1919–1969).* Seoul: Kyungseong Bangjik, 1969.

Lee Byung-Chul. *Hoamjajeon [Autobiograpny of Lee Byung-Chul].* Seoul: Joong-Ang Ilbo, 1986.

Lee, Ga-Won, et al. "Characterization of Polymer Gate Transistors with Low-Temperature Atomic-Layer-Deposition-Grown Oxide Spacer." *IEEE Electron Device Letters, 30:2* (2009), 181–184.

Lee, Heon-Bok, Ju-Hyung We, Hyun-Jeong Yang, Kukjoo Kim, Kyung-Cheol Choi, and Byung-Jin Cho. "Thermoelectric Properties of Screen-Printed ZnSb Film." *Thin Solid Films, 519* (20100), 5441–5443.

Lee, Hyunjin, et al. "Fully Integrated and Functioned 44nm DRAM Technology for 1Gb DRAM." *2008 Symposium on VLSI Technology Digest of Technical Papers* (2008), 86–87.

Lee Jae-Hoon, Woo-Jin Nam, Hee-Sun Shin, Min-Koo Han, Yong-Min Ha, Chang-Hwan Lee, Hong-Seok Choi, and Soon-Kwang Hong. "Highly Efficient Current Scaling AMOLED Panel Employing A New Current Mirror Pixel Circuit Fabricated by Excimer Laser Annealed Poly-Si TFT." *IEEE International Electron Device Meeting (IEDM) Technical Digest* (2005), 931–934.

Lee Jae-Hyun. *Samsung iraeseo ganghada: Samsungjeonja Bandoche Memorisaeopbu Iyagi 1.* Seoul: Barunbooks, 2017.

Lee, Seok-Hee, Byung-Jin Cho, Jong-Chul Kim, and Soo-Han Choi. "Quasi-Breakdown of Ultrathin Gate Oxide under High Field Stress." *IEEE International Electron Devices Meeting (IEDM) Technical Digest* (December 1994), 605–608.

LG Display. *2009 Annual Report* (in Korean). March 17, 2010.

————. *2010 Annual Report* (in Korean). March 28, 2011.

———. *2012 Annual Report* (in Korean). March 21, 2013.

———. *2013 Annual Report* (in Korean). March 21, 2014.

———. *2014 Annual Report* (in Korean). March 30, 2015.

———. *2017 Annual Report* (in Korean). April 2, 2018.

———. *2018 Annual Report* (in Korean). April 1, 2019.

———. *2019 Annual Report* (in Korean). March 30, 2020.

———. *2020 Annual Report* (in Korean). March 15, 2021.

LG Electronics. *LGjeonja 50nyeonsa* [*LG Electronics 50-Year History*], 4 volumes. Seoul: LG Electronics, 2008.

LG Philips LCD. *2002 Annual Report* (in Korean). March 31, 2003.

———. *2004 Annual Report* (in Korean). March 31, 2005.

———. *2005 Annual Report* (in Korean). March 31, 2006.

———. *2006 Annual Report* (in Korean). March 30, 2007.

———. *2008 Annual Report* (in Korean). March 31, 2009.

LG Sciencepark. "President Message." www.lgsciencepark.com/EN/about.php.

Lim, Hyung-Kyu. "Charge-Based Modeling of Thin-Film Silicon-on-Insulator MOS Field-Effect Transistors." Ph.D. thesis, University of Florida, 1984.

———. "Design and Fabrication of a Seven-Segment Decoder/Drive with PMOS Technology." Master's thesis. Seoul: KAIS, 1978. Later published in *Jeonjagonghakhoeji, 15:3* (July 1978), 11–17.

Lim, Hyung-Kyu, and Jerry G. Fossum. "An Analytic Characterization of Weak-Inversion Drift Current in a Long-Channel MOSFET." *IEEE Transactions on Electron Devices, 30:6* (1983), 713–715.

———. "A Charge-Based Large-Signal Model for Thin-Film SOI MOSFET's." *IEEE Transactions on Electron Devices, 32:2* (1985), 446–457.

———. "Current Voltage Characteristics of Thin-Film SOI MOSFET's Strong Inversion." *IEEE Transactions on Electron Devices, 31:4* (1984), 401–408.

———. "Threshold Voltage of Thin-Film Silicon-on-Insulator (SOI) MOSFET's." *IEEE Transactions on Electron Devices, 30:10* (1983), 1244–1251.

Lim, Youngil. *Technology and Productivity: The Korean Way of Learning and Catching Up.* Cambridge, Mass: The MIT Press, 1999.

Lin, Youn-Long, Chong-Min Kyung, Hiroto Yasuura, and Yongpan Liu, eds. *Smart Sensors and Systems.* Dordrecht: Springer, 2015.

LNF-Wiki. "Low Pressure Chemical Vapor Deposition." https://lnf-wiki.eecs.umich.edu/wiki/Low_pressure_chemical_vapor_deposition.

Macrotrends. "South Korea (Economy)." www.macrotrends.net/countries/KOR/south-korea/gnp-gross-national-product.

Maeil Kyungje. "Bandoche Wigigam keojineunde Teukbyeolbeop mung-gaegoitneun Gukhoe-eu Musa-anil." *Maeil Kyungje* (September 7, 2022). www.mk.co.kr/opinion/editorial/view/2022/09/792545/.

———. "Bimemorido Samsung apjang … Chin Dae-Jedaepyo." *Maeil Kyungje* (February 27, 1997). www.mk.co.kr/news/it/view/1997/02/11540/.

———. "'Chakyonghyeong Jeonjagigi' Jejak Pilsumuljilin Gobunja Jeolyeonmak Gaebal." *Maeil Kyungje* (March 12, 2015). www.mk.co.kr/news/it/view/2015/03/227417/.

———. "Chey Tae-Won, Imwonbandae-edo Gyeoldan … Hynix 12jo beoleodeulineun Hoesaro Daebyeonsin." *Maeil Kyungje* (January 20, 2022). www.mk.co.kr/news/business/view/2022/01/57791/.

———. "Dr. Chung Han." *Maeil Kyungje* (February 4, 2003). www.mk.co.kr/news/home/view/2003/02/38971/.

———. "Hynixsajang Kim Jong-Gabssi Naejeong." *Maeil Kyungje* (February 27, 2007). www.mk.co.kr/news/home/view/2007/02/102236/.

———. "Hyundaijeonja Daeduke Goseongneugjikjeopsistem Yeonkucenter Gigong." *Maeil Kyungje* (November 23, 1994). www.mk.co.kr/news/home/view/1994/11/60172/.

———. "iPhone 4 tteuteoboni CPU-Samsung, LCD-LGjepum." *Maeil Kyungje* (June 9, 2010). www.mk.co.kr/news/business/view/2010/06/298328/.

———. "iPhone X Bujin Jikgyeoktan … Samsung Disply Gadongyul Bantomak Uryeo." *Maeil Kyungje* (February 2, 2018). http://vip.mk.co.kr/news/view/21/20/1571953.html.

———. "Jeongjinki Eonronmunhwasang Sisang." *Maeil Kyungje* (July 13, 1989). http://m.mk.co.kr/onews/1989/999213#mkmain.

———. "K Bandoche, Mieopsineun Jonripbulga … Jung Nunchibol Pilyoeopseo." *Maeil Kyungje* (April 20, 2021). www.mk.co.kr/news/economy/view/2021/04/380924/.

———. "LG Display, 10nyeon Georaecheo Apple notchyeotda." *Maeil Kyungje* (January 26, 2017). www.mk.co.kr/news/business/view/2017/01/60988/.

———. "Samsung Saesaup Yeonkubijung 50%ro." *Maeil Kyungje* (July 12, 2007). www.mk.co.kr/news/business/view/2007/07/368243/.

———. "Samsung-group 468myeong Choedaegyumo Imwoninsa Danhaeng." *Maeil Kyungje* (December 8, 1995). www.mk.co.kr/news/business/view/1995/12/56489/.

———. "Samsungjeonja 'Dream Team'e Haeksimproject matgyeo." *Maeil Kyungje* (September 23, 2004). www.mk.co.kr/news/economy/view/2004/09/333443/.

———. "Sege Choeso Capsule Naesigyeong Gaebal." *Maeil Kyungje* (January 28, 2003). www.mk.co.kr/news/it/view/2003/01/31687/.

———. "Segye Choecho, Cheoneuro Jeonkireul Saengsanhaneun 'Wearable Baljeonsoja.' " *Maeil Kyungje* (February 21, 2015). www.mk.co.kr/news/home/view/2015/02/167666/.

Magazine Hankyung. "LG Display, 2018 Imwon Insa … 'Yeokdae choedae 26myeong Seungjin.' " *Magazine Hankyung* (November 30, 2017). https://magazine.hankyung.com/business/article/201711307543b.

Marshable. "Apple Is Trying to Rely Less on Samsung by Having LG Make Some of Its iPhone OLED Screens." *Marshable* (June 28, 2018). https://mashable.com/article/apple-lg-samsung-oled-screens.

Masuoka, Fujio. "Technology Trend of Flash-EEPROM. Can Flash-EEPROM Overcome DRAM?" *1992 Symposium on VLSI Technology Digest of Technical Papers* (1992), 6–9.

Matthews, John A., and Dong-Sung Cho. *Tiger Technology: The Creation of a Semiconductor Industry in East Asia.* Cambridge: Cambridge University Press, 2000.

Media SK. "Park Sung-Kye at SK Hynix, Who Lives in the Nanometer World." *Media SK* (January 18, 2013). http://mediask.co.kr/1115.

Michigan Medicine. University of Michigan "Capsule Endoscopy." www.uofmhealth.org/conditions-treatments/digestive-and-liver-health/capsule-endoscopy.

Ministry of Education, Science and Technology. "2011 'Global Frontier-Saeop' Gongo." June 17, 2011 press release.

Ministry of Government Administration. *The Official Gazette, No. 8740* (in Korean). Seoul: Ministry of Government Administration, January 14, 1981.

Ministry of Industry and Commerce. "Ten New Technologies in 2004" (in Korean). January 20, 2004 press release.

Ministry of Knowledge and Economy. "Soje-Bupum Seongkwabogo Daehoe." November 1, 2012 press release.

Ministry of Science and Technology. "Chosohyeong Gogineung Capsulhyeong Naesigyeong 'MIRO' Gaebal." January 28, 2003 press release.

———. "Policy and Strategy for Science and Technology" (1975), 40, in Frederick E. Terman Papers, Department of Special Collections, Stanford University, Stanford, CA, SC 160 Vi, 20/7.

Ministry of Science, ICT, and Future Planning, "Wearable Che-on Jeonryeoksaengsan Gisul, 2015 UNSECO 10dae Gisul 1-wi Grand Prix Susang." February 4, 2015 press release.

Ministry of Trade, Industry and Energy. "21segi Frontier Gisulgaebalsaeop Project Sujuhyeonhwang." press release (circa 2005, exact date unknown).

———. "Bandoche, Display Saneop Donghyang." *e-Narajipyo* (September 14, 2022). www.index.go.kr/unity/potal/main/EachDtlPageDetail.do?idx_cd=1155 (searched on October 6, 2022).

———. "The Fifth Display Day." October 6, 2014 press release.

Momodomi, Masaki, et al. "New Device Technologies for 5 V-only 4Mb EEPROM with NAND Structure Cell." *International Electron Devices Meeting (IEDM) Technical Digest* (1988), 412–415.

Money Today. "iPhone 4 Bupumjung LG Displayga gajang bissa." *Money Today* (June 29, 2010). https://news.mt.co.kr/mtview.php?no=2010062910394529760.

Moon, Chan. "Fabrication of $Hg_{0.7}Cd_{0.3}Te$ IR Photodiode and Investigation on the Reverse Leakage Current." Master's thesis. Seoul: KAIST, 1990.

Moon, Hanul, et al. "Synthesis of Ultrathin Polymer Insulating Layers by Initiated Chemical Vapour Deposition for Low-Powered Soft Electronics." *Nature Materials, 14* (2015), 628–635.

Moon Man-Yong. "Hankukgwahakgisulyeonkuwon (KIST) ui Byeoncheon-gwa Yeonkuhwaldong." *Hankuk Gwahaksa Hakhoeji, 28:1* (2006), 81–115.

Munhwa Broadcasting Corporation (MBC). "100-minutes Discussion" (May 9, 2002).

———. "Jeongbuchaegwondan Hynix Heolgapmaegak Nonran." *MBC News* (April 19, 2002). https://imnews.imbc.com/replay/2002/nwdesk/article/1889228_30761.html.

Nam, Choong-Mo, Sung-Kye Park, Sang-Ho Lee, Jai-Bum Suh, Gyu-Han Yoon, and Sung-Ho Jang. "A New Extraction Method of Retention from the Leakage Current in 0.23 μm DRAM Memory Cell." *Proceedings of the 2000 International Conference on Microelectronic Test Structures* (2000), 102–105.

Nanoelectronic And Neuromorphic Device Lab (NAND Lab). "Culture Column" in Culture Activity. https://nand.kaist.ac.kr:54856/bbs/board.php?bo_table=sub6_1&page=1.

———. "Monolithic 3D Integration." https://nand.kaist.ac.kr:54856/sub2_1_b.php.

———. "Research Topics." https://nand.kaist.ac.kr:54856.

National Archives of Korea. "Girokeuro boneun Gyeongjaegaebal 5gaenyeon Gyehoek." https://theme.archives.go.kr//next/economicDevelopment/overview.do.

National Chiao Tung University. "Simon M. Sze." https://eenctu.nctu.edu.tw/en/teacher/p1.php?num=127&page=1.

National Inventors Hall of Fame. "Dawon Kahang." www.invent.org/inductees/dawon-kahng.

National NanoFab Center. "Mission and Vision." www.nnfc.re.kr/eng/pageView/322.

National Research Foundation of Korea. "Engineering Research Center." www.nrf.re.kr/cms/page/main?menu_no=131.

National University of Singapore (NUS) Faculty of Engineering. *A Vision for Tomorrow: Annual Report 2004–2005.* Singapore: NUS, 2005. www.eng.nus.edu.sg/wp-content/uploads/2019/02/NUS_AR_2004.pdf.

The New York Times. "Losses Mount at Hynix as Chip Prices Slump." *The New York Times* (February 5, 2009). www.nytimes.com/2009/02/05/business/worldbusiness/05iht-chip.1.19949158.html.

———. "The Silicon Godfather; The Man Behind Taiwan's Rise in the Chip Industry." *The New York Times* (February 1, 2000). www.nytimes.com/2000/02/01/business/the-silicon-godfather-the-man-behind-taiwan-s-rise-in-the-chip-industry.html.

———. "With Turnaround, Hynix Comes Full Circle." *The New York Times* (November 28, 2006). www.nytimes.com/2006/11/28/business/worldbusiness/28iht-hynix.3696087.html.

Newspim. "LGga Hynix Insue Chamyeo anhaetdeon Iyu." *Newspim* (July 13, 2011). www.newspim.com/news/view/20110713000206.

———. "SK Hynix, Segye Choego 96dan NAND Flash Gaebal bonkyeokhwa." *Newspim* (December 7, 2017)., www.newspim.com/news/view/20171207000187.

OBELAB. "Application Areas." www.obelab.com/product/product_nirsit.php.

Oh, Choon-Sik. "MOSFET Source and Drain Structures for High-Density CMOS Integrated Circuits." Doctoral thesis. Seoul: KAIST, 1986.

Oh, Choon-Sik, Yo-Hwan Koh, and Choong-Ki Kim. "A New P-Channel MOSFET Structure with Schottky-Clamped Source and Drain." *International Electron Devices Meeting (IEDM)*, San Francisco (December 1984), 609–613.

Oh, Myung, and James F. Larson. *Digital Development in Korea*. New York: Routledge, 2011.

Okimoto, Daniel I., Takuo Sugano, and Franklin B. Weinstein, eds. *Competitive Edge: The Semiconductor Industry in the U.S. and Japan*. Stanford, Calif.: Stanford University Press, 1984.

OLED-info. "DSCC Details its 2020 OLED Market Estimates." *OLED-info* (March 24, 2020). www.oled-info.com/dscc-details-their-2020-oled-market-estimates.

———. "DSCC: OLED to Overtake LCD Production Capacity for Mobile Applications in 2020." *OLED-info* (September 23, 2019). www.oled-info.com/dscc-oled-overtake-lcd-production-mobile-applications-2020.

———. "An Introduction to OLED Displays." www.oled-info.com/oled-introduction.

———. "UBI: The OLED Market Grew Only 0.7% in 2020, Details Revenue by Company." *OLED-info* (March 10, 2021). www.oled-info.com/ubi-oled-market-grew-only-07-2020-details-revenue-company.

Omdia. "BOE Becomes World's Largest Flat-Panel Display Manufacturer in 2019 as China Continues Rise to Global Market Dominance." *Omdia* (June 4, 2019). https://omdia.tech.informa.com/OM003804/BOE-Becomes-Worlds-Largest-Flat-Panel-Display-Manufacturer-in-2019-as-China-Continues-Rise-to-Global-Market-Dominance.

Opinion News. "1995nyeon cheot Bandoche Hohwang eottaetna." *Opinion News* (December 22, 2020). www.opinionnews.co.kr/news/articleView.html?idxno=44364.

Optris. "Focal Plane Array (FPA)." www.optris.com/focal-plane-array.

Pang Pyoug-Son et al. *Hanbandoui Heuktojagiro taeonada*. Seoul: Kyeongjin Munhwasa, 2010.

Park Bang-Ju. "Kyung Chong-Min Miraechangjokwahakbu ITyunghapsistemsaeopdanjang." *Science and Technology*, 528 (2013), 66–69.

Park Jae-Woo, Park Sung-Kye, and Kim Choong-Ki. "LOCOS Process to Reduce the Edge Effect on SOI NMOSFET." *Daehanjeonjagonghakhoe Haksuldaehoe*, 17:1 (1994), 209–210.

Park, Jong-Kyung, Seung-Min Song, Jeong-Hun Mun, and Byung-Jin Cho. "Graphene Gate Electrode for MOS Structure-Based Electronic Devices." *NANO Letters* (November 2011), 5383–5386.

Park Jong-Man. "Technological Trend for Wireless Ingestible Capsule Design." *Hankuktongsh inhakhoenonmunjip*, 34:12 (2009), 1524–1534.

Park No-Chun. "Munchebanjeong." *Encyclopedia of Korean Culture*. http://encykorea.aks.ac.kr/Contents/Item/E0019697.

Park Sang-In et al. *Uri Kim Choong-Ki Seonsaengnim* [*Our Teacher, Kim Choong-Ki*]. Daejeon: Privately printed, 2002.

Park Seong-Rae et al. *A Study of the Formation of Modern Scientists and Engineers in Korea* (in Korean). Daejeon: Korea Science and Engineering Foundation Report, 1995.

———. *A Study of the Foundation of Modern Scientists and Engineers in Korea II: Studying in America* (in Korean). Daejeon: Korea Science and Engineering Foundation Report, 1998.

Park Song-Bae, Kyung Chong-Min, Im In-Chil, Cha Gyun-Hyun, and Kim Hyung-Gon. "Research on the Development of CAD Software." *Teukjeongyeonku Gyeolgwa Balpyohoe Nonmunjip* (1987), 69–72.

Park Sung-Kye. "Design of Metal Gate CMOS Process Using Rapid Thermal Process." Master's thesis. Seoul: KAIST, 1990.

———. "Device Design for Suppression of Floating Body Effect in Fully Depleted SOI MOSFET's." Doctoral thesis. Daejeon: KAIST, 1994.

Park Sung-Kye and Kim Choong-Ki. "A Device Parameter Extraction Method for Thin Film SOI MOSFET's." *Daehanjeonkihakhoe Haksuldaehoe Nonmunjip*, 7 (1992), 820–824.

Park Sung-Kye, Young-Chul Lee, Moon-Sik Suh, Sang-Ho Lee, Hyung-Jae Lee, and Gyu-Han Yoon. "Moisture Induced Hump Characteristics of Shallow Trench-Isolated Sub 1/4 μm nMOSFET." *IEICE Technical Report*, 99:234 (1999), 21–25.

Park Sung-Kye, Moon-Sik Suh, Jae-Young Kim, and Sung-Ho Jang. "CMOSFET Characteristics Induced by Moisture Diffusion from Inter-Layer Dielectric in 0.23 μm DRAM Technology with Shallow Trench Isolation." *2000 IEEE International Reliability Physics Symposium Proceedings* (2000), 164–168.

Park Young-Jun. *VLSI Soja Iron* [*Theory of VLSI Device*] (in Korean). Seoul: Kyohaksa, 1995.

People of Kagoshima. "Chin Jukan Kiln." https://peopleofkagoshima.com/arts-and-crafts/sats uma-ware-chin-jukan-kiln/.

Person of Distinguished Service to Science and Technology. "Godfather of Semiconductor Industry Who Led the Technological Development by Problem-Solving Research and Education." www.koreascientists.kr/eng/merit/merit-list/?boardId=bbs_000000000 0000051&mode=view&cntId=58&category=2019&pageIdx=.

Phys Org. "Hynix Develops World's Fastest and Highest Density Graphic Memory." *Phys Org* (December 5, 2005). https://phys.org/news/2005-12-hynix-worlds-fastest-highest-dens ity.html.

Pico SERS. "Core Technology." www.picofd.com/technology/core-technology/.

Point2 Tech. "About Us." www.point2tech.com/product_ETube.php.

POSCO TJ Park Foundation. "2021 POSCO TJ Park Prize Awardees." www.postf.org/en/page/ award/history.do.

Pulse News. "5-Month-Old-Start-up TEGway wins Grand Prix at UNESCO's Netexplo Award." *Pulse News* (February 5, 2015). https://pulsenews.co.kr/view.php?year= 2015&no=118570.

Ramakrishana, S., and T. C. Lim. "Overview of the NUS Nanoscience and Nanotechnology Initiative and Its Available Facilities," presented at South East Asia Materials Network Meeting, IMRE, Singapore (November 14–16, 2005). www.icmr.ucsb.edu/programs/ archive/documents/Kim.pdf.

Reference for Business. "Concurrent Engineering." www.referenceforbusiness.com/managem ent/Comp-De/Concurrent-Engineering.html.

Reuters. "China's Memory Upstart YMTC Edges Closer to Rivals with 232-layer Chip." *Reuters* (August 4, 2022). www.reuters.com/technology/chinas-memory-upstart-ymtc-edges-closer-rivals-with-232-layer-chip-2022-08-04/.

———. "LG Signs LCD Supply Deal with Apple." *Reuters* (January 11, 2009). www.reut ers.com/article/us-lgdisplay-apple/lg-signs-lcd-supply-deal-with-apple-idUSTRE50B 0FW20090112.

———. "SK Hynix Says Has Developed Its Most Advanced 238-layer Storage Chip." *Reuters* (August 2, 2022). www.reuters.com/technology/sk-hynix-says-has-developed-its-most-advanced-238-layer-storage-chip-2022-08-02/.

Sakakibara, Kiyonori. *From Imitation to Innovation: The Very Large Scale Integrated (VLSI) Semiconductor Project in Japan.* Cambridge, Mass.: MIT Press, 1983.

Sakui, Koji. "Professor Fujio Masuoka's Passion and Patience Toward Flash Memory." *IEEE Solid-State Circuits Magazine, 5:4* (2013), 30–33.

Samsung. "Samsung's System LSI Business Introduced Strategic Business Opportunities with Local Design Houses" (July 30, 2001). www.samsung.com/semiconductor/newsroom/ news-events/samsungs-system-lsi-business-introduces-strategic-business-opportunit ies-with-local-design-houses/.

Samsung Advanced Institute of Technology (SAIT). "About SAIT." www.sait.samsung.co.kr/ saithome/about/who.do.

Samsung Display. "History." www.samsungdisplay.com/eng/intro/history/2000s.jsp#anchor.

Samsung Electronics. *2002 Annual Report* (in Korean). Suwon, Kyeongkido: Samsung Electronics, 2003.

———. *2004 Annual Report* (in Korean). Suwon, Kyeongkido: Samsung Electronics, 2005.

———. *2005 Annual Report* (in Korean). Suwon, Kyeongkido: Samsung Electronics, 2006.

———. *Samsungjeonja 30nyeonsa* [*Thirty-Year History of Samsung Electronics*]. Seoul: Samsung Electronics 1999.

———. *Samsungjeonja 40nyeon: Dojeonkwa Chongjoui Yeoksa* [*Forty-year History of Samsung Electronics: The Legacy of the Challenge and Creativity*]. Suwon, Kyeongkido: Samsung Electronics, 2010.

Samsung Newsroom. "Samsungjeonja 2022nyeon Jeongki Imwon Insa" (December 9, 2021). https://news.samsung.com/kr/삼성전자-2022년-정기-임원-인사.

Samsung Semiconductor and Telecommunications. *Samsung Bandochetongsin 10nyeonsa* [*Ten-Year History of Samsung Semiconductor and Telecommunications*]. Seoul: Samsung Bandochetongsin,1987.

Semicon China. "Dr. Simon M. Sze." www.semiconchina.org/en/788.

Seo, Kwang-Seok, and Choong-Ki Kim. "On the Geometrical Factor of Lateral p-n-p Transistors." *IEEE Transactions on Electron Devices, ED-27* (January 1980), 295–297.

Seo Moon-Seok. "Iljeha Gogeupseomyugisuljadeuleu Yanseong-gwa Sahoejinchule gwanhan Yeongu." *Kyungjesahak, 34* (2003), 83–116.

Seoul Kyungje. "Jeong Kyeong-Do gukbang, '4chsanseophyeokmyeong Haekshimgisul Hwalyonghae Bangsansuchul Gyeongjaengryeok Ganghwa.'" *Seoul Kyungje* (December 14, 2018). www.yna.co.kr/view/AKR20181214119400503.

Seoul National University (SNU). "Minnesota Project, Seouldaehakgyo Jaegeoneul wihan Noryeok." www.snu.ac.kr/about/history/history_record?md=v&bbsidx=131605.

———. *Seouldeahakgyo 50nyeonsa* [*Fifty-Year History of Seoul National University*], 2 volumes. Seoul: Seoul National University Press, 1998.

Seth, Michael J. *A Concise History of Korea*, 2nd edition. Lanham, Md.: Rowman & Littlefield, 2016.

Shulaker, Max M., Tony F. Wu, Mohamed M. Sabry, Hai Wei, H.S. Philip Wong, and Subhasish Mitra. "Monolithic 3D Integration: A Path from Concept to Reality," *2015 Design, Automation & Test in Europe Conference & Exhibition* (2015), 1197–1202.

SK. "History." https://eng.sk.com/history.

———. "Our Companies." https://eng.sk.com/companies#industry.

SK Hynix. *2012 Annual Report* (in Korean). March 28, 2013.

———. *2013 Annual Report* (in Korean). March 31, 2014.

———. *2014 Annual Report* (in Korean). March 31, 2015.

———. *2015 Annual Report* (in Korean). March 20, 2016.

———. *2016 Annual Report* (in Korean). March 31, 2017.

———. *2017 Annual Report* (in Korean). April 2, 2018.

———. "Hyundai Electronics-Hyundai Semiconductor, Tonghapbeopineuro Gongsikchul-beom" (October 15, 1999). https://news.skhynix.co.kr/presscenter/officially-launched-as-an-integrated-corporation.

———. "SK Hynix Develops World's Highest 238-Layer 4D NAND Flash" (August 2, 2022). https://news.skhynix.com/sk-hynix-develops-worlds-highest-238-layer-4d-nand-flash/.

———. "SK Hynix Inc. Launches the World's First 'CTF-based 4D NAND Flash' (96-Layer 512Gb TLC)." *SK Hynix News* (November 4, 2018). https://news.skhynix.com/sk-hynix-inc-launches-the-worlds-first-ctf-based-4d-nand-flash-96-layer-512gb-tlc/.

———. "SK Hynix Starts Mass-Producing World's first 128-Layer 4-D NAND." *SK Hynix News Newsroom* (June 26, 2019). https://news.skhynix.com/sk-hynix-starts-mass-producing-worlds-first-128-layer-4d-nand/.

———. "SK Hynix Unveils the Industry's Most Multilayered 176-Layer 4-D NAND Flash." *SK Hynix Newsroom* (December 4, 2020). https://news.skhynix.com/sk-hynix-unveils-the-industrys-highest-layer-176-layer-4d-nand-flash/.

Sisa Journal. "Bandochesaeop Big Deal mueoteul namgyeotna." *Sisa Journal* (January 7, 1999). www.sisajournal.com/news/articleView.html?idxno=81583.

Son Seung-Chul. "Imjinwaeran Piroin." *Encylopedia of Korean Culture*. http://encykorea.aks.ac.kr/Contents/Item/E0073566.

———. "Joseontongshinsa Piroin Swaehwankwa geuhanke." *Jeonbuk Sahak*, *42* (2013), 167–200.

Song Sungsoo. "The Growth of Samsung's Semiconductor Sector and the Development of Technological Capabilities" (in Korean). *Hangukgwahasa Hakhoeji*, *20:2* (1998), 151–188.

Statista. "Global LCD TV Panel Unit Shipments from H1 2016 to H1 2020, by Vendor." *Statista* (March 9, 2022). www.statista.com/statistics/760270/global-market-share-of-led-lcd-tv-vendors/.

———. "NAND flash Manufacturers Revenue Share Worldwide from 2010 to 2021 by Quarter." *Statista* (August 31, 2021). www.statista.com/statistics/275886/market-share-held-by-leading-nand-flash-memory-manufacturers-worldwide/.

Streetman, Ben G., and Sanjay Kumar Banerjee. *Solid State Electronic Devices*, 7th edition. Harlow, Essex: Pearson, 2016.

Suh, Kang-Deog, et al. "A 3.3V 32Mb NAND Flash Memory with Incremental Step Pulse Programming Scheme." *Proceedings of 1995 International Solid-State Circuits Conference* (1995), 128–129.

Taiwan Semiconductor Manufacturing Company (TSMC). "Dedicated IC Foundry." www.tsmc.com/english/dedicatedFoundry.

Techspot. "SK Hynix Launches 96-layer 4D NAND Flash." *Techspot* (November 6, 2018). www.techspot.com/news/77269-sk-hynix-launches-96-layer-4d-nand-ssd.html.

Terman, Frederick E., et al. "Survey Report on the Establishment of the Korea Advanced Institute of Science," in Frederick E. Terman Papers, Department of Special Collections, Stanford University, Stanford, CA, SC 160. X313, 9.

Technology and Innovation. "Choegogisulgyeongyeongin Interview—SK Hynix Miraegisulwon R&D Gongjeongdamdang Pi Seung-Hobusajang." *Technology and Innovation*, *437* (January 2020). http://azine.kr/m/_webzine/wz.php?c=71&b=103404&g=.

Technology Business Incubation Center (TBIC). https://tbic.kaist.ac.kr/#01_02_sect.

TechTarget. "3D NAND Flash." www.techtarget.com/searchstorage/definition/3D-NAND-flash.

TEGway. "Company Overview." http://tegway.co/tegway/.

The 30th Korean Conference on Semiconductors, "Shisang Gaeyo." http://kcs.cosar.or.kr/2023/about_award1.jsp.

Tongsan "Bandochesaneop: dashi oneun Chohohwangki, gyeonjohan Heureumse Jeonmang." *Tongsang, 104* (January 2021). https://tongsangnews.kr/webzine/2101/sub2_2.html.

Top Class. "Segye Choesohyung Microcapsule Naesigyeong 'Miro' mandeun Kim Tae-Songbaksa." *Top Class* (March 2008). http://topclass.chosun.com/mobile/board/view.asp?catecode=K&tnu=200803100005.

Toshiba. *Flash Memory: Semiconductor Catalog March 2016.* (2016).

Turley, Jim. "Masked ROM," in Jim Turley, ed., *The Essential Guide to Semiconductors.* Upper Saddle River, N.J.: Prentice Hall, 2003.

The University of Manchester. "Graphene: Applications." www.graphene.manchester.ac.uk/learn/applications/ (searched on March 11, 2021).

US Chamber of Commerce. "Made in China 2025." (March 16, 2017). www.uschamber.com/international/made-china-2025-global-ambitions-built-local-protections-0.

Utah Arch. "A DRAM Refresh Tutorial" (November 27, 2013). http://utaharch.blogspot.com/2013/11/a-dram-refresh-tutorial.html.

The Verge. "Apple Reportedly Adds LG as Second OLED Display Supplier for iPhone XS and XS Max." *The Verge* (September 14, 2018). www.theverge.com/circuitbreaker/2018/9/14/17860688/apple-iphone-xs-max-lg-oled-display-samsung.

Vogel, Ezra F. *The Four Little Dragons: The Spread of Industrialization in East Asia.* Cambridge, Mass.: Harvard University Press, 1991.

The Wall Street Journal. "Apple Can't Cut Its Dependence on Rival Samsung's Screens." *The Wall Street Journal* (April 20, 2018). www.wsj.com/articles/apple-struggles-with-effort-to-diversify-screen-suppliers-1524216606?mod=Searchresults_pos7&page=3.

———. "SK Telecom to Buy Hynix Stake for $3.04 Billion." *The Wall Street Journal* (November 14, 2011). www.wsj.com/articles/SB1000142405297020419050457703786405 9713458.

WARP Solution. "What We Do." https://warpsolution.com.

Weber, Max. "The Principal Characteristics of Charismatic Authority and Its Relation to Forms of Communal Organization," in idem, *The Theory of Social and Economic Organization*, translated by A. M. Henderson and Talcott Parsons, edited with an introduction by Talcott Parsons. London: The Fee Press of Glancoe, Collier-Macmillan Ltd., 1947, 364–369.

Weekly Chosun. "Sege Choebinguk Hankuk eotteoke Suchul Ogangi doeeotna?" *Weekly Chosun* (November 4, 2015). http://weekly.chosun.com/client/col/col_view.asp?Idx=244&Newsnumb=20151118668.

Western Digital. "Western Digital Announces Industry's First 96-Layer 3-D NAND Technology" (June 27, 2017). www.westerndigital.com/company/newsroom/press-releases/2017/2017-06-27-western-digital-announces-industrys-first-96-layer-3d-nand-technology.

Wikipedia. "Fairchild Camera and Instrument." https://en.wikipedia.org/wiki/Fairchild_Camera_and_Instrument.

Wong, Dawn. "Creating the Future." *The Straits Times* (July 1, 2005), 3.

World Ranking Guide. "Helsinki School of Economics (HSE)." https://worldranking.blogspot.com/2010/02/helsinki-school-of-economics-hse.html (searched on October 20, 2021).

Wu, Muh-Cherng. "Chapter 5: IC Foundries: A Booming Industry," in Chun-Yen Chang and Po-Lung Yu, eds., *Made by Taiwan: Booming in the Information Technology Era.* Singapore: World Scientific, 2001.

Xu, Kevin. "Morris Chang's Last Speech." *Interconnected* (September 12, 2021). https://int erconnected.blog/morris-changs-last-speech/.

Xunzi. *Xunzi: The Complete Text*, translated and with an introduction by Eric L. Hutton. Princeton, N.J.: Princeton University Press, 2014.

Yi, Jeong-Hyong, Sung-Kye Park, Young-June Park, and Hong Shick Min. "Numerical Analysis of Deep-Trap Behaviors on Retention Time Distribution of DRAMs with Negative Worldline Bias." *IEEE Transactions on Electron Devices, 52:4* (2005), 554–560.

Yonhap News. "Cheoneuro Jeonkisaengsan … 'Wearable Battery' Gisulgaebal." *Yonhap News* (April 7, 2014). www.yna.co.kr/view/AKR20140407081500017.

———. "Hynix Bandoche Maegak Ilji." *Yonhap News* (November 11, 2011). www.yna.co.kr/view/AKR20111111149800002.

———. "Koo Bon-Moohoejang Pyeongsaengui Han, Bandoche." *Yonhap News* (May 20, 2018). www.yna.co.kr/view/AKR20180518082300003.

———. "LG Olhaebuteo Apple-e OLED Display Gongkeun Gidae." *Yonhap News* (May 25, 2019). www.yna.co.kr/view/AKR20190525020800091.

———. "Minju, Jeon Samsungjeonjasajang Buleo 'Global Sanup' Yeolgong." *Yonhap News* (June 30, 2020). www.yna.co.kr/view/AKR20200630191400001.

———. "Smart IT Yunghapsistemyeonkudan Gaeso." *Yonhap News* (May 30, 2012). www.yna.co.kr/view/AKR20120529018700017.

Yoo, Jei-Hwan, et al. "A 32-bank 1 Gb DRAM with 1 GB/s Bandwidth." *1996 IEEE International Solid-State Circuits Conference, Digest of Technical Papers* (1996), 378–379.

Yoo, Sangwoon. "Innovation in Practice: The 'Technological Drive Policy' and the 4Mb DRAM R&D Consortium in South Korea in the 1980s." *Technology and Culture, 61:2* (2020), 385–415.

Yoon, Jun-Bo, Chul-Hi Han, Euisik Yoon, and Choong-Ki Kim. "High-Performance Three-Dimensional On-Chip Inductors Fabricated by Novel Micromachining Technology for RF MMIC." *IEEE MTT-S International Microwave Symposium Digest, 4* (1999), 1523–1526.

Yoon Taesik, Woo-Cheol Shin, Taek-Young Kim, Jeong-Hun Mun, Taek-Soo Kim, and Byung-Jin Cho. "Direct Measurement of Adhesion Energy of Monolayer Graphene As-Grown on Copper and Its Application to Renewable Transfer Process." *NANO Letters* (February 2012), 1448–1452.

Yoon Yong-I. *Uri yet Dojagiui Arumdaum.* Kyeonggi-do, Paju-si: Dolbegae, 2007.

Youtube. "Apple WWDC 2010—iPhone 4 Introduction." www.youtube.com/watch?v=z__jxoczNWc.

YTN. "Kangsogieopi Himida: Dokbojeokin Gisulro mandeun Cheomdan Mugiui Nun, i3system." *YTN* (March 26, 2017). www.youtube.com/watch?v=5Pub-BdT3ck.

ZD Net Korea. "SK Hynix, 96dan 3D NAND Gaebal Sidong." *ZD Net Korea* (December 18, 2017). https://zdnet.co.kr/view/?no=20171218140855.

Index

Note: Page numbers in *italics* refer to figures.

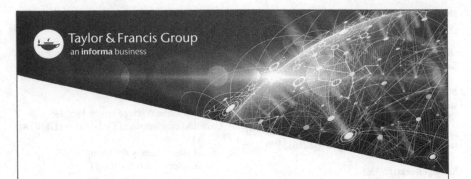

Printed in the United States
by Baker & Taylor Publisher Services